INFRARED SPECTRA OF CELLULOSE AND ITS DERIVATIVES

INFRAKRASNYE SPEKTRY TSELLYULOZY I EE PROIZVODNYKH

ИНФРАКРАСНЫЕ СПЕКТРЫ ЦЕЛЛЮЛОЗЫ И ЕЕ ПРОИЗВОДНЫХ

Edited by
Academician B. I. Stepanov
Director, Institute of Physics
Academy of Sciences of the Belorussian SSR

Translated from Russian by
A. B. Densham
The Gas Council
London, England

INFRARED SPECTRA OF CELLULOSE
AND ITS DERIVATIVES

By

Rostislav Georgievich Zhbankov
Institute of Physics
Academy of Sciences of the Belorussian SSR
Minsk

Springer Science+Business Media, LLC 1966

Rostislav Georgievich Zhbankov, senior scientist at the Institute of
Physics of the Academy of Sciences of the Belorussian SSR, was born
in 1930. A graduate of the Belorussian State University and a lecturer
in the Physics Department of the Belorussian Polytechnical Institute,
his research concerns cellulose spectroscopy and other polycarbohydrates. Zhbankov is the author of more than 50 scientific papers and
patents and is a co-worker of the State Prize winner Academician
B. I. Stepanov.

The Russian text, which the author has corrected and updated
for this edition, was originally published for the Institute of
Physics of the Academy of Sciences of the Belorussian SSR
by Nauka i Tekhnika Press in 1964.

Library of Congress Catalog Card Number 65-25268

ISBN 978-1-4899-2734-7   ISBN 978-1-4899-2732-3 (eBook)
DOI 10.1007/978-1-4899-2732-3

FOREWORD

This monograph is concerned with systematization of the infrared spectra of an important natural polymer, cellulose, and its derivatives.

The infrared spectra of the main classes of cellulose derivatives are described and interpreted and those of such model compounds as mono-, di-, and trisaccharides are considered. Considerable attention is given to problems of technique in obtaining infrared spectra of fibrous cellulose materials, and to the analytical possibilities of infrared spectroscopy in studies of the properties of cellulose and its derivatives.

The book will be of use to scientific and plant workers interested in the study and treatment of cellulose, compounds related to cellulose (carbohydrates and polycarbohydrates), and other polymers.

INTRODUCTION

Spectroscopy has nowadays acquired great scientific and practical importance. Its possibilities are based on the specificity of the emission and absorption spectra of all types of material, from elements to complex natural products. Most widely used are the methods of emission spectral analysis (analysis of the emission spectrum from an incandescent body). The rapidity and availability of these methods, together with their high sensitivity and selectivity, has made them indispensable in the practice of plant and scientific laboratories for establishing the presence of specific elements in a substance under investigation.

Molecular spectral analysis has been introduced to a lesser extent owing to technical difficulties. However, these technical difficulties in investigation of the spectra of a number of specific materials are compensated for by the wide possibilities of this method for investigating physical and chemical structures, identifying materials, etc. The methods of molecular spectroscopy can be applied to a great variety of products in the course of scientific research in the fields of chemistry, biology, medicine, etc. No chemical product can be manufactured without determining the chemical structure of the starting material, knowing what impurities it contains, and checking the quality of the finished product. Without adequate methods of control, it is impossible to improve production or introduce changes in technology.

Methods of molecular spectral analysis are in many cases highly selective, rather precise, rapid, and economical. Naturally, they are not universally applicable. It would be absurd to suggest that the methods of chemical and spectral analysis are mutually exclusive. The correct procedure is to make use of the advantages of both methods.

Infrared spectroscopy is the most effective method of molecular spectral analysis. It provides one of the main means for structural analysis of macromolecules, has been used to elucidate many of the properties of polymers, and has made it possible to solve such fundamental scientific problems as the study of macromolecule configuration, reaction mechanisms, the nature of intermolecular forces, etc.

Infrared spectroscopy is a promising method for investigating the properties of cellulose, an important natural high polymer. Cellulose is widely used in the production of artificial silk, films, varnishes, and plastics, in the textile and paper industries, etc. Its value to the national economy is continuously increasing because cellulose products can ensure the development of new rapidly progressing branches of technology. Reserves of cellulose are continuously replaced by nature, which ensures that it will remain cheap and readily available. Cellulose is indeed the most widely distributed natural organic material.

In the last few years infrared spectral methods have become more and more widely used in the practice of scientific and industrial analysis of cellulose. They can be applied with advantage to analysis of the various structural and chemical modifications of cellulose, and are beginning to be used to elucidate cellulose transformations in the process of viscose silk production, to determine the nature of hydrogen bonding in cellulose, etc. It should be noted, however, that the application of these methods to the study of cellulose has hitherto been inadequate and, indeed, much less than their application to the study of other high polymers. We believe that the following reasons are responsible for this: until recently, there were no satisfactory techniques for obtaining the infrared spectra of cellulose fibers; analysis of cellulose in the form of a film could not give adequate information on the special features of the structure, and this restricted the field of investigation; there was no generally accepted interpretation of the main frequencies in the cellulose spectrum and there were errors and lack of precision; it was difficult to collect literature data on the infrared spectrum of cellulose; it was difficult to compare experimental data, obtained by various authors with different equipment and under different conditions.

Satisfactory methods have now been developed for obtaining the infrared spectra of fibers of cellulose and its derivatives: pressing ground-up fibers with KBr or KCl, making films by pressing cellulose fibers sprayed on to a metallic mirror surface, without an immersion medium, etc. The main absorption bands in the spectra of cellulose and its important derivatives have now been interpreted, but this interpretation requires further checking and must be made more precise in the course of research and analytical work on the infrared spectrum of cellulose.

The present work correlates, to a definite degree, the results of investigations on the infrared spectra of cellulose and its derivatives, carried out in the Belorussian State University and at the Institute of Physics of the Academy of Sciences of the Belorussian SSR.

When investigating such a complex high polymer as cellulose, it was necessary to study the spectra of several model compounds, cellulose derivatives, cellulose products with known structural alterations, etc. The present book shows the infrared spectra of numerous cellulose materials, differing in origin and treatment, and also of several sugars and polyhydric alcohols.

The monograph consists of seven chapters and a special appendix, which consists of a catalog of the infrared spectra of cellulose, its derivatives, and a number of model compounds, with tables of characteristic frequencies.

Special attention is devoted in this book to problems of technique in obtaining the infrared spectra of fibers of cellulose materials, and to the analytical possibilities of infrared spectroscopy in studies on the properties of cellulose and its derivatives. Consideration is given to the special features of the infrared spectra of important classes of cellulose derivatives, and the main frequencies are interpreted.

The spectra given in the catalog are the characteristics of compounds obtained from various sources, and by various methods of isolation and synthesis. Hence, these spectra cannot be used as standards, but they do illustrate specific special features for the given class of compounds.

It is not impossible that, depending on the conditions for obtaining a particular product, some distinguishing features of the spectrum may appear while the general qualitative picture remains the same.

Chemical analyses were not available for some of the compounds shown in the catalog. The spectra of such compounds are in general only useful for showing the region of the most intense bands of new functional groups introduced into the cellulose molecule. The order of arrangement of the spectra in the catalog is similar to that of their description in the main part of the book.

All the spectra shown in the catalog were obtained with a UR-10 infrared spectrometer (German Democratic Republic, the people's firm of Carl Zeiss, Jena), with prisms of KBr ($400-700$ cm^{-1}), NaCl ($700-2000$ cm^{-1}), and LiF ($2000-3800$ cm^{-1}).

The purpose of the table is to assist the investigator in a rapid preliminary appraisal of the origin of individual absorption bands in the infrared spectra of cellulose materials, and in the choice of spectral region for the determination of existing atomic groups.

The table shows the vibration frequencies (in the most convenient range for investigation, 400-3600 cm^{-1}) of a number of the main structural groups which determine the specific properties of cellulosic materials.

The frequencies given by various authors for a particular absorption band may vary over a range of 10 cm^{-1}, so that, in order to facilitate comparison, the frequencies given in the table are rounded off to the nearest 10 cm^{-1}.

It is not claimed that the table is exhaustive; it needs to be made more precise and requires supplementation.

The cellulose samples were kindly supplied for recording of their infrared spectra by Professors Z. A. Rogovin, P. V. Kozlov, A. A. Konkin, O.P. Golova, S. N. Danilov, N. N. Shorygina, and V.I. Sharkov, and by Doctors of Chemical Sciences A. I. Skrigan, O.P. Koz'mina, N.V. Shulyatikova, A. M. Shishko, F. N. Kaputskii, and N. Ya. Lenshina.

The recording of the spectra of the large number of cellulose derivatives and model compounds, shown in the book, was carried out by R. Marupov (most of the modified celluloses and the mono-, di-, and polysaccharides), N. I. Grabuz (unmodified cellulose, cellulose hydrated, products of partial hydrolysis of cellulose), and N. V. Ivanova (alkaline cellulose and cellulose esters). The very laborious work of preparing the figures was done by M. M. Isafarova, and the large amount of photographic work was carried out by S. A. Vankovich.

The author wishes to express his deepest gratitude to Academician of the Academy of Sciences of the Belorussian SSR B. I. Stepanov for his initiative and for directing this work, and to Doctor of Physico-Mathematical Sciences V. M. Chulanovskii, Doctor of Technical Sciences Z. A. Rogovin, Candidate in Technical Sciences L. S. Gal'braikh, Candidate in Physico-Mathematical Sciences N. A. Borisevich, and the collective leadership of their laboratory for valuable comments and a great deal of help in this work.

CONTENTS

CHAPTER I. Methods for Obtaining the Infrared Spectra of
Cellulose and Related Materials 1
Investigation of Unchanged Fibrous Material 18
The Suspension Method. 20
The Solid Immersion Media Method 20
Production of Thin Fiber Films Without Use of an Immersion
Medium 23
Production of Infrared Spectra of Cellulose from Alkaline
Aqueous Solutions 31
Production of Thin Films of Water-Soluble Compounds from
Their Aqueous Solutions 31

CHAPTER II. Model Cellulose Compounds. Unmodified
Celluloses 35
Monosaccharides. 37
Disaccharides. 51
Polysaccharides. 53
Cellulose from One-Year-Old Plants 57
Cellulose from Young Wood Shoots 63
Cellulose from Woods of Various Ages. 68
Samples of Technical Cellulose 71

CHAPTER III. Hydrocellulose. Products of the Partial Hydrolysis of
Cellulose. 73
Cellulose Regenerated After Treatment with Alkali of
Various Concentrations. 80
Cellulose Subjected to Grinding 84
Viscose Fibers. 86
Effect of Hydrogen Bonds on the Structure of Cellulose . . . 86
Alkaline Cellulose. 89
Products of the Partial Hydrolysis of Cellulose 90

CHAPTER IV. Cellulose Esters and Ethers.   93
    Cellulose Acetates Produced under Homogeneous
        Conditions of Acetylation and Partial Hydrolysis.   100
    Cellulose Acetates Produced under Heterogeneous
        Conditions of Acetylation and Partial Hydrolysis.   103
    Comparison of Spectral Data with the Results of Viscometric
        Analyses of Cellulose Acetates.   105
    Acetylcellulose Fibers with Different Acetyl Group Contents .   107
    Special Features of the Spectra of Ethers and Esters of
        Cellulose.   111

CHAPTER V. Oxidation Products of Cellulose. Salts of Oxidation
        Products of Cellulose   117
    Monocarboxycellulose   119
    Dialdehydrocellulose   122
    Dialdehydrocellulose Oxidized by Nitrogen Oxides   123
    Dicarboxycellulose   124
    Dialcoholcellulose   124
    Salts of Oxidation Products of Cellulose   125
    Salts of Cellulose Oxidation Products with Inorganic Cations .   127
    Salts of Cellulose Oxidation Products with Organic Cations. .   129

CHAPTER VI. New Types of Cellulose Derivative.   131
    Esters of Cellulose with Phosphorus-Containing Acids.   132
    Esters of Cellulose with Chlorinated Aliphatic Acids   137
    Stable Xanthate Derivatives of Cellulose   140
    Investigation of the Products of Thermal Decomposition of
        Cellulose Methylxanthate   145
    Cellulose Compounds Containing New Functional Groups . . .   148
    Graft Copolymers of Cellulose with Poly-2-methyl-5-
        vinylpyridine.   150
    Graft Copolymers of Cellulose and Carbon Chain Polymers,
        Obtained by Initiating the Graft Polymerization with
        Pentavalent Vanadium Compounds   157
    Products of the Ion-Exchange of Graft Copolymers of
        Cellulose and Polyacrylhydroxamic Acid with Ions of
        Fe^{3+} and Cu^{2+}.   160

CHAPTER VII. The Possibilities of the Infrared Spectroscopic
    Method for Investigation of the Properties of Cellulose and
    Its Derivatives   167
    The 2000-4000 cm^{-1} Region   167
    The 1500-2000 cm^{-1} Region   174
    The 1200-1500 cm^{-1} Region   177

The 950-1200 cm^{-1} Region 181

The 700-950 cm^{-1} Region. 182

The 400-700 cm^{-1} Region. 184

APPENDICES* 187

    Appendix I. Spectra of Mono-, Di-, and Polysaccharides
and their Derivatives, and Spectra of Polyhydric
Alcohols 189

    Appendix II. Spectra of Unmodified Celluloses 219

    Appendix III. Spectra of Celluloses Regenerated After
Treatment with Aqueous Alkali of Various Concentrations. Spectra of Viscose Silk and of Alkaline Cellulose 235

    Appendix IV. Spectra of Products of Partial Hydrolysis and
Ethanolysis of Cellulose 244

    Appendix V. Spectra of Cellulose Ethers and Esters. 249

    Appendix VI. Spectra of Oxidation Products of Cellulose. . . 263

    Appendix VII. Spectra of Salts of Cellulose Oxidation
Products 273

    Appendix VIII. Spectra of New Types of Cellulose
Derivatives and of Modified Celluloses. 277

    Appendix IX. Characteristic Vibration Frequencies of
Groups and Bonds in the Molecules of Cellulose and
Its Derivatives 304

LITERATURE CITED. 325

* A detailed listing of all the spectra presented in each Appendix appears on the first page of that Appendix.

PUBLISHER'S NOTE

The following Soviet journals cited in this book
are available in cover-to-cover English translation:

Russian title	English title	Publisher
Izvestiya Akademii Nauk SSSR: Otdelenie Khimicheskikh Nauk	Bulletin of the Academy of Sciences of the USSR: Division of Chemical Sciences	Consultants Bureau
Izvestiya Akademii Nauk SSSR: Seriya Fizicheskaya	Bulletin of the Academy of Sciences of the USSR: Physical Series	Columbia Technical Translations
Kolloidnyi Zhurnal	Colloid Journal	Consultants Bureau
Optika i Spektroskopiya	Optics and Spectroscopy	American Institute of Physics
Uspekhi Khimii	Russian Chemical Reviews	The Chemical Society (London)
Zhurnal Fizicheskoi Khimii	Russian Journal of Physical Chemistry	The Chemical Society (London)
Zhurnal Obshchei Khimii	Journal of General Chemistry of the USSR	Consultants Bureau
Zhurnal Prikladnoi Khimii	Journal of Applied Chemistry of the USSR	Consultants Bureau
Zhurnal Tekhnicheskoi Fiziki	Soviet Physics — Technical Physics	American Institute of Physics

CHAPTER I

METHODS FOR OBTAINING THE INFRARED SPECTRA OF CELLULOSE AND RELATED MATERIALS

Spectroscopic methods are based on the study of characteristic absorption or emission of electromagnetic radiation by a substance. No media exist which absorb all electromagnetic radiations; any medium strongly absorbs radiation only at definite wavelengths.

If no account is taken of the energy of reciprocating movement of the molecule as a whole, then the energy of a molecule can be considered, to a first approximation, as the sum of the energies of movement of electrons, of vibration of the atomic nuclei, and of rotation of the molecule. This division is based on the considerable differences in energy, emitted or absorbed by a molecule, corresponding to transitions between energy levels of the types considered. For example, the difference between successive electronic levels is of the order 2-10 eV, between vibration levels it is 0.05-0.5 eV, and between rotation levels it is 0.005-0.025 eV. Thus, for a given electronic state, a molecule may show various vibrational and rotational states, and for given electronic and vibrational states it may show various rotational states (Fig. 1). Interaction between a molecule and an electromagnetic field may lead to a transfer of energy from the field to the molecule, as a result of which the molecule changes from one quantum state to another (excitation) and acquires energy ΔE, related to the frequency ν' of the absorbed radiation by the equation

$$\Delta E = h\nu', \tag{1}$$

where h is Planck's constant (h = $1.623 \cdot 10^{-27}$ erg $\cdot$ sec).

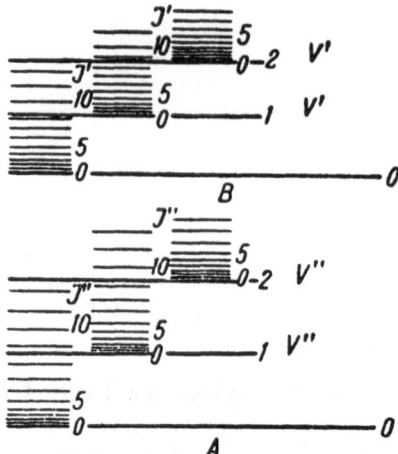

Fig. 1. Diagram of energy levels of di-
atomic molecule: A and B are electronic
levels; V' and V" are the quantum numbers of the vibration levels; J' and J" are
the quantum numbers of the rotation levels.

If the molecule returns to its original state it emits the same quantum
of energy ΔE of frequency ν'. It follows from equation (1) that the highest
frequency (i.e., the lowest wavelength, since $\lambda = C/\nu'$, where C is the rate
of propagation of radiation) is associated with radiation absorbed or emitted
by a change in electronic state, while the lowest frequency corresponds to a
change in rotation state.

The electronic states of molecules can be investigated by means of
the spectrum in the ultraviolet, visible, and near-infrared regions ($\lambda \approx 1000$
to 12,000 A; 1 A = 10^{-8} cm), vibration states in the medium-infrared region
($\lambda \approx 1.2$-40 μ, 1μ = 10^{-4} cm), and rotational states in the far-infrared and
short radio-frequency wave regions ($\lambda = 4 \cdot 10^{-3}$ to 1.5 cm).

As a spectroscopic characteristic it is convenient to use the wave number, i.e., the reciprocal of the wavelength. For example, if $\lambda = 1\mu$ = 10^{-4}
cm, then $\nu = 1/\lambda = 1/(10^{-4}$ cm$) = 10^4$ cm^{-1}. This value is equal to the
number of wavelengths in 1 cm and is normally also called the frequency,
although the dimensions of frequency [ν] are sec^{-1}.

Investigation of the vibrations and rotations of molecules can also be
carried out in the visible region of the spectrum by means of the Raman effect, discovered by the Soviet scientists Mandel'shtam and Landsberg, and by
the Indian scientist Raman (1928). The basis of the Raman effect is the

modulation of light scattered from a substance by the vibrational and rotational movements of its molecules. Because of this modulation, the Raman spectrum shows, in addition to the frequency of the incident radiation, other bands which differ from it in frequency by frequencies corresponding to vibrational and rotational transitions.

Electronic spectra are particularly important for investigating various questions in theoretical chemistry and greatly assist in the solution of many photochemical problems, such as the colors of organic compounds. The movement of the outer electrons, which is directly reflected in the electronic spectrum, largely determines the chemical properties of a substance, the reactivity of its atoms, bond strengths, etc.

However, electronic spectra are not always sufficiently selective, and this restricts their use for analytical purposes. Purely rotational spectra provide the most precise means for determining bond lengths and valence angles. However, it is very difficult to obtain these spectra experimentally. The lines are normally of low intensity and are difficult to measure.

Of all spectroscopic methods for investigating molecules (particularly complex molecules), the one most widely used nowadays is infrared vibration spectroscopy. This is not surprising when consideration is given to the rich source of information provided by the vibration spectrum of a molecule. Indeed, each molecule has its own specific set of vibrations, depending on the geometry of the molecule, the nature of the vibrating atoms, the values of the interatomic distances, the types of inter- and intramolecular interactions, etc. Infrared spectra reveal quite small changes in the structures of molecules and are therefore very effective in the investigation of rotational isomers, stereoisomers, branched and unbranched isomers, etc. The vibration spectrum provides an unambiguous characteristic of a substance, like a "passport."

A nonlinear molecule, consisting of N atoms, has $3N-6$ vibrational degrees of freedom. According to theory, only those vibrational movements which correspond to changes in dipole moment can lead to emission or absorption of radiation. The dipole moment of a molecule $\vec{\mu}$ can be expressed as a Taylor series:

$$\vec{\mu} = \vec{\mu_0} \sum_i^k \left(\frac{\partial \vec{\mu}}{\partial q_i} \right)_0 \cos(\omega_i t + \varphi_i) + \frac{1}{4} \sum_{ij}^k c_i c_j \frac{\partial^2 \vec{\mu}}{\partial q_i \partial q_j} \times$$

$$\{ \cos[(\omega_i + \omega_j)t + (\varphi_i + \varphi_j)] + \cos[(\omega_i - \omega_j)t + (\varphi_i - \varphi_j)] + \dots \}, \quad (2)$$

where the suffix "0" denotes the equilibrium state; c_i and φ_i are constants;

$i = 1, 2, \ldots, k;$ and q_i are so-called normal coordinates. By the use of these coordinates all the atoms in a molecule can be described as vibrating with one frequency and phase. The vibration energy of the system can be determined by adding the energies corresponding to the normal vibrations.

It follows from equation (2) that the dipole moment of a molecule varies with time at frequencies equal to the frequencies of the normal vibrations, their sums, and their differences. Consequently, the molecule will selectively emit or absorb radiation only of these frequencies.

Frequencies corresponding to normal vibrations are called fundamental frequencies; double, treble frequencies, and so on, are called the first, second overtones, and so on; frequencies corresponding to sums or differences of normal vibrations are called component frequencies.

Fundamental frequencies in absorption and emission are forbidden if $\partial \vec{\mu} / \partial q_i = 0$. A similar conclusion is reached by selection rules for vibration absorption spectra, based on quantum-mechanical calculations.

Since absorption or emission of radiation can only occur when there is a change in dipole moment, it follows that the frequencies of the absorbed or emitted radiation will be determined by the symmetry of the molecule.

Figure 2 shows examples of active and inactive vibrations in infrared spectra.

In principle, all the atoms participate in the vibrations of a molecule, and their bond angles and bond lengths alter simultaneously. However, the parts played by various atoms in a particular type of movement may vary considerably. Very often the vibration is confined to a particular part of the molecule and, in separate cases, may lead only to changes in bond angle or to changes in bond length. This is the basis for the classification of normal vibrations into stretching and deformation types. In the case of stretching vibrations, the main variation is in the bond length, while the angles between the bonds remain approximately constant, whereas in deformation vibrations the main variation is in the angles between the bonds.

We must also consider an additional classification of stretching and deformation vibrations of groups XY_2 and XY_3, since structural elements of this type will often be encountered in the spectra of the compounds discussed in this book. Stretching and deformation vibrations can be classified as symmetrical or asymmetrical. The significance of this terminology should be obvious from Figs. 3 and 4. It is further necessary to introduce the terms external and internal to characterize deformation vibrations. An internal deformation vibration is associated with changes in the bond angles within a group, whereas external vibrations are associated with a change in the spatial

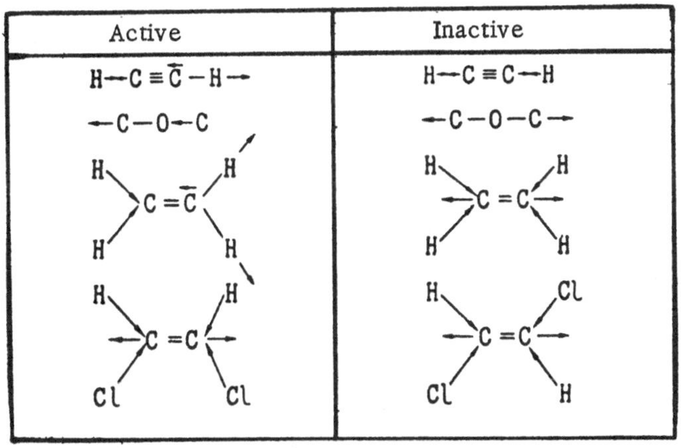

Active	Inactive
$H—C \equiv \overline{C}—H \rightarrow$	$H—C \equiv C—H$
$\leftarrow C—O \leftarrow C$	$\leftarrow C—O—C \rightarrow$

Fig. 2. Types of vibrations which are active and inactive in infrared spectra.

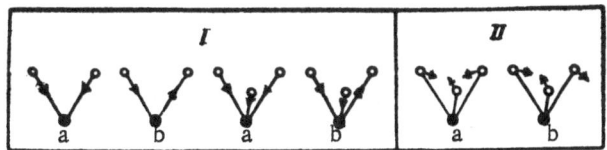

Fig. 3. Stretching and deformation vibrations: a) symmetrical; b) asymmetrical. I) Stretching vibrations; II) deformation vibrations.

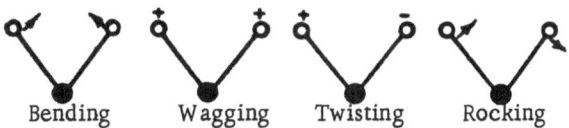

Bending   Wagging   Twisting   Rocking

Fig. 4. Types of deformation vibrations of the group XY_2.

positioning of the group as a whole (Figs. 3 and 4). Internal deformation vibrations of the group XY_2 are often referred to in the literature as bending or scissoring, while external deformation vibrations, depending on their type, are referred to as wagging, twisting, or rocking.

It is well known that molecules which differ in structure may contain common structural groups, such as OH, CH_2, $C=O$, various types of rings, etc. A molecule can be subject to a vibration in which only a definite structural group participates to an appreciable extent. It has been shown experimentally that some frequencies can be attributed respectively to the vibrations of definite atomic groupings. These are called characteristic frequencies and play an important role in the structural analysis of molecules by means of their vibration spectra.

The characteristic frequencies of a group of atoms vary slightly, depending on the nature of the compound. This gives useful additional information as to the special features of the environment of this or that functional group and the nature of inter- and intramolecular interactions. Thus, the literature does not give one value of the characteristic frequency for a particular group, but a frequency range in which it can appear, and this range may be considerable. For example, vibration frequencies of alcoholic OH groups are in the region 2500-3700 cm^{-1}, of CH groups in the region 2800-3100 cm^{-1}, of C $=$ O in the region 1600-1900 cm^{-1}, and of C $\equiv$ N in the region 2100-2300 cm^{-1}, while internal deformation frequencies of CH_2 and CH_3 groups are in the region 1340-1480 cm^{-1}, etc. It is possible to make definite suggestions as to the nature of a group from the frequency of its absorption band and the shape of the band contour.

The understanding of frequency characteristics has developed from the comparison of the spectra of different molecules, so that the determination of these frequencies is purely experimental in character. The frequency, characterizing the vibration of any particular bond, would only have precisely the same value in all molecules if there were no interaction with other angular and bond vibrations. This occurs very seldom in practice. For this reason, the presence of repeating frequencies in the spectra of various molecules indicates that there is the same interaction on the given vibration by the vibrations of adjacent angles and bonds.

Tables of the characteristic frequencies of many of the important structural elements of molecules have now been compiled, and these have provided a basis for the introduction of infrared spectroscopic methods into the practice of scientific and plant laboratories. However, it must be appreciated that the detailed interpretation of frequencies in infrared spectra requires particular skills which only may be acquired by experience. The task is especially complicated in the analysis of compounds with strongly interacting basic structural elements. It is then nearly always necessary to make use of the spectra of various model substances. In many cases the object of analysis is simply to establish the presence of definite groupings, such as OH, CH, C $=$ O, C $\equiv$ C, ONO_2, etc., or, indeed, to establish the identity of a compound. The infrared spectroscopic method solves these problems simply and unambiguously.

As an example, let us consider, omitting details, the spectroscopic appearance of vibrations of the CHOH (CH_2OH) group, which is one of the main structural elements of the cellulose macromolecule. The vibrations of this group may be divided into the stretching and deformation vibrations of hydroxyl, the stretching and deformation vibrations of CH, and the stretching vibrations of C—O.

The stretching vibrations of hydroxyl groups lead to the appearance of intense bands in the region 2500-3700 cm^{-1}. If the hydroxyl groups are free, their bands appear within the narrow range 3590-3670 cm^{-1}. However, hydrogen bonding in the hydroxyl groups leads to the emergence of stronger diffuse bands in the region 2500-3550 cm^{-1}, and the appearance of bands in the interval 2500-3200 cm^{-1} indicates the existence of very strong hydrogen bonding. It is normally possible to distinguish between hydrogen bonds of the dimer type OH . . . O, which give narrower bands in the region 3480-3530 cm^{-1}, and polymeric OH . . . OH . . . OH bonds, which give bands in the range 3200-3400 cm^{-1}. Bands due to intermolecular hydrogen bonding can be distinguished by the fact that they disappear in a nonpolar solvent or, as is sometimes possible, when the material is investigated in the gas phase. Theoretical calculations, confirmed by experiments, have shown that the frequencies of the stretching vibrations of hydroxyl groups, participating in hydrogen bonding, depend on the chain lengths of the oxygen bridges and reflect the energy of the hydrogen bonds. When there is a complex system of hydroxyl association in the material, there may exist an assembly of energetically nonequivalent hydrogen bonds, and these may not be resolvable spectroscopically owing to band overlap.

Much less characteristic are the frequencies of the deformation vibrations of hydroxyl groups (angular vibrations of C—O—H), whose absorption bands are not so strong and are located at longer wavelengths. The probable frequency region for planar deformation angular vibrations of C—O—H is 1100-1400 cm^{-1}. For example, Stepanov has calculated that the frequency of the angular vibration band of C—O—H in the spectrum of methyl alcohol is at 1340 cm^{-1}.

The frequencies of the stretching vibrations of C—O are not characteristic when there are C—C bonds present. However, the high intensity of the bands makes it possible to establish the presence of C—O bonds by means of a strong absorption band in the range 1000-1250 cm^{-1}.

The stretching vibrations of the CH group, not associated with an aromatic ring and not accompanied by a double bond, are located in the range 2800-3000 cm^{-1}. The stretching vibration bands of $\succ$CH are considerably weaker than those of > CH_2, but determination of methylene groups becomes difficult when there are relatively many CH groups in the molecule. The most characteristic frequencies are the internal deformation vibrations of methylene groups in the region 1420-1470 cm^{-1}, and this is the most convenient band for determining the presence of these groups. The frequencies of the wagging and twisting deformation vibrations of the methylene group coincide with the frequency range of the deformation vibrations of hydroxyl at 1200-1400 cm^{-1}, which also includes the deformation vibration frequency of $\geqslant$ CH. When all these types of vibrations occur together it becomes practically impossible to interpret frequencies in the range 1200-1400 cm^{-1}, without the assistance of model compounds or the use of deuteration methods.

The rocking vibrations of methylene groups are at a still lower frequency, in the region 700-900 cm^{-1}. This spectral region is covered by the vibration frequencies of a number of other elements of the cellulose macromolecule (particularly pyran ring frequencies), so that the attribution of bands in this region also requires special care.

In principle, it is possible to investigate the infrared spectrum of a substance by emission, reflection, or absorption of radiation. However, the infrared radiation, emitted by a substance under normal conditions, is very weak, and increasing the radiation by heating the substance may lead to a change in molecular structure or dissociation. In some cases it is possible to use for analysis the thermal emission from a substance at room temperature. To do this, the normal hot source of radiation is replaced by a cold source, such as a strongly cooled body. * Organic substances have, as a rule, a weak reflecting power; moreover, the properties of a substance are not always the same at the surface and in the interior. Most of the investigations of the vibration infrared spectra of organic molecules have been carried out by means of absorption spectra. Infrared spectroscopy has largely become synonymous with absorption spectroscopy.

Infrared spectrometers differ from the analogous instruments used in the ultraviolet and visible regions of the spectrum mainly because of the optical properties of infrared radiation.

Most optical materials transmit infrared radiation within restricted limits (Figs. 5 and 6). For example, glass prisms do not transmit radiation of wavelength longer than 2.5 μ, quartz not beyond 3.5 μ, etc. In infrared

---

*B.I. Stepanov. The Principles of Spectroscopy with Negative Light Beams, V.I. Lenin Press of the Belorussian State University, Minsk (1961).

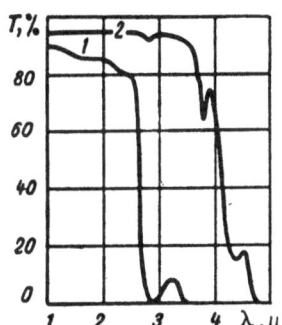

Fig. 5. Curves for the spectral transmission of: 1) crown glass, 3 mm thick; 2) fused quartz, 2.5 mm thick.

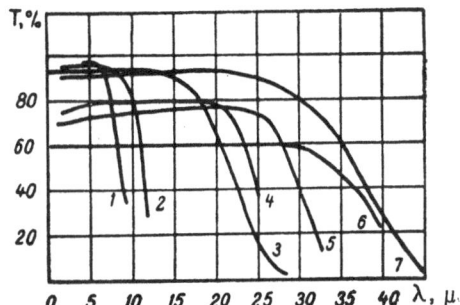

Fig. 6. Curves for the spectral transmission of: 1) lithium fluoride, 1 mm thick; 2) calcium fluoride, 1 mm thick; 3) rocksalt, 1 mm thick; 4) silver chloride, 5 mm thick; 5) KRS-6, 3.5 mm thick; 6) KRS-5, 4 mm thick; 7) potassium bromide, 1 mm thick.

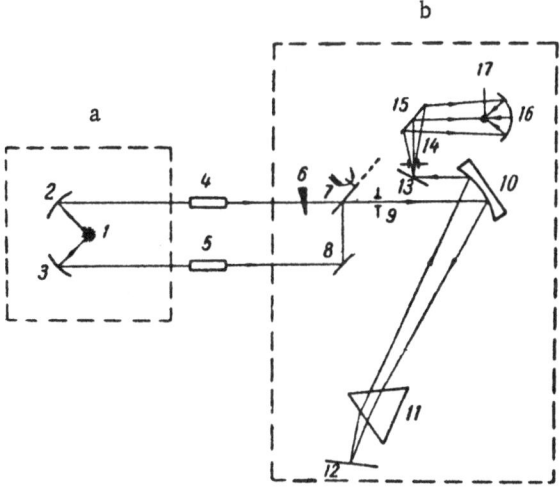

Fig. 7. Simplified optical diagram of double-beam infrared spectrometer. a) Radiation source with focusing arrangements; b) monochromator and radiation detector. 1) Radiation source; 2,3) spherical mirrors; 4,5) reference and sample cells; 6) compensating attenuator; 7,8, 13,15) plane mirrors; 12) Littrow mirror; 9,14) inlet and outlet slits of monochromator, respectively; 10) parabolic collimating mirror; 11) prism; 16) elliptical mirror; 17) detector.

spectrometers it is normal to use mirrors instead of lenses, since mirrors have a high reflecting power at all wavelengths and are not subject to chromatic aberration.

The operating principles of any infrared spectrometer are as follows. Continuous infrared radiation from a radiation source is focused onto the sample to be investigated. After passage through the sample, the radiation is dispersed by the monochromator to give a spectrum. The individual parts of the spectrum are successively directed through the exit slit of the monochromator onto a detector, and the output from the latter is amplified and fed to a recorder.

The absorption spectrum of the atmosphere is superimposed on that of the sample when the spectrum is recorded with a single-beam infrared spectrometer. In this case, the absorption spectra of the sample and of atmospheric vapors are recorded against the background of the radiant source. These difficulties are increased when the substance is investigated in solution, since most known solvents have their own intense absorption bands.

However, all these difficulties can be avoided by using a double-beam infrared spectrometer (Fig. 7). With this instrument, the radiation from the source is split into two beams, one of which passes through a cell containing the sample to be investigated, and the other through a similar cell which does not contain the sample. The spectrum of the radiation in the sample channel differs from that of the radiation in the reference channel by the amount of selective absorption of radiation by the sample. The radiation detector is so made that it records only the difference in intensity of the two radiation beams. The pulsating signal from the detector is amplified and directed to a special device (the attenuator), which compensates for the difference between the two light beams. The movement of the attenuator is synchronized with the operation of a recorder, which shows the difference in energy between the light beams (the out-of-balance signal) as a curve relating transmission by the sample and wavelength (or frequency).

The radiation sources most widely used with infrared spectrometers are a Nernst filament, consisting of compressed rare-earth oxides, thoria, ceria, and zirconia, or a Silit resistor (Globar). The maximum energy of the radiation from a Globar lies at a longer wavelength (1.8-$2\ \mu$) than that from a Nernst filament ($1.4\ \mu$). Over the range 600-$5000\ cm^{-1}$, the energy of the radiation falls by a factor of 600 for a Globar, or by a factor of more than 1000 for a Nernst filament. Thus, it is better to use a Globar to investigate the long wavelength region of the spectrum.

Prisms are usually employed for dispersion in infrared spectrometry, because they are easier to fabricate than diffraction gratings and are satisfactory

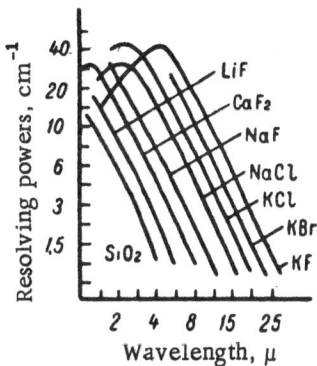

Fig. 8. The dispersions of various materials used for making infrared spectrometer prisms [81].

for present analytical requirements. The prisms are constructed from monocrystals of the salts of alkali metals and of alkaline-earth metals. Table 1 shows the transparency regions for a number of prisms and the most suitable spectral ranges for their use when dispersion is considered (see also Fig. 8).

After passage through the prism, the spectrum of the radiation is directed successively onto the exit slit of the monochromator by the Littrow mirror (see Fig. 7) and rotated by a special cam, which is shaped to give a linear scale of frequency or wavelength.

The detector used in an infrared spectrometer is normally a thermocouple or a bolometer, which changes in resistance on heating. An optico-acoustic device is also sometimes used to measure modulated infrared radiation; this gives a measure of energy absorption and operates on the principle that the pressure of a gas changes on heating.

Qualitative analyses by infrared spectroscopy can solve various experimental problems:

1) identification of a substance;

2) establishment of correspondence between a proposed chemical formula and the structure of a substance under investigation;

3) establishment of the structural formula of a substance when its elementary composition and certain characteristics (melting point, boiling point, etc.) are known;

4) obtaining additional data which is difficult or practically impossible to obtain by chemical means (types of hydrogen bond, rotational isomers, the presence of certain functional groups, etc.);

5) identification of mixture if the range of possible components is known approximately.

For most purposes concerned with structural analysis, it is sufficient to investigate the frequency region 400-4000 cm^{-1} and, for this reason, most infrared spectral measurements are carried out in this range.

It is easiest to interpret spectral bands lying between 1400 and 3700 cm^{-1}. This region includes the stretching vibration frequencies of the groups

Table 1. The Transparency to Infrared Radiation of Prism Materials
and the Most Suitable Regions for Their Use

Prism material	Transparency range, μ	Most suitable range for use, μ
Flint glass	Up to 2.5	Up to 2.0
Quartz	3.5	1.5- 3.5
Lithium fluoride	6.0	2.5- 5.0
Calcium fluoride (fluorite)	8.0	5.5- 8.5
Sodium fluoride	10.0	4.0-10.0
Sodium chloride	16.0	8.5-15.5
Potassium chloride	22.0	11.0-20.0
Potassium bromide	28.0	15.0-28.0
Potassium iodide	32.0	22.0-32.0
Cesium iodide	55.0	25.0-55.0

XH, OD, and triple and double bonds, and the deformation vibration frequencies of XH. The number of different frequencies is relatively small, so that identification is quite certain. It is more difficult to give attributions to bands in the 700-1400 cm^{-1} region, which includes vibrations associated with angular deformation of CCH and COH, stretching vibrations of C—O and C—C, skeletal vibrations of ring structures, etc. Many of these vibrations have frequencies which are very close together and interact strongly, so that it may be difficult to relate individual frequencies to particular types of vibration.

There has been little investigation of carbohydrate spectra in the 400-700 cm^{-1} region, which contains the overtone frequencies of hydrogen bonds and the out-of-planar deformation vibrations of hydroxyl groups. The still lower frequency region offers new possibilities for the study of inter- and intramolecular interaction and, hence, of structural details. However, there has been practically no work on the spectra of carbohydrates in this region. The small number of analytical problems to be solved here and considerable experimental difficulties have delayed developments in this direction.

The use of polarized infrared radiation has been of great assistance to the experimenter. The polarizers are normally constructed of selenium or silver chloride plates. By passing polarized radiation through a sample film or fiber, oriented in a particular direction, it is possible to determine the spatial distribution of individual vibrating dipoles. The maximum absorption is obtained when the direction of the vibrating dipole coincides with that of the electric vector of the incident radiation. If these directions do not coincide, which is usually the case in practice, then it is possible to determine the so-called coefficient of dichroism by passing through the sample radiation polarized in different directions. It is normal to use two directions of polarization, perpendicular and parallel to the orientation of the sample. From a knowledge of the dichroism coefficient it is possible to draw conclusions as to the angular spatial distribution of particular structural elements relative to the orientation axis of the molecule. In this case, infrared spectra can supplement the information obtainable by x-ray structural analysis, particularly in the case of light atom groups (OH, NH, CH), where it is very difficult to determine orientations by x-ray structural methods.

It has already been mentioned that in many cases there is no need to interpret all the frequencies observed in the spectrum, where it is sufficient to establish the presence of definite groups or bonds.

Before recording the infrared spectrum of a substance, it is desirable for the operator to be familiar with the structural formula of the material, or the range of possible structural elements, and to select the requisite spectral regions for carrying out the analysis and the best method of preparing the sample for investigation.

Studies on the infrared spectra of cellulose and related compounds are conveniently carried out in the following order.

1. The $2700\text{-}3700$ cm^{-1} region is investigated. The presence of bands in the range $3200\text{-}3700$ cm^{-1} is conclusive evidence of the presence of OH and/or NH groups. It is then necessary to look for overtone bands and component frequencies, which are usually of low intensity. From the character of the bands (frequency, intensity, contour) it is possible to draw conclusions as to the nature of the association of these groups, the degree of esterification of the hydroxyl groups, etc.

Absorption bands in the range $2700\text{-}3200$ cm^{-1} are due to stretching vibrations of CH, CH_2, and CH_3 groups. Narrow bands in the range $3000\text{-}3200$ cm^{-1} indicate the presence of CH groups, linked to double bonds or in aromatic structures.

2. The $2000\text{-}2600$ cm^{-1} region is investigated for samples containing the groups OD, SH, and the treble bonds $C \equiv N$ and $C \equiv C$. Attribution of the bands does not usually present any difficulty here, since very few vibration frequencies are known in this range. Investigation of this region is not of great interest in the absence of the groups listed above.

3. Investigation of the $1500\text{-}1800$ cm^{-1} region can show the presence of the double bonds $C = O$, $C = C$, and $N = O$, the deformation vibrations of amino groups, etc. The double bonds $C = O$ and $N = O$ are the most easily detected; they give intense bands at about 1750 cm^{-1} for $C = O$, 1650 cm^{-1}

for ONO_2, 1550 cm^{-1} for CNO_2, and 1600 and 1500 cm^{-1} for aromatics. There are no special difficulties in establishing the presence of the NH_2 band, except in aromatic derivatives when it becomes more complicated. Determination of the presence of the $C=C$ group is more difficult because the bands are of low intensity.

4. Absorption bands in the 1000-1500 cm^{-1} region can be used to show the presence of methylene and methyl groups and ester bonds of the groups $P=O$ and SO_2 or SO_3; they also give additional information as to the presence of the groups ONO_2, COOH, $C-OH$, etc.

Identification of methylene and methyl groups, bonded to carbon atoms, is most conveniently carried out using the absorption band corresponding to internal deformation vibrations at 1370-1470 cm^{-1}. Their detection in the stretching frequency region is complicated by absorption by CH groups.

Investigation of the 1000 to 1300 cm^{-1} region is particularly interesting when there is a need for direct demonstration of the presence of ester bonds. For example, the spectra of cellulose esters are characterized by the presence of two strong bands in the regions 1150-1300 and 1000-1100 cm^{-1}. Attribution of the frequencies in this region is conveniently carried out in association with analysis of other parts of the spectrum.

5. The frequency spectrum in the 700-1000 cm^{-1} region gives relatively little independent information on the presence of this or that functional group. The region is generally used at present for identification, because of the specificity of the spectra of cellulose materials in this range. However, it is possible that greater precision in the attribution of bands in this region might lead to valuable additional information as to the structure of cellulose and similar compounds, their conformation, etc.

When analyzing the spectra of cellulose and its derivatives it is necessary to take into account any special structural features. Cellulose materials can differ in the orderliness of their macromolecules, degree of asymmetry, ratio between various types of rotational isomer, etc. For example, a lower degree of structural orderliness in the cellulose macromolecules of the sample can be responsible for a more diffuse general appearance of the spectrum. Cellulose can exist in various structural modifications which are reflected in their spectra. Unless all these factors are allowed for, it is possible to make mistakes in the attribution of the infrared absorption bands of cellulose materials.

In order to determine the intensity of an absorption band, it is necessary, in quantitative spectral analysis, to make use of the Bouguer—Lambert—Beer law:

$$I_v = I_{0v} e^{-k_v cd}, \tag{3}$$

where k_v is the absorption coefficient for radiation of frequency v, calculated for unit concentration of the substance in a layer of unit thickness; I_{0v} and I_v are the intensities of the incident and transmitted radiation of frequency v.*

If we use logarithms to the base ten instead of natural logarithms, then the absorption coefficient is called the decimal absorption coefficient for unit concentration of the substance. The absorption coefficient for unit concentration is also called the extinction coefficient or absorptivity. If it is impossible to determine the concentration, then the absorption is given by the Bouguer–Lambert law:

$$I_v = I_{0v} e^{-k_v d}, \tag{4}$$

where k_v is now called the absorption coefficient for a layer of unit thickness.

The ratio of the transmitted to the incident beam, $T = I/I_0$, is usually called the transmission. The ratio of the radiation absorbed to the incident radiation, $A = (I_0 - I)/I_0 = 1 - T$, is called the absorption. Both these are often expressed as percentages.

The logarithm of the reciprocal of the transmission, $D = -\log T = \log I_0/I = k_v cd$, is called the optical density.

The intensity of the beam is reduced, not only by absorption, but also by scattering. In this case, the total attenuation of the beam is given by the equation

$$I_v = I_{0v} e^{-(k_v + k_{vp})d},$$

where k_{vs} is the scattering coefficient.

In the case of fibrous materials, particularly those with a rough surface, the scattering of radiation may be considerable, and this can greatly distort the spectral characteristics. Consequently, the main difficulty in obtaining the infrared spectra of these products is to develop a technique for reducing scattering of radiation as far as possible.

The spectral characteristics of a substance, as obtained with a double-beam infrared spectrometer, are usually recorded as the relation between

---

*The definition of d was omitted in the original Russian text; it is, of course, the thickness of the sample layer — Translator.

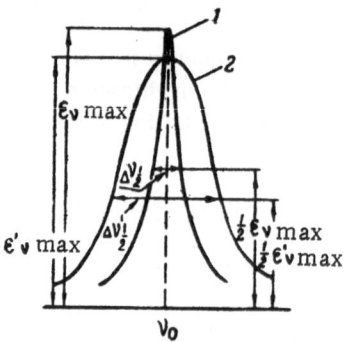

Fig. 9. Values characterizing the shape and intensity of an infrared absorption band. 1) True absorption band; 2) absorption band altered by finite slit width. $\varepsilon_{\nu\,max}$ = intensity at absorption band maximum; $\Delta\nu_{\frac{1}{2}}$ = half slit width.

transmission or absorption and wavelength or wave number. The use of calibrated paper, with a logarithmic ordinate scale, makes it possible to measure either percentage transmission or optical density.

Deviations from the Bouguer—Lambert—Beer law can arise from the nature of the substance itself (intermolecular reactions, dissociation processes, etc.), or from purely instrumental causes, the most important of which is the finite slit width. The Bouguer—Lambert—Beer law is only strictly correct for monochromatic radiation. However, the use in infrared spectrometers of a radiation source with a continuous spectrum makes it impossible to obtain strictly monochromatic radiation. When the spectrometer is set to a frequency ν, the radiation passing through the exit slit of the monochromator will cover some range of frequency $\nu \pm \Delta\nu$.

When the slit width is comparable with the half bandwidth, then the observed band will differ from its true form by about 20% in optical density and by 2-3% in integral absorption intensity. Thus, in order to make precise comparisons of absorption-band intensities, it is preferable to use the band area (integral intensity) rather than the band intensity at the absorption maximum (Fig. 9). The distorting action of the instrument may be characterized by its apparatus function.

When carrying out quantitative analyses on the basis of the Bouguer—Lambert—Beer law, it is necessary to determine whether there are deviations from this law, and to measure precisely concentrations and the thickness of the absorbing layer. In many analytical applications it is convenient to use a calibration curve relating relative absorption intensity and concentration of component to be determined, constructed on the basis of standard samples at known concentration. This makes it possible to carry out quantitative analyses without depending on the Bouguer—Lambert—Beer law.

Generalizing the above, we may say that infrared spectra can be widely used for determining the structure of molecules and for the qualitative and quantitative analysis of mixtures. The infrared spectrum shows a number of features which are closely related to the structure of the substance: the vibration frequencies, the band intensities, the band contours, including the diffusiveness and the presence of a structure, etc., the relation between an absorption band and the state of aggregation of the sample and the temperature, and band dichroism.

A simple calculation based on the number of absorption bands gives definite information as to the degree of symmetry of the molecule and the presence of isomeric forms. The observed frequencies establish the presence in the molecule of definite groups or bonds, and the frequency shifts from one compound to another give details of special features in the environments of these groups and of their interaction with other structural elements. A valuable characteristic of an absorption band is its intensity, which indicates the polarity of the molecule or of an individual element, and makes it possible to carry out quantitative spectral analysis. It should be noted that, in some cases, the intensity of a band can reveal the presence of a group or bond, even when its frequency is not a characteristic of the group or bond. The contour of an absorption band is very sensitive to the inter- and intramolecular interaction of vibrating groups and can provide a typical indication of a definite type of interaction. Large-intensity changes in vibration spectra with change of temperature or solvent indicate the existence of strong intermolecular interaction, or the presence of several isomeric forms of the substance under investigation. The use of polarized infrared radiation for the analysis of oriented samples makes it possible to determine the spatial distribution of vibrating dipoles.

The use of infrared spectroscopic methods makes it possible to obtain much valuable data on the properties of high polymers, as determined by the conformational features of their macromolecules. The presence of numerous independent units, showing a certain independence, gives polymeric materials a number of specific properties. The need arises to develop methods for analyzing the structural differences of polymers which are chemically identical. Infrared spectroscopy at present provides the most effective of all spectroscopic methods for investigating high polymers.

The following obvious advantages of this method should be noted: high selectivity, directness, reliability, very high sensitivity, rapidity of analysis, and the small amount of material required.

Infrared spectroscopic methods can provide the answers to many important chemical and physicochemical problems. However, it must be appreciated that the possibilities of these methods are limited, and that the best results are obtained by combining spectroscopic, chemical, and other methods of investigation.

The study of cellulose in the form of a regenerated film cannot give adequate information as to its properties. The inherent fibrous structure of cellulose is destroyed in the process of film formation, and the structural change is accompanied by changes in inter- and intramolecular interactions. The sample always contains some impurity or other, degradation products, etc. Moreover, it is not possible to obtain a film from all types of cellulose preparation. It is also difficult to investigate cellulose spectra in solution because cellulose is insoluble in most of the acceptable solvents. Again, the dissolution of cellulose is associated with formation of complexes with the solvent, and the structures of these are not at all obvious.

Thus, the problem is to develop methods which are of general application and which do not involve any serious changes in the morphological constitution of cellulose, or in the frequencies corresponding to its fibrous structure. In spite of the obvious difficulties of this task, the practical importance of fibrous materials demands a continuous search for new methods and the improvement of existing methods for solving various analytical problems.

We will now consider the methods based on different ways of preparing the fibrous material: 1) investigation of unchanged fibrous material; 2) the use of ground-up material in an immersion medium; 3) production of thin fiber films by pressing without an immersion medium.

Investigation of Unchanged Fibrous Material

The essence of the method is to obtain a layer of parallel closely packed fibers by a winding process [105, 106]. It is convenient to use one or more long threads of thickness $10-40 \mu$ [2]. The threads are usually wound under a microscope on bushes or frames. In this way, Ruscher and Schmolke obtained satisfactory spectra of polyamide and polyacrylic fibers [106]. An immersion medium is required in addition for fibers with a rough surface. N.I. Makarevich, a colleague of the Institute of Physics of the Academy of Sciences of the Belorussian SSR, has proposed a convenient modification of this method. The fiber was wound on a bush of KCl or KBr, with slotted edges, and placed directly in a cell of constant thickness. In this way he obtained good spectra of polypropylene, fluorlon, and certain other synthetic fibers. The winding method has made it possible to reveal a number of the structural features of viscose cord fibers. Satisfactory immersion media for this are a mixture of CCl_4 and CS_2, or tetrachloroethylene (Fig. 10). When there is a need for a thicker absorptive layer (e.g., in an analysis in the overtone region, etc.), it is possible to make direct use of a bundle of fibers placed in an immersion medium. This was done successfully by Nikitin [1].

In spite of its advantages, the winding method is difficult and requires the existence of long fibers of definite thickness and adequate strength. These factors restrict its use, and the need arose to develop other methods of analysis.

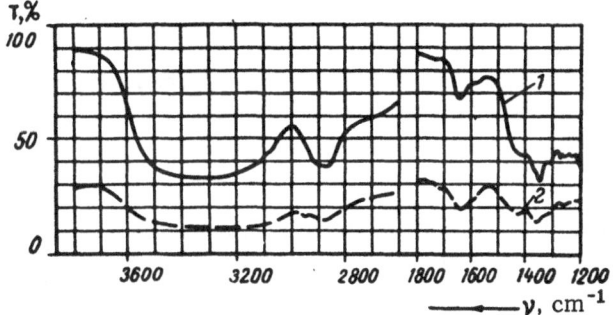

Fig. 10. Spectra of viscose silk fibers, obtained by the winding method. 1) In a CCl$_4$ immersion medium; 2) without an immersion medium.

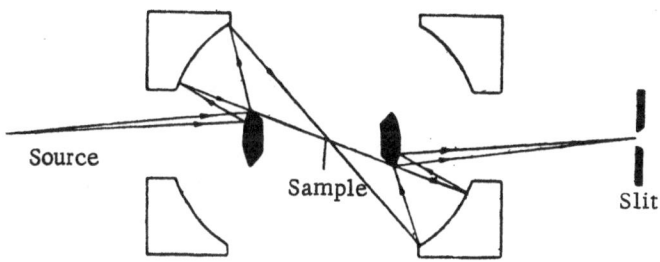

Fig. 11. Diagram of reflecting microscope [109].

In many cases, it is necessary to investigate samples whose width is less than the slit width of the spectrometer. This requires the use of a special projecting attachment which gives an enlarged image of the sample on the spectrometer slit. A reflecting microscope is preferred because optical materials absorb strongly in the infrared region. Tests with an immersion lens did not give encouraging results [107].

The first reflecting microscope for a spectrometer was described in 1949 [108], and since then the design has been continuously improved [109]. Figure 11 shows the principle of a reflecting microscope with aspherical mirrors.

If the magnification is more than 10, the collimator of a normal spectrometer can hardly ever be filled completely. In every case, passage through an enlarging system involves loss of a considerable part of the energy, sometimes 70%. The loss of radiant energy may be compensated by working with a considerably larger slit, but this greatly reduces the resolving power of the instrument.

It becomes very difficult experimentally to measure the ratio of the intensities of the incident and transmitted radiation. A simple removal of the sample to measure I_0 is not really effective, because it involves some defocusing of the beam and does not take into account the loss in energy by reflection at the sample surface. The latter effect, it is true, can be eliminated by using an immersion medium.

The possibility of using reflecting microscopes in double-beam infrared spectrometers has been considered [2, 110]. The matter is complicated by the need for the same adjustment and focusing for two systems and, if only one microscope is used, there is a need to compensate for differences in the optical paths and in atmospheric absorption. An original and convenient method of obtaining differential records, for the sample with respect to the source, is to move the sample in and out of the beam, by means of a solenoid [2]. This technique is distinguished by its simplicity, precision, and reproducibility.

In spite of the considerable experimental difficulties, satisfactory spectra have been obtained for a number of micro-objects by means of reflecting microscopes. Further improvements in technique should open up new possibilities in the study of the spectra of fibers. Possible applications of micro-attachments to the study of polymeric materials have been reviewed [2].

The Suspension Method

This is one of the earliest methods for analyzing fibrous materials. The fiber is carefully ground up and is immersed, normally, in a paraffin medium (a high-boiling petroleum fraction, the so-called "Nujol"). Forziati and Rowen [111] described the following technique for obtaining infrared spectra of cellulose fibers: 50 mg of the sample was ground up with 0.25 ml of oil. This technique reduces light scattering at adjacent surfaces, but it involves considerable grinding of the sample, which is accompanied by serious degradation and structural changes. Moreover, grinding of fibrous material is not always possible, and then only to a certain extent. Attention must be given to the strong Nujol bands in the region of the stretching and deformation vibrations of CH_3 and CH_2 groups; these make it very difficult to interpret the spectra in these regions.

The Solid Immersion Media Method

Extensive use has been made, in recent years, of the technique of pressing the material to be investigated in a solid immersion medium. This medium normally consists of alkali metal salts, KBr, KCl, AgCl, NaCl, etc., which are quite transparent in the infrared region. A sample of the ground-up

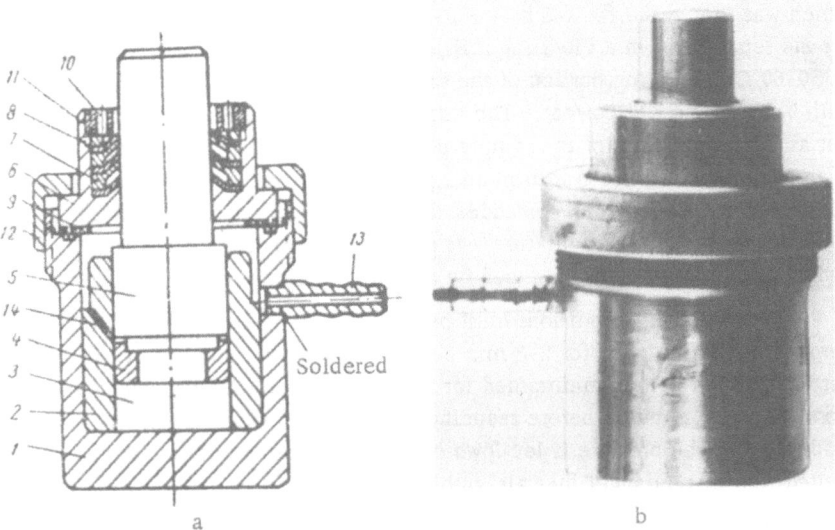

Fig. 12. a) Diagram of vacuum press, as developed in the Institute of Physics of the Academy of Sciences of the Belorussian SSR, for pressing discs of sample material with an alkali metal halide; b) external appearance of press [4]. 1) Body; 2) guide sleeve; 3) anvil with highly polished upper surface; 4) ring holding sample; 5) piston with highly polished lower surface; 6-9) vacuum sealing system; 6) sleeve; 7) conical ring; 8,9) rubber vacuum packings; 10) nut compressing the packing (8) and conical ring (7); 11) collar; 12) nut connecting the sleeve to the body; 13) connection to backing pump; 14) opening to remove air from working part of press.

fiber powder is well mixed with dry KBr (etc.) powder, normally in the ratio of 1 : 100 to 1 : 1000 [106], and pressed under vacuum in a special press. O'Connor, Du Pre, and McCall [112] first used KBr as a solid immersion medium for cellulose fibers and pointed out its advantages over the Nujol emulsion technique. The method proposed by O'Connor and his co-workers for preparing the samples for analysis was as follows: a 0.5-1 g sample of the material to be investigated was ground up in a vibrating mill and passed through a 20-mesh sieve; 2 mg of the ground-up fiber was well mixed with 350 g of KBr, and 300 mg of the mixture was pressed for 10 min, at a pressure of 2500 lb/in^2, in a vacuum press at 3 mm Hg. The technique of pressing the sample with an alkali metal salt gives satisfactory spectra and does not involve excessive grinding of the fibers. R. Marupov and the author proposed the following modification of the method, by means of which they obtained highly reproducible spectra of cellulose fibers. Small KBr crystals were thoroughly ground in an agate mortar, so as to give a uniform powder, which was then dried for 4-5 h at 200-300°C. The sample material was ground separately, in a vibrating mill or agate mortar, and dried for 3-5 h at 50-60°C. A 12-mg portion of the sample powder was then well mixed with 0.2 g of the KBr powder. The mixing process was conveniently carried out as follows: 12 mg of the sample powder was first mixed with 30-50 mg of the KBr powder for 4-5 min in an agate mortar; then the remaining 170-150 mg of the KBr powder was added, and mixing was continued for another 5-10 min. The resulting mixture was pressed in a special press (Fig. 12), under vacuum conditions produced by a normal backing pump.

The following conditions must be observed when pressing: the air is pumped out of the press for 2-3 min before applying the pressure of 5-7 tons per cm^2, which is then maintained for 2-3 min; the press is disconnected from the vacuum pump before reducing the pressure, which is then let down gradually. If the pressure is let down before the vacuum pump is disconnected, there is a danger that air, rushing violently into the space between the plunger and the ring, may cause cloudiness or cracking of the disc. Light scattering by the discs is found to be within permissible limits. It is more difficult to carry out quantitative measurements with Nujol than with these halide discs, which are therefore to be preferred for spectroscopy of fibers. The possibilities of the KBr pressed disc method have been discussed in detail in the literature [3, 4].

Higgins [113] recommended using KCl because it is less reactive and more stable at high temperature. It is also less hygroscopic and more easily ground into particles of the requisite dimensions, but it requires a higher operating pressure. According to certain authors, the lowest working pressure for potassium chloride is 6.5 tons/cm^2 [114]. Higgins proposed the following technique for the preparation and analysis of cellulose fiber samples: the sample was first ground very thoroughly (to pass a 60 openings/inch sieve), and 0.75 mg of the product (weighed to the nearest 0.01 mg) was mixed with enough KCl powder to give a total weight of 0.15 g. The final grinding and mixing were carried out by hand in an evaporating basin with a slightly roughened bottom, using a glass pestle with a roughened surface. In the author's view, this was more convenient than using a mechanical homogenizer. Pressing was carried out under vacuum (0.5 mm Hg) for 3-5 min. Higgins obtained satisfactory spectra with a pressure of 8 tons/cm^2.

The main disadvantages of the alkali halide pressed disc method are the need to disperse the fibers and their contact with alkali halides, which are not in general inert. Surface compounds are formed in some cases and give rise to so-called anomalies in the spectra. Examples are known of chemical interaction between the material to be analyzed and basic

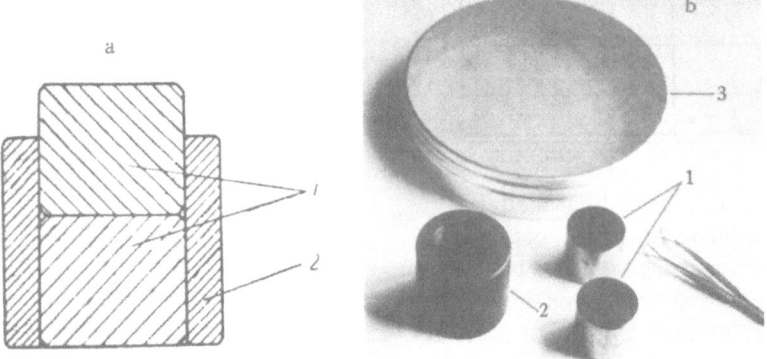

Fig. 13. a) Diagram of press for pressing out fiber films, without an immersion medium, by the direct-pressing technique; b) external view with
accessories. 1) Dies; 2) guide; 3) sieve.

material of the disc, of changes in the sample structure during grinding,etc.
[3, 5, 115-120]. The halides of alkali metals are appreciably hygroscopic,
and this complicates analysis in the hydroxyl group region. It should also
be noted that the pressed-disc method is not very effective when there is a
large difference in refractive index between the immersion medium and the
sample, or if these are difficult to mix.

Great care is needed when the technique used involves drying an
aqueous solution of the sample containing added alkali metal halide. This
method is used, for example, in biochemical analyses. Our investigations
have shown that there are differences in the spectra of the powders of some
water-soluble polymers, obtained from aqueous solutions which do or do not
contain NaCl.

Production of Thin Fiber Films Without Use of an Immersion Medium

In 1956, Zhbankov and Ermolenko developed a technique which greatly reduced light scattering by cellulose fibers, without use of an immersion
medium [6]. The basis of the method was to press fibers, which had been
sifted through a sieve, without adding anything to them. The sifted fibers
were uniformly distributed over the surface of a die, which had been highly
polished until it shone like a mirror. The press for making these fiber films
(Fig. 13) was very simple. It consisted of two cylindrical dies 1 and a guide
2. The die surfaces were lapped with a precision of not less than second
class, and the fineness of the surface was in the range of 10th to 14th class.

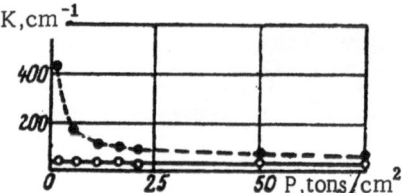

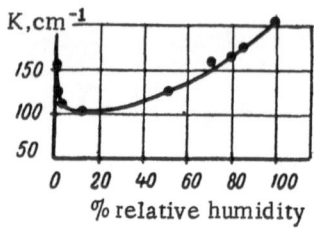

Fig. 14. Change in light scattering by a cellulose fiber film, obtained by direct pressing, as a function of the pressure applied. Top curve — without immersion medium; lower curve — in a medium of tetrachloroethylene.

Fig. 15. Changes in the light scattering of cellulose fiber films, made by direct pressing, as a function of the relative humidity.

Steels of grades Kh-40 or U-8 were suitable for making the dies. It was important that the dies should move freely in the guide. If they did not, there was a possibility that the upper die, when placed in the guide, would disturb the edge of the fiber layer sprayed onto the lower die, or that a flow of air might affect the uniformity of the fiber layer.

A normal soil sieve, with a gauze of metal or synthetic material, was used for sifting the fibers. The working range of the press was 5-10 tons/cm². Any further increase in pressure did not lead to any substantial reduction in light scattering (Fig. 14).

The working surfaces of the dies could be rectangular in shape, thus making it possible to reduce the total pressing load for a given value of the pressure.

Spraying of fibers onto the die was achieved by uniform movement with forceps of a tuft of fibers over the surface of a sieve, located above the die. The thickness of the fiber layer was controlled visually. After the pressure has been applied, the dies were removed from the guide and parted by movement in a horizontal plane from adjacent surfaces. The fiber film produced could conveniently be stripped off with a safety razor blade. The pressing process gave a mechanically stable semitransparent disc with flat lustrous surfaces. The morphological structure of cellulose fibers was not broken down within the limits of pressure applied, 5-10 tons/cm² [6].

It was possible to show, by exposing the fiber films to atmospheres containing various different water contents, that light scattering by the samples was a minimum at 10-12% relative humidity (Fig. 15). The films gradually became porous if the humidity was further increased, and were then almost useless for analysis. Thus, in any investigation, it is necessary to take into account the humidity of the film. In many cases, when studying qualitative changes in spectra, it is possible to use samples which are in equilibrium with atmospheric moisture (with the natural exception of products especially sensitive to moisture, or in some special investigations).

By making use of the direct-pressing technique, it is possible to obtain valuable information on the interaction between fibers in fibrous materials and on the effect on this interaction of adsorbed moisture.

The relation between light scattering and pressure used was investigated by recording the spectra of a given fiber film, produced at pressures gradually increasing from 1 to 75 tons/cm^2. Figure 14 shows that light scattering by the fiber film was reduced as the pressure was increased from 1 to 10 tons per cm^2, but that any further considerable increase in pressure had very little additional effect.

It should be noted that even a relatively high pressure (75 tons/cm^2) had very little effect on the spectrum of the resulting film.

The proposed method makes it possible to obtain more optically uniform systems than by pressing with an alkali metal salt, whose refractive index does not coincide with that of the fiber. Pressing the fibers alone ensures that they are in contact with each other and thus largely eliminates air/fiber interfaces.

Film thickness can be estimated with the aid of the usual length gauges (IZV 1, IZV 2, etc.), or by weighing. For semiquantitative or quantitative investigations, it is possible to use comparison with an unchanged absorption band, an internal standard, or measurement of thickness by a radiometric method, etc.

In the case of direct measurement of the thickness of fiber films, obtained by straightforward pressing, it is necessary to consider differences in fiber density, the nature of the packing, and interactions between fibers. For example, with oxidized cellulose films [7], the dialdehydocelluloses are characterized by the most open packing and, consequently, are the thickest for a given quantity of material by weight.

Ermolenko and Gusev [8] investigated methods for measuring the thickness of fiber films, obtained by direct pressing. They showed that quantitative work, based on simple measurement of the sample thickness, could be carried out for comparative measurements on materials which were similar in structure.

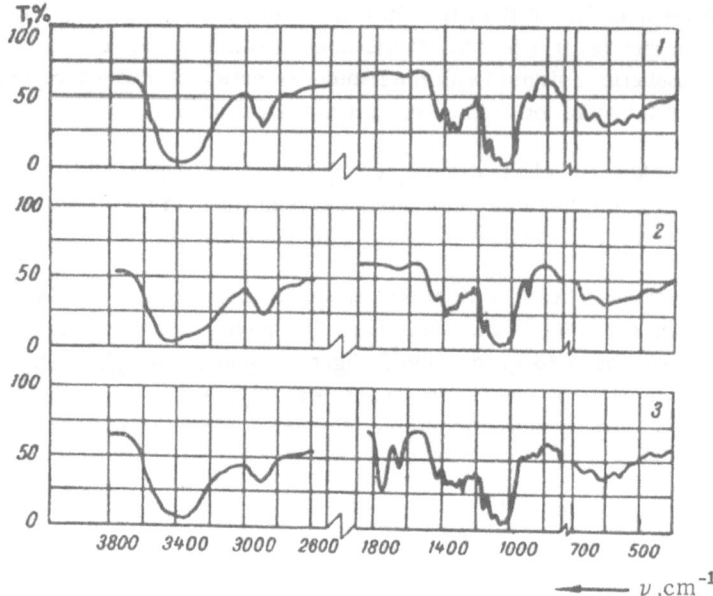

Fig. 16. Spectra of cellulose fibers, obtained by the direct pressing technique. 1) Cotton cellulose; 2) mercerized cotton cellulose; 3) cotton cellulose oxidized by nitrogen oxides.

Fig. 17. A device for moving a fiber film of varying thickness in a vertical plane.

These authors concluded that the radiometric method gave the most reliable results. However, because of its specificity, it has not been used very widely in practice. In any particular case, there is a need to analyze the possibilities and to use the most suitable method for measuring the thickness.

The direct pressing technique is very simple and is used in scientific and industrial practice for elucidating various chemical and structural changes in cellulose [9-15]. It should be noted that, with this method, it is possible to study spectroscopically the interaction of fibers with chemical reagents, directly during the course of the reaction [16]. This gives it a distinct advantage over the Nujol or KBr disc techniques. A disadvantage of direct pressing is the difficulty of obtaining very thin specimens (of the order of microns) for analyzing strongly absorptive groups. Moreover, it is not possible to reduce light scattering by the fiber film to a minimum, particularly at the sample surface. The background, due to light scattering, is considerable when the sample has a rough surface (Fig. 16), and this makes it difficult to determine small amounts of functional groups, or various contaminants and impurities in fiber samples. However, these disadvantages can be largely overcome by recording differential spectra. A simple device for doing this is shown in Fig. 17.

The fibers to be investigated are made into a film of varying thickness by the direct pressing method, and this film is supported by a guiding tape of robust material. The guiding tape is wound on the roller of the device, so that rotation of the rollers moves the film. The film of varying thickness can thus be moved until it compensates for the absorption of a standard sample, placed in the reference beam of the spectrometer. Differential recording of spectra can also be used with the pressed KBr disc technique. In this case, as shown by Garbuz, it is convenient to press a disc with sectoral changes in concentration of the material to be investigated.

When preparing fiber films of varying thickness, it is preferable to use a press with dies of rectangular section (about 15 × 60 mm).

Burgess and Spedding [121] investigated the possibilities of the direct pressing technique [6] for obtaining spectra of cellulose and a variety of synthetic fibers (nylon, terylene, orlon, etc.). These authors noted the advantages of the technique for identification of fibers, since it does not involve grinding or dissolution, which can lead to structural changes. According to Burgess and Spedding, the method is particularly useful for investigating samples of low hydroxyl content. The KBr disc method is then unsuitable owing to the hygroscopicity of the matrix material.

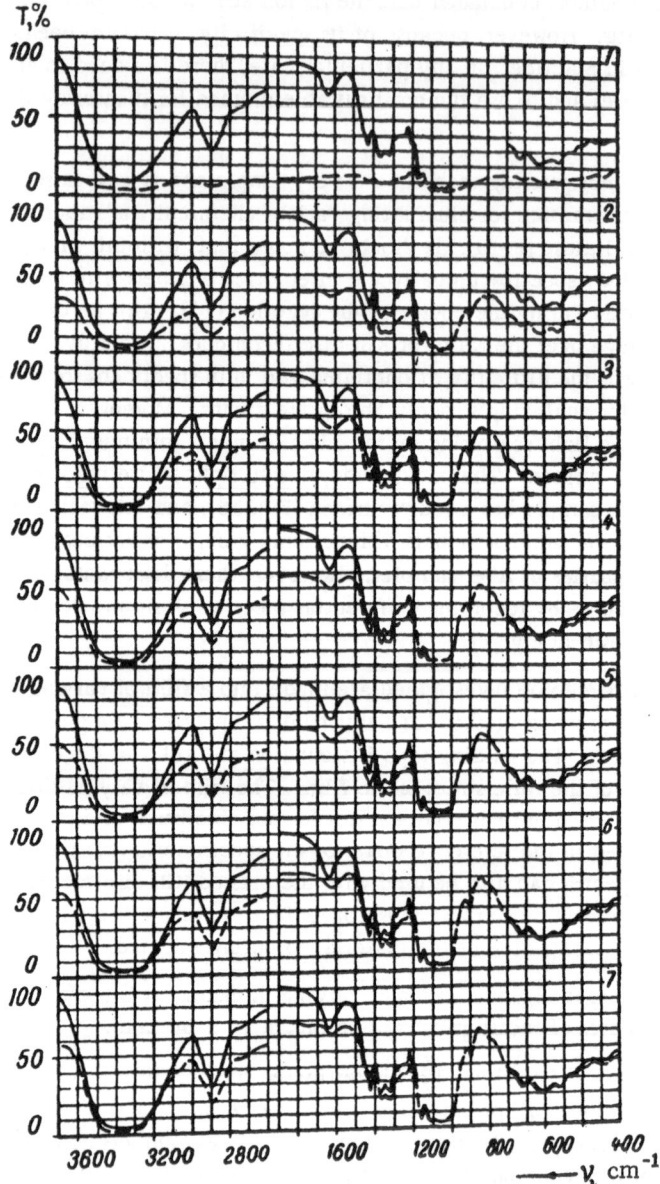

Fig. 18. Spectra of films of the same cellulose fiber, obtained by direct pressing at various pressures, before (dotted lines) and after (continuous lines) immersion in a liquid medium (tetrachloroethylene). 1, 2, 3, 4, 5, 6, 7) Pressed at 1, 5, 10, 15, 20, 50, and 75 tons/cm^2, respectively.

a

b

Fig. 19. General view of directly pressed cellulose fiber films, placed between windows, before (a) and after impregnation with tetrachloroethylene (b). 1) 50 μ thick; 2) 20 μ thick.

Stepanov, Zhbankov, and Marupov [17] proposed a way for considerably reducing light scattering still further by cellulose fibers. They showed that the greatest effect in reducing light scattering by cellulose fibers was achieved by placing thin films [6] in a liquid immersion medium, such as tetrachloroethylene, carbon tetrachloride, etc. This gave spectra which as a rule were more clearly defined than those obtained from KBr discs. Placing the fiber film in an immersion medium meant that the pressure applied in making the film could be greatly reduced while maintaining the same reduction in light scattering. For example, satisfactory cellulose spectra were obtained by immersing a fiber film, pressed at only 1 ton/cm^2, in a medium of carbon tetrachloride (Fig. 18). However, the use of such low pressures is generally undesirable, because the films produced are very loose and have open gaps in them.

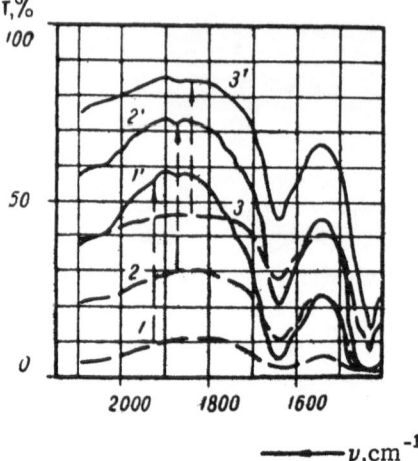

Fig. 20. Spectra of cellulose fiber films of various
thickness, before (dotted lines) and after (continuous
lines) steeping in an immersion medium (tetrachloroethylene). 1,2,3) 20, 50, and 80 μ, respectively.

The process of preparing the sample for analysis is extremely simple,
provided that the immersion medium is not too volatile: The fiber film is
steeped in the liquid and pressed between two windows, transparent to infrared radiation (Fig. 19). Provided that they do not attack the material under
investigation, the solvents normally used in infrared spectroscopy are, under
otherwise similar conditions, very suitable for use as immersion media,since
they are sufficiently transparent in thin layers. The proposed method [17]
for reducing light scattering by cellulose fibers considerably extends the possibilities of infrared spectroscopy in the study of fine spectroscopic effects in
cellulose materials, such as the presence in the structure of various impurities, small numbers of new functional groups, etc. By this means the thickness of a cellulose fiber film can be increased considerably with only a relatively small increase in light scattering. For example, when the film thickness was increased from 20 to 50 and 80 μ, the background due to scattering
of radiation increased by about 15-30% (Fig. 20).

The deficiencies of the KBr pressed disc technique were pointed out
[122] and efforts were made to find a more convenient method for analyzing
cellulose fibers. The authors noted that, even with careful grinding of the
fibers, it was difficult to avoid conglomeration, which in turn made it difficult to obtain an optically uniform medium. They concluded that a more
rational procedure was to prepare fiber films, 20-30 μ thick, and to press

these between two layers of KBr, in order to reduce scattering of radiation — the so-called "sandwich" technique. However, considerable scattering persisted, although there was some improvement. The inadequacy of the experimental material did not enable them to draw any definite conclusions as to the effectiveness of this technique. It is not at all clear that a solid immersion medium has any advantages over a liquid in reducing light scattering.

Production of Infrared Spectra of Cellulose from Alkaline Aqueous Solutions

It is well known that treatment of cellulose with alkali is one of the main steps in technological processes for producing viscose silk, many technically important cellulose derivatives, etc. [19]. Treatment of cellulose with concentrated alkali gives a new structural modification — hydrocellulose. The nature of the interaction of cellulose with alkali is still a subject for discussion, so that considerable interest attaches to the investigation of this interaction by infrared spectroscopy. However, an appreciable water content in the fibers makes it practically impossible to analyze the spectrum of cellulose in the hydroxyl group region. A study of dried samples gives an incomplete picture of the structure, since drying can give rise to a number of additional processes.

Zhbankov, Ivanova, and Rozenberg [16] developed a technique for obtaining infrared spectra of cellulose in contact with aqueous alkali, which enabled them to study the cellulose spectrum, in the hydroxyl group region, while the cellulose was being treated with alkali. They used the direct pressing method [6] to obtain a fiber film 9-11 μ thick, placed this between two windows of lithium fluoride, or any other nonhygroscopic material transparent to infrared radiation, and applied a few drops of a solution of NaOD in D_2O. It was an advantage to isolate the samples from the atmosphere. The stretching vibrations of the OD group are considerably displaced toward longer wavelength, as compared with those of the OH group, so that it was possible to analyze the spectra in the region of the hydroxyl groups of cellulose. However, when using this technique, it was necessary to allow for the process of deuterium exchange, and this involved carrying out control experiments with model substances.

Production of Thin Films of Water-Soluble Compounds from Their Aqueous Solutions

In many cases, considerable interest attaches to obtaining the infrared spectra of water-soluble saccharides and their polymers directly from aqueous solutions. An example is the biologically synthesized polymer of d-glucose,

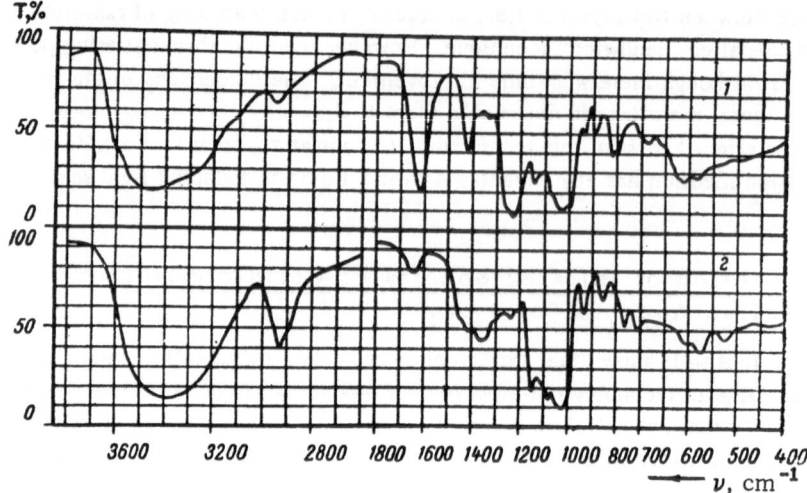

Fig. 21. Spectra of heparin (1) and glycogen (2), obtained by the applied thin
film method.

dextran, which is widely used, in the form of a 6% solution in water (polyglucin), as an effective artificial plasma in cases of shock or heavy loss of
blood [20].

Until now, pharmacological and chemical methods of analysis have
been used mainly for investigating polyglucin. Infrared spectroscopic methods, which have proved very useful in the study of similar compounds, have
hardly been used at all. This is due to the considerable experimental difficulties in obtaining infrared spectra of polyglucin. This particular polysaccharide is not soluble in any of the solvents normally used in infrared spectroscopy. Application of the pressed disc method (KCl, KBr, etc.) to polyglucin powder obtained from aqueous solution does not give satisfactory results. Experiments have shown that the scattering of radiation by such discs
is considerable, and that the spectral bands are broad and diffuse, without any
well-defined structure. This is particularly true of powder samples obtained
directly from finished products, which contain added common salt. On the
other hand, it is often of interest to analyze precisely these samples. Again,
the technique of pressing without an immersion medium is not very effective
for this purpose; it is impossible to obtain thin enough films and, because the
material is very hygroscopic, the films tend to be sticky and are difficult to
separate from the die. Production of thin films of dextran, directly from its
aqueous solutions, also gives unsatisfactory results; dextran films are very
brittle, adhere tightly to the die surface, and are thus very difficult to separate in practice. It should also be noted that polyglucin is not readily wetted
by most known media.

The various methods proposed have been considered, and it appears that the most effective and at the same time the simplest method for obtaining infrared spectra of water-soluble plasma-substitute materials is that which has been used for studying the spectra of polyvinyl alcohol [123]. The essence of the method [21] is to obtain, directly from the solution to be analyzed, a thin film (3-5 μ) on a plate of KRS-5, or of any other material which is transparent over a wide enough spectral range. The thickness of the film can be controlled by the concentration of the solution and the number of drops taken. For example, in the case of the 6% aqueous solution normally used in clinics, a good spectrum can be obtained by applying 2-3 drops to a KRS-5 window, 25-30 mm in diameter. The process of drying the applied layer of solution gives a transparent film with a lustrous uniform surface. In order to achieve a film of uniform thickness, the window surface should be level during drying. The drying should be carried out under vacuum at room temperature. This method gives minimum light scattering by the sample, in contrast to the KBr disc technique, when the background, due to scattering, may be quite large. The same method is applicable to the analysis of films of other water-soluble carbohydrates, such as d-glucose, glycogen, heparin, and other biologically important polysaccharides, starting with aqueous solutions. It is clear from Fig. 21 that the spectra of heparin and glycogen, obtained by this method, show well-defined spectral features.

CHAPTER II

MODEL CELLULOSE COMPOUNDS. UNMODIFIED CELLULOSES

In order to apply infrared spectroscopy effectively to the study of the structures of cellulose and its derivatives, and of the mechanisms of their re- actions, it is first necessary to interpret the main infrared absorption bands. For this purpose it is convenient to begin with an investigation of the infra- red spectra of various types of mono-, di-, and polysaccharides. These compounds have structural elements which are very similar to those of cellu- lose and may therefore be considered as model compounds for cellulose and its derivatives.

Several investigations have been published, dealing with the infrared spectra of saccharides and their derivatives. However, the published data are incomplete, and the interpretation of band frequencies has either not been considered, or is based on inadequate foundations.

Kuhn [124] used the Nujol emulsion method to obtain the infrared spectra of various saccharides but, except in the case of D-glucose, did not state the types of anomers. Barker and others [125] used the same method to investigate the spectra of α- and β-glucopyranoses, and of many of their de- rivatives, in the region 730-973 cm^{-1}. They noted the following frequency ranges (cm^{-1}) characteristic of α- and β-anomers:

Type 1 — α-anomer 917 ± 13, β-anomer 920 ± 5
Type 2 — α-anomer 844 ± 8, β-anomer 891 ± 7
Type 3 — α-anomer 766 ± 10, β-anomer 774 ± 9.

The authors considered that only the type 2 frequency range could be used for differentiation. However, the presence of an absorption band in the 891 ± 7 cm^{-1} is not in itself evidence of the presence of a β-anomer owing to the proximity and possibility of overlap of the frequency ranges of various anomers. Barker and his colleagues considered that the bands in the type 1 region corresponded mainly to the antisymmetric stretching vibrations of the $C-O-C$ ring, while those in the type 2 region were due essentially to deformation vibrations of hydrogen atoms at the C_1 atom. The frequency variations between the α- and β-anomers were attributed to differences in location (equatorial or axial) of a hydrogen atom at C_1, which altered the interaction of H_1 with the other hydrogen atoms. The type 3 vibrations were attributed to pulsation vibrations of the ring, by analogy with the interpretation of the vibration spectrum of tetrahydropyran [126].

In another paper [127], Barker and his colleagues considered the spectra of the D-glucopyranose acetates in the 750-850 cm^{-1} region, and also of derivatives of galactopyranose, mannopyranose, xylopyranose, and arabinopyranose. The bands of frequency previously classed [125] as of type 1 were not considered, because all the acetates showed absorption in this region. The following frequency ranges (cm^{-1}) were used for distinguishing the anomers of glucose, galactose, and mannose:

	Type 2a	Type 2b	Type 2c	Type 3
α-D-glucose	843 ± 4	—		
β-D-glucose	—	890 ± 8	—	753 ± 17
α-D-galactose	825 ± 11	—		
β-D-galactose	—	895 ± 9	871 ± 7	752 ± 20
α-D-mannose	833 ± 8	—		
β-D-mannose	—	893 ± 6	876 ± 9	791 ± 18

The higher vibration frequencies of type 2b for the β-anomers, as compared with those of type 2a for the α-anomers, can be explained by the additional Van der Waals interaction of the hydrogen atoms at C_1 and C_5 located axially. However, there are no absorption bands of type 2a frequency for derivatives of α-D-xylose. The authors' explanation for the failure of these frequencies to appear lay in the reduction in the derivative dipole moment of the deformation vibration of C_1-H, owing to the absence of a CH_2OH group at the C_5 atom. This explanation seems rather artificial.

Absorption bands of type 2c [127], characteristic of D-galactose and D-mannose derivatives (α- and β-anomers), and absent from the spectra of D-glucose derivatives, were attributed by the authors [127] to equatorial CH groups at C_2 and C_4. The spectroscopic data of [11, 119] is incomplete, so that it is impossible to make a comparison over a wide spectral range with the results of our own work. Independently of the investigations of Barker and his colleagues, Whistler and House [128] studied the differences in the

spectra of anomers of acetyl, methyl, and phenyl derivatives of monosaccharides, and also of a number of mixed ethers, in the region 700-1600
cm^{-1}. These authors observed the greatest differences in the spectra of the
α- and β-anomers of the phenylglucosides, as compared with the other compounds investigated. It should be noted that it is hardly feasible to construct
correlations for anomers by comparing the spectra of saccharoses of various
types, since, in view of the strong interaction between structural elements,
it is impossible to attribute vibrations to individual bonds or valence angles.

Konkin, Shigorin, and Novikova [22] obtained the infrared spectra of
a number of saccharoses in the region 2700-3600 cm^{-1} and investigated the
character of the hydrogen bonds between hydroxyl groups. Farmer [117] obtained the spectrum of D-glucose over the wide range of 625-5000 cm^{-1},
using the KBr disc technique, but did not compare frequencies. Urbanski
and his colleagues [129] used the Nujol emulsion method to obtain the spectra of nine mono-, di-, and trisaccharides, isolated by crystallization from
water, but it was not possible to separate the anomers. Tipson and Isbell
[130] proposed a classification of saccharide derivatives, based on a structural criterion, and obtained the spectra of 24 aldopyranosides with a partial
and provisional interpretation of the spectral frequencies; they attributed absorption bands in the 1235-1267 cm^{-1} region to C—O stretching vibrations,
but it will be seen below that this is incorrect. Tipson and his co-workers
[131] also studied a large number of infrared spectra of cyclic saccharose
acetals, aldopyranosides, and their aceto-derivatives in the region 250-
5000 cm^{-1}.

The spectra of some aqueous sugar solutions have been published [132].
This work is of considerable interest with respect to the technical possibilities of obtaining the spectra of water-soluble carbohydrates in aqueous solution.

Presented below are some general data on the infrared spectra of
mono-, di-, and polysaccharides, and of some polyhydric alcohols, obtained
with the help of A. M. Prim, R. Marupov, and N. V. Ivanova. These results
will assist in the consideration of subsequent material. The spectra of most
of the compounds were recorded with a UR-10 spectrometer, over the range
400-3700 cm^{-1}, by the KBr pressed disc technique. They are shown in Figs.
22-25 and 27-30, and in Appendix I, Nos. 1-53.

Monosaccharides

A study of the spectra of the simplest compounds, the monosaccharides,
is of particular interest, since these compounds only differ in the stereochemical location of their hydroxyl groups.

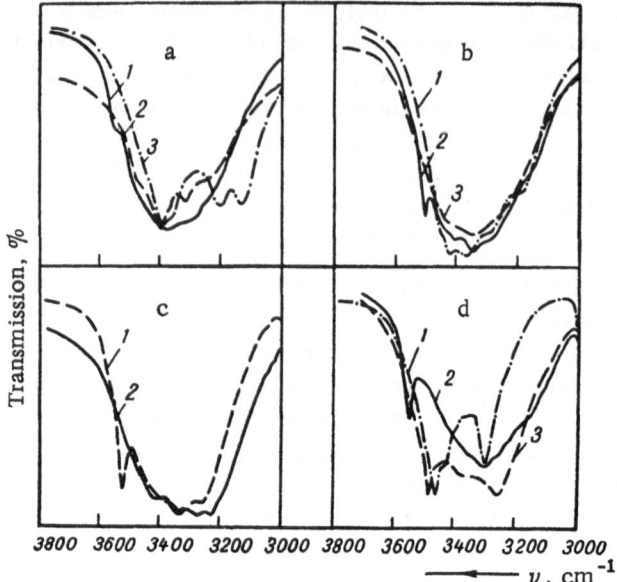

Fig. 22. Spectra of mono- and disaccharides. a. 1) α-D-talose; 2) α-D-glucose; 3) α-D-galactose. b. 1) Cellobiose; 2) trehalose; 3) maltose. c. 1) β-D-lyxose; 2) α-D-xylose. d. 1) α-Methyl-D-mannoside; 2) α-Methyl-D-glucoside; 3) rhamnose.

The spectra of all the sugars considered are characterized by the intense broad bands, in the 3250-3550 cm^{-1} region, of the stretching vibrations of OH groups involved in hydrogen bonds (Fig. 22). The spectrum of β-D-lyxose shows a sharp band at about 3520 cm^{-1}, which indicates the presence of an OH group with weaker hydrogen bonding. This must be largely attributed to the axially located OH group at atom C_2. In fact, there is no similar feature in the spectrum of α-D-xylose.

Figure 22 shows that the spectra of monosaccharides contain a number of distinct absorption bands within the 3200-3600 cm^{-1} range used for analysis: 3130, 3210, 3250, 3300, 3340, 3400, 3430, 3490, and 3540 cm^{-1}.

Sokolov [23] established a correlation between the energy of a hydrogen bond and the vibration frequency of the group participating in the hydrogen bond. It must, therefore, be accepted that in compounds of similar type there is an assembly of nonequivalent hydrogen bonds. Comparison of the spectra of these compounds shows that a change in the spatial location of even individual hydroxyl groups, or their replacement, can generally lead to a considerable change in the whole system of hydrogen bonds (Fig. 22). Thus,

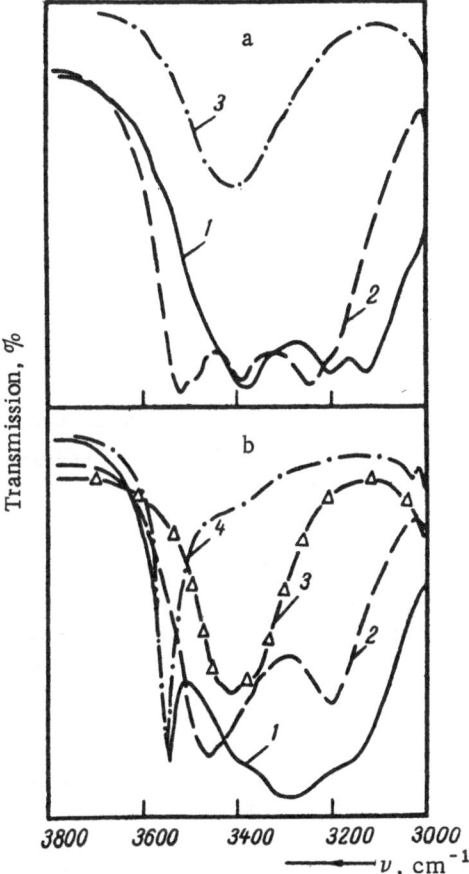

Fig. 23. Spectra of selectively substituted monosac-
carides in the hydroxyl hydrogen bond region. a. 1) α-
D-galactose; 2) α-methyl-D-galactoside; 3) 2,3,4,6-
tetra-O-methyl-D-galactose. b. 1) α-Methyl-D-
glucoside; 2) 2,3-di-O-methyl-β-D-glucose; 3) 2,3,4,6-
tetra-O-methyl-D-glucose; 4) 1,3-dimethyl-2,6-tosyl-
D-glucose.

the spectrum of α-D-glucose shows two bands at about 3410 and 3310 cm^{-1},
while that of α-D-galactose shows three separate bands, of approximately
equal intensity, at 3390, 3210, and 3130 cm^{-1}; the 3390 cm^{-1} band in the
latter is more diffuse than the 3410 cm^{-1} band in the former. The spectrum
of α-methyl-D-glucoside shows a sharp band at 3550 cm^{-1} and a broad diffuse band in the region 3100-3450 cm^{-1}, with its main maximum at 3300
and inflections at 3400 and 3150 cm^{-1}. A sharp change in the system of

hydrogen bonds is very marked in the transition from α-D-galactose to α-methyl-D-galactoside; new bands appear at 3520, 3490, and 3250 cm^{-1}, and the bands at 3210 and 3130 cm^{-1} disappear. It may be concluded from this that, in compounds of a similar type, the hydrogen bonds of the hydroxyl groups form a complete system, and that a change in one part of the system affects the nature of the remaining linkages.

Because of the strong interaction between hydroxyl groups in the structure of the compounds under consideration, the frequencies of OH groups concerned in hydrogen bonding are not only determined by the lengths of the oxygen bridges and the bond angles O--------O, but also by the mutual spatial locations of the hydroxyl groups. Without allowance for the configuration of OH groups, and if only one frequency is considered, it is possible to reach erroneous conclusions as to the energy of the hydrogen bonds of OH groups.

The highly labile nature of the hydrogen bonds in these compounds and their ability to regroup may be responsible for the specific structure of hydrogen bonding as a function of substituent location, and for changes in location of hydroxyl groups during the course of chemical reactions.

This view is, indeed, confirmed by analysis of the spectra of selectively methylated samples of monosaccharides (see Figs. 22 and 23). The spectrum of α-D-glucose shows three diffuse bands at about 3410, 3310, and 3260 cm^{-1}, but there is a marked change in the structure of the hydroxyl group bands when the hydroxyl group at C_1 is substituted to give α-methyl-D-glucoside: a sharp band appears at 3550 cm^{-1}, together with a broad band having its main maximum at about 3300 and inflections at 3400 and 3150 cm^{-1}. The spectrum of β-D-glucose with substituted hydroxyl groups at C_2 and C_3 shows two separate bands at 3460 and 3200 cm^{-1}, that of 2,4,3,6-tetra-O-methyl-D-glucose shows a relatively sharp band with its main maximum at about 3420 and an inflection at 3370 cm^{-1}, while that of 1,3-dimethyl-2,6-tosyl-D-glucose shows one sharp band at about 3550 cm^{-1}. There are marked changes in the hydrogen bonding of hydroxyl groups with the change from α-D-galactose to α-methyl-D-galactoside: the spectrum of α-D-galactose shows three distinct OH bands at 3390, 3210, and 3130 cm^{-1}, while that of α-methyl-D-galactoside shows bands at 3520, 3490, 3400, and 3250 cm^{-1}, etc.

Many sugars also show absorption bands in the region 2600-2800 cm^{-1} (Fig. 24). These bands are diffuse in character and cannot be attributed to overtones or compound frequencies. Absorption in this region is characteristic of compounds containing hydroxyl groups with very strong hydrogen

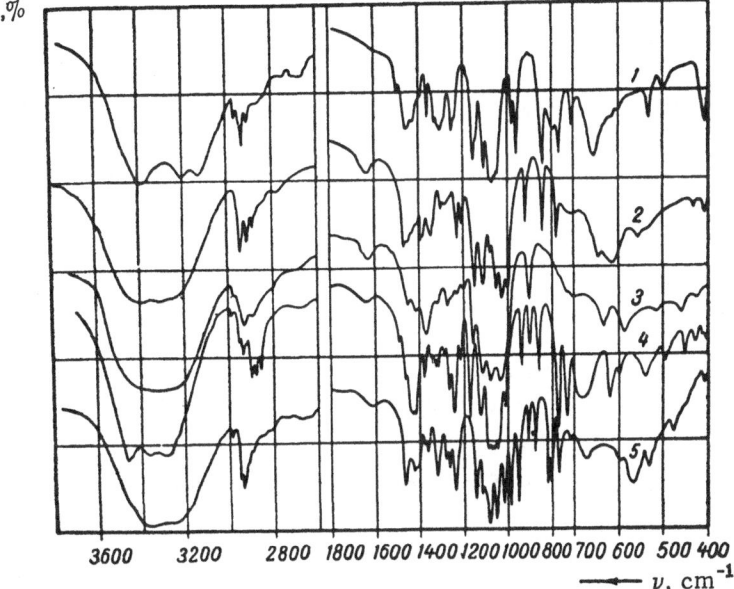

Fig. 24. Spectra of monosaccharides. 1) α-D-galactose; 2) α-D-
glucose; 3) β-D-glucose; 4) β-D-mannose; 5) α-D-talose.

bonding of the chelate type. Similar very strong hydrogen bonding has been
observed, for example, in β-diketones and in carboxylic acids [24], where
the following type of structure has been assigned:

$$R-C\diagdown\begin{smallmatrix}O...H-O\\O-H...O\end{smallmatrix}\diagup C-R \cdot \qquad \text{(a)}$$

Formation of a cyclic structure by means of hydrogen bonds is possible in the
case of sugars, as follows:

$$R-O\diagdown\begin{smallmatrix}H\\H\end{smallmatrix}\diagup O-R \cdot \qquad \text{(b)}$$

Formation of intramolecular hydrogen bonds of a similar type can occur when
adjacent hydroxyl groups are located equatorially:

As noted above, the possibility of establishing various types of hydrogen bonding in mono- and polysaccharides has been considered [22]. The authors suggested five variants of hydrogen bond formation and gave frequency intervals for these. However, they did not consider the possibility of formation of very strong hydrogen bonds in accordance with scheme b. Moreover, the frequencies given for definite types of hydrogen bonding were not conclusively established. The stretching vibrations of OH can vary considerably for a given type of hydrogen bonding, depending on the surroundings of the hydrogen bond, the geometry of the ring, and the location of side groups. The most clearly visible, very strong hydrogen bonds are revealed in the spectra of α-D-galactose (with maxima at 2670 and 2870 cm^{-1}), but three other types of hydrogen bonds are clearly characterized by maxima at 3130, 3210, and 3390 cm^{-1}: α-D-xylose, β-D-lyxose, and polyhydric alcohols.

The absorption bands of the stronger hydrogen bonds overlap significantly with the absorption bands of CH, CH$_2$, and CH$_3$ groups, located in the range 2800-3000 cm^{-1}, and this increases the absorption intensity in this region of the spectrum.

The presence of a large number of strongly polar hydroxyl groups and the interdependence of the hydrogen bond system largely determine the special features of the spectra of this class of compound.

Recent investigations, carried out in cooperation with N. V. Ivanova, have shown that it is easiest to investigate special features in the hydrogen bonding of OH groups in carbohydrates by obtaining the spectra at low temperatures, such as liquid nitrogen temperature. Under these conditions, the OH bands split up into sharper components and there are fairly marked shifts to longer wavelengths. The latter give useful additional information on the nature of the hydrogen bonding.

It is clear from the spectra shown in Fig. 24 (see also Appendix I, 1-9) that a change in the stereochemical positioning of even one of the hydroxyl groups leads to a marked change in the spectrum over the whole analytical range of 400-3600 cm^{-1}. For example, the spectrum of α-D-glucose differs from that of β-D-glucose in the appearance of sharp bands at 2910, 2890, 1460, 1340, 1220, 840, and 770 cm^{-1}, etc., and a redistribution of intensity in a number of other absorption bands. Particularly noticeable are the differences in the spectra of the α- and β-anomers of D-glucose in the regions 1200-1500 and 700-950 cm^{-1}. The spectra of all the monosaccharides considered are also quite specific, thus indicating that a change in the spatial location of individual hydroxyl groups leads to a change in the interaction between the main structural elements. This is in all probability due to the presence in these compounds of numerous strongly polar hydroxyl groups.

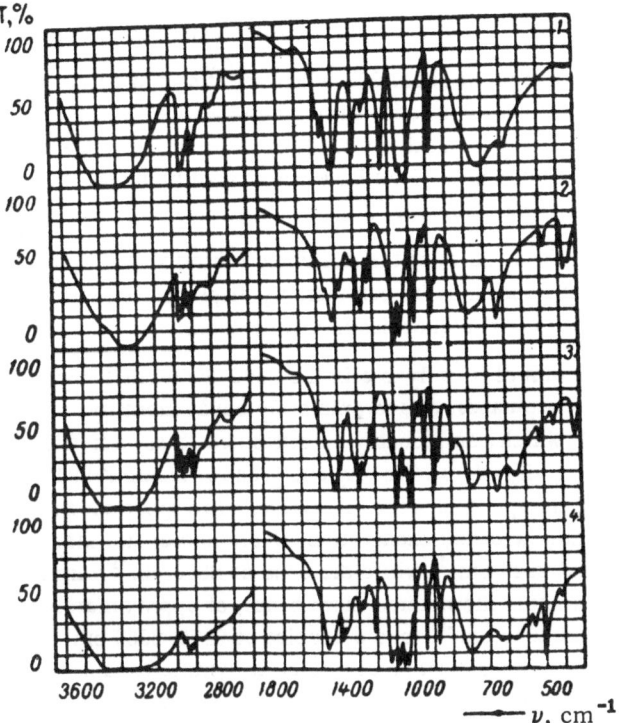

Fig. 25. Spectra of polyhydric alcohols. 1) Pentaerythritol;
2) i-erythritol; 3) mannitol; 4) dulcitol.

Highly characteristic structural bands, specific for each compound, are to be observed in the spectral region corresponding to the stretching vibrations of the CH groups. For example, the spectrum of α-D-glucose shows sharp bands at 2940, 2910, and 2890 cm^{-1}, and weaker bands at 2880 and 2850 cm^{-1}; the spectrum of α-D-galactose shows bands of different intensity at 2970, 2930, and 2910 cm^{-1}, while the spectrum of α-D-mannose shows a strong doublet at 2920-2930 cm^{-1} and much weaker bands at 2850, 2950, and 2970 cm^{-1}.

In order to determine the region corresponding to the stretching vibration frequencies of the CH_2 groups in the structure of polyhydric alcohols, spectra were obtained of the following: pentaerythritol $C(CH_2OH)_4$, i-erythritol $CH_2OH-CHOH-CHOH-CHOH-CH_2OH$, mannitol $CH_2OH-CHOH-CHOH-CHOH-CHOH-CH_2OH$, etc. (Fig. 25).

The spectrum of pentaerythritol is found to contain a strong doublet at 2940-2950, an inflection at 2925, and a strong band at 2890 cm^{-1}.

The band system at about 2950-2890 cm^{-1} should certainly be assigned to the asymmetric and symmetric stretching vibrations of the CH$_2$ groups. The spectrum of i-erythritol shows bands of approximately the same intensity at 2970, 2960, 2930, and 2910 cm^{-1}, while that of mannitol shows bands at 2980, 2970, 2910, and 2900 cm^{-1}.

Thus, the frequencies of the stretching vibration of CH$_2$ in carbohydrates and related compounds may be considerably displaced to shorter wavelengths, as compared with their frequencies in hydrocarbons.

The spectrum of β-D-lyxose in the 2800-3000 cm^{-1} region consists mainly of two sharp bands, of approximately equal intensity, at about 2930 and 2890 cm^{-1}, which can probably be attributed to asymmetric and symmetric stretching vibrations of methylene groups. Bands at 2940 and 2890 cm^{-1} appear in the spectrum of α-D-xylose. The spectrum in this region alters considerably with the change to β-D-arabinose; two sharp bands appear at 2950 and 2940 cm^{-1}, while the band at 2890 cm^{-1} vanishes. The spectrum of rhamnose shows three sharp bands, of approximately equal intensity, at 2890 and 2940 cm^{-1}, and a doublet at 2970-2980 cm^{-1}. Two, if not all three, of these bands must be attributed to symmetric and asymmetric stretching vibrations of methyl groups. The spectrum of D-glucose-6-phosphate (barium salt) shows only one weak band, at about 2920 cm^{-1}, in the 2800-3000 cm^{-1} region. These results all indicate that methylene and methyl groups, present in the structure of model compounds for cellulose, give strong bands in the region under consideration. The CH$_2$ groups are of great importance, even if they do not determine the total intensity and the character of the 2800-3000 cm^{-1} bands. It is known that the CH$_2$ group's bands are much stronger than those of CH bands [24]. In view of the strong interactions of these groups with adjacent structural elements, it is usually difficult to separate the stretching vibration frequencies of CH$_2$ and CH, which can vary within quite wide limits, depending on the material.

The possible existence of rotational isomers must also be considered, arising from turning or rotation of CH$_2$OH about the C_5-C_6 bond; this may be responsible for an increase in the number of absorption bands. Spectral analysis in the 2800-3000 cm^{-1} should allow for this form of isomerism.

There is no appreciable absorption by carbonyl in the 1600-1800 cm^{-1} region, which covers the absorption band of water of crystallization (about 1650 cm^{-1}). The possibility of existence of an aldehyde form is accordingly excluded for the case of monosaccharides in the crystalline state.

The $1300-1500$ cm^{-1} region contains the deformation vibration frequencies of CH_3, CH_2, and CH groups, and of the valence angles of $C-O-H$. It is well known [24] that, in hydrocarbon spectra, the internal deformation vibrations of CH_2 are characteristic and are located in the region 1460 ± 10 cm^{-1}. In the case of mono-, di-, and polysaccharide spectra, it is impossible to distinguish a similar narrow spectral range for the scissoring vibrations of methylene groups. In fact, the spectra of these compounds, in the region of the CH_2 deformation vibrations, show groups of bands which are specific for each compound and, in order of decreasing intensity, are 1415, 1430, 1440, 1445, 1465, 1490 cm^{-1} (β-D-mannose); 1460, 1450, 1430, 1410 cm^{-1} (α-D-glucose); 1425, 1485, 1450, 1405 cm^{-1} (cellobiose); 1435, 1460 cm^{-1} (maltose); 1460, 1480 cm^{-1} (α-D-xylose); 1425, 1465, 1400 cm^{-1} (β-D-lyxose), etc. This may be attributed particularly to interaction between CH_2 and accompanying groups, which can lead to substantial changes in the methylene group frequencies, depending on the conformation of the CH_2OH group [17]. Confirmation of this view is provided by the considerable differences, in the spectral region under consideration, between the spectra of α-D-glucose and α-D-galactose, whose structures differ only in the location of the hydroxyl group at C_4 (i.e., in the immediate vicinity of the CH_2OH group). The character of the spectrum in the range covering deformation vibrations of the CH_2 group is not only affected by changes in the surroundings of the C_6 atom, but also by changes at more remote carbon atoms (see, for example, the spectra of the α and β anomers of D-glucose in Appendix I, 1-2, and the spectra in [17]), thus establishing that there is interaction between the structural elements. In all probability the strongly polar hydroxyl groups play an important role here.

There are at present conflicting views concerning the angular deformation frequencies of $C-O-H$ in the spectra of sugars and related compounds. Some authors do not distinguish between these vibrations and the stretching vibrations of $C-O$, or else consider that the deformation vibration frequencies of $C-O-H$ appear in the region $1100-1150$ cm^{-1} [24, 134-136]. Higgins and his colleagues, and some other investigators [133, 137], on the basis of work with deuterated cellulose, attribute bands at 1450 and 1335 cm^{-1} to deformation vibrations of OH. Falk and Whally [138] showed that, in the case of methyl alcohol, OH deformation vibrations are responsible for the frequencies in the $1420-1506$ cm^{-1} region in the liquid and solid states, and at 1346 cm^{-1} in the gaseous state. Consideration of the spectra of sugars reveals that, in spite of there being only one CH_2 group in the molecule, there are always several bands or inflections in the range $1390-1480$ cm^{-1}. Absorption in this region can be explained by the deformation vibrations of $C-O-H$ together with the possible existence of rotation isomers.

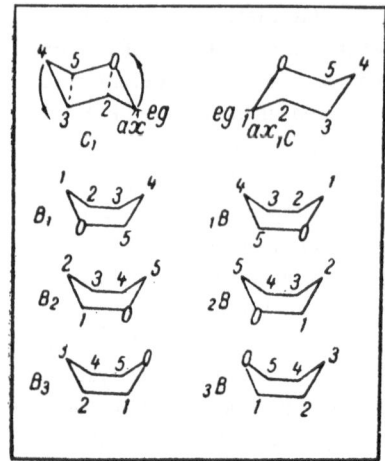

Fig. 26. The most stable configurations of pyranose rings.

There are several bands of medium intensity in the 1200-1400 cm^{-1} region, which are sharper in the case of monosaccharides containing CH$_2$OH groups. In all probability the vibration frequencies of CH$_2$OH accompany the frequencies of other groups in this range. For example, pentaerythritol has a strong band here. The character of the spectrum in the 1200-1400 cm^{-1} region is very sensitive to any change in the location of hydroxyl groups. This is shown very clearly by monosaccharides containing primary hydroxyl groups (compare the spectra of D-glucose, D-mannose, and D-galactose), and indicates that there is relatively strong interaction between the structural elements and the CH$_2$OH groups in these compounds.

All sugars are characterized by very strong bands in the region 1000-1200 cm^{-1}. A comparison of the spectra of cyclic monosaccharides with those of polyhydric alcohols, whose molecules have a linear structure, indicates that the presence of strong absorption bands in the range 1120-1170 cm^{-1} is usually characteristic of the cyclic monosaccharide structure.

In the spectra of all the polyhydric alcohols which we investigated (dulcitol, i-erythritol, mannitol, sorbitol), the frequencies of the strong bands in this region did not exceed 1120 cm^{-1} (Fig. 25). The range 1000-1200 cm^{-1} may contain frequencies corresponding to stretching vibrations of C−O, C−C, and ring structures, external deformation vibrations of CH$_2$ groups, etc. The strongest bands in this region are attributed to stretching vibrations of C−O bonds.

The spectra of monosaccharides show diffuse complex bands in the 1000-1100 cm^{-1} region and two well-defined sharp bands in the region 1120-1170 cm^{-1}. There is a band at 1150 cm^{-1}, which remains constant for the α and β anomers of D-glucose, and for α-D-galactose, and shifts to 1170 cm^{-1} with β-D-lyxose, and to 1130 cm^{-1} with α-D-mannose. Comparison of the structures of these sugars indicates that the 1150 cm^{-1} band may be largely attributed to stretching vibrations of the C$_2$−O bond. However, it must be appreciated that attribution of frequencies in this region to particular bonds or angles can only be an approximation, since in any vibration there is strong interaction with other elements of the structure.

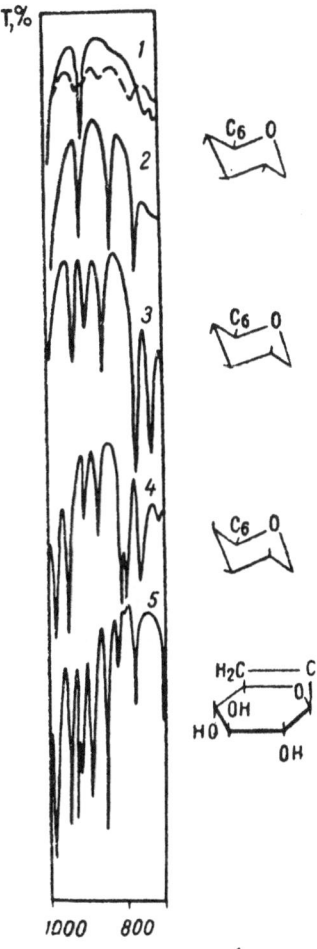

v, cm^{-1}

Fig. 27. Spectra of: 1) β-D-glucose; 2) α-D-glucose; 3) β-D-mannose; 4) α-D-talose; 5) levo-glucosan. The dotted line shows the spectrum of β-D-glucose after thermal treatment.

The spectra of monosaccharides and of polyhydric alcohols show several strong sharp bands in the ranges 700-1000 cm^{-1} for monosaccharides, and 850-1000 cm^{-1} for polyhydric alcohols. The strong bands at 700-800 cm^{-1} in the spectra of monosaccharides can probably be attributed to pulsation vibrations of the pyranose ring skeleton.

Strong bands in the range 800-950 cm^{-1} are observed in the spectra of monosaccharides and of polyhydric alcohols. This region may cover the external deformation vibration frequencies of the CH groups. A doublet band can be observed at about 900 cm^{-1} in the spectrum of ethylene glycol and has been attributed to rocking vibrations of CH$_2$ groups in the gauche and trans configurations [139]. It has been noted [140] that new bands in the range 850-900 cm^{-1} appear with the transition from tetrahydrofuran to tetrahydrofuryl alcohol. It should also be noted that the band at about 900 cm^{-1} in the spectrum of ethyl alcohol vanishes when the alcohol is deuterated to CH$_3$CD$_2$OH [141]. There are bands at 903 and 920 cm^{-1} in the spectrum of cyclobutane, and these have been attributed to deformation vibrations of CH$_2$ [24]. The spectrum of pentaerythritol shows a strong band at about 870 cm^{-1}, which must be attributed to the rocking vibrations of the methylene groups. Consequently, the spectra of sugars and related compounds may show strong bands between 800 and 950 cm^{-1}, attributable to external deformation vibrations of the methylene groups. It is possible that the relative increase in the number of bands in this region, observed in the spectra of some monosaccharides containing a CH$_2$OH group, may be due to rotational isomerism associated with this grouping.

Table 2. Compounds Containing Combined Hydrofuranol and Pyranose Rings

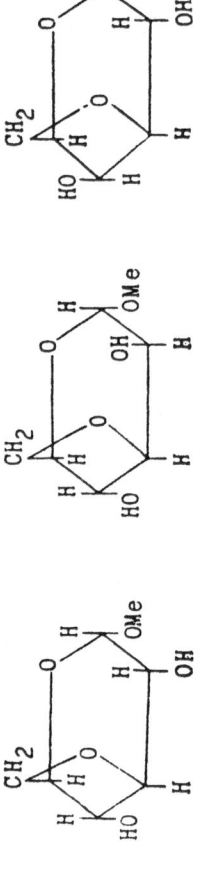

Methyl-3-6-anhydro-α-D-glucopyranoside

Methyl-3,6-anhydro-α-D-mannopyranoside

Methyl-3,6-anhydro-α-D-galactopyranoside

Compound and possible form of pyranose	Includes C–O stretch vibrations of methyl ethers	Types A, 1 and 2b	Type B (or 2c)	Other rings	Type C (or 2a)	Type 2a	Type D	Type 3
Methyl-α-D-glucopyranoside (CI)	992 v.s*	896 s	–	–	–	840 s	–	745 s
3,6-Anhydro derivative (IC)	987 v.s / 942 s	909 v.s / 892 v.s	868 s	–	842 s	–	815 v.s	742 v.s
2-Deoxy derivative (CI)	966 s / 914 s	896 s	–	870 s†	–	837 v.s	–	760 s
3,6-Anhydro-2-deoxy derivative (IC)	997 s / 978 s	916 v.s / 909 v.s	868 v.s	–	843 v.s	–	826 s / 818 s / 790 vs	737 s
Methyl-α-D-mannopyranoside (CI)	972 s	914 m	888 w	–	–	843 m	–	808 s

Compound								
3,6-Anhydro derivative (IC)	990 s 967 m 942 s	919 m 913 s	878 s	849 m	—	—	801	757 s
3,6-Anhydro-2-O-methyl derivative (IC)	(987,973) v.s 963 v.s	929 v.s 907 s	873 v.s	853 s	—	—	823 v.s 810 v.s	767 s 754 s
3,6-Anhydro-2,4-di-O-methyl derivative (IC)	(984,965) m	935 s 901 m	878 m	854 w	—	—	811 m	752 m
Methyl-β-D-mannopyranoside (CI)	947 m	928 m 896 m	875 s	—	830 m	—	—	795 v.s
3,6-Anhydro derivative (IC or IB)	(997,984) v.s (963,956) v.s	897 m	870 s	(814 v.s)	814 v.s	—	778 v.s	739 m
3,6-Anhydro-2,4-di-O-methyl derivative (IC or IB)	(997,955) v.s	901 m 892 m	876 m	859 m	826 s	—	775 s	753 m
Methyl-α-D-galactopyranoside (CI)	966 s	923 s	868 s	—	—	818 v.s	—	784 v.s
3,6-Anhydro derivative (IC)	987 m 966 v.s	924 v.s 903 v.s	865 w	835 m	847 s	—	768 s	734 v.s
3,6-Anhydro-2-deoxy derivative (IC)	951 v.s	914 v.s	860 v.s	834 v.s	850 v.s*	—	797 v.s	754 s
Methyl-α-D-galactopyranoside (CI)	981 s	940 s 887 s	868 s	—	821 v.w	—	—	782 v.s
3,6-Anhydro-2,4-di-O-methyl derivative (IB)	(970,944) v.s	907 v.s	861 s	(811 s)	811 s	—	767 s	733 s
Values and standard deviations for 3,6-anhydro derivatives	—	—	870 ± 7	838 ± 16	—	—	798 ± 21	747 ± 11

*v.s = very strong; s = strong; m = medium; w = weak.
†CH_2 swinging (deoxy group).

In the analysis of carbohydrate spectra, account must be taken of the possibility of conformational changes in structure. Because of the hetero-atom in a pyranose there are 8 more stable conformations (Fig. 26) and nu-merous intermediate (skew) forms. On the basis of the instability factors, in-troduced by Reeves [142], the CI conformation of β-D-glucose is the most stable, having no instability factors. In fact, the spectrum of β-D-glucose shows the fewest bands and only one sharp band in the frequency range 700-950 cm^{-1} (Fig. 27). At the same time, the spectra of β-D-mannose and α-D-talose, for which both CI and IC conformations are possible, show more bands in this frequency region. The region under consideration is the most interesting, because it corresponds to pyran ring frequencies. The individual bands in this part of the spectrum are sensitive to the spatial location of the C$-$H groups [125, 127], and the fact that the bands are relatively few and sharp means that they can be resolved sufficiently to facilitate comparisons.

It is known that raising the temperature can lead to the appearance of unstable conformations. Our experiments showed that new bands at about 920, 840, and 770 cm^{-1} (Fig. 27) appear in the spectrum of β-D-glucose when the temperature is raised. This conflicts with the conclusion of Barker and others [125, 127] that the 840 cm^{-1} band is specific for α-D-gluco-pyranosides, only. On the other hand, on the basis of studies of numerous anomers of D-glucopyranosides, it has been shown that the 840 cm^{-1} band appears in the spectrum when the C_1-H group is in the equatorial position. A transition of the C_1-H group in the structure of β-D-glucose to the equa-torial position is possible if there is a change in ring conformation.

It cannot be excluded that the spectrum in the 700-950 cm^{-1} region may also reflect unusual conformational changes in carbohydrate structures, and may therefore be of considerable interest for further investigation.

In the development of conformational analysis it is important to study the modification products of sugars, since the modifications can lead to con-formational changes in the pyran rings. An investigation has been published [140] comparing the spectra of various sugars and their derivatives contain-ing 3,6-anhydro rings, for which the pyran rings change from conformation CI to IC and IB. Table 2, taken from this work [140], shows the band fre-quencies of these compounds in the range 700-1000 cm^{-1}. Without going in-to the interpretation of these bands (some of which may be attributed to vi-brations of 3,6-anhydro rings, types A, B, C, and D [140]), it will be seen that there are very marked changes in the spectra, alterations in intensity and frequency shifts of various bands, accompanying conformational changes in the pyran rings. This is particularly evident for the bands corresponding to the pulsation vibrations of the pyranose rings (the so-called type 3 [125,127]).

All the compounds considered have specific spectra in the region 400-700 cm^{-1}. The bands in this range may be sharp or broad, and differ in intensity, depending on the nature of the compound. For most compounds the strongest absorption is between 600 and 700 cm^{-1} (α-D-glucose, α-D-galactose, i-erythritol, pentaerythritol, dulcitol). On changing from the spectrum of D-glucose to that of glucose penta-acetate, there is a general reduction in absorption and several new sharp bands appear. A reduced absorption in this region is characteristic of cellulose derivatives [25-27]. It should be noted that the spectra of α-D-glucose and α-D-mannose, as compared with those of α-methyl-D-glucoside and α-methyl-D-mannoside, are characterized by an increase in the number of sharp strong bands in the 400-700 cm^{-1} region. The structure of the hydroxyl band at 3200-3600 cm^{-1} also becomes more clearly defined. A study of the vibration spectrum of methanol [138] has shown that the nonplanar vibrations of OH give rise to absorption in the range 650-790 cm^{-1}. The supposition that hydroxyl groups absorb in this region is confirmed by the general appearance of the spectrum: A background or separate broad bands can normally be observed in this region. In this part of the spectrum may be found the frequencies of out-of-planar deformation vibrations of hydroxyl, as well as the overtones of related hydrogen bonds. As discussed above, the hydroxyl groups in these compounds are concerned mainly in hydrogen bonding, and this may be the reason for the diffuse character of the bands.

Disaccharides

With the transition from monosaccharides to disaccharides there is a considerable change in the whole spectrum. Many of the absorption bands shift and change in intensity, and there is a reduction in the number of sharp strong bands (Fig. 28). As already explained, the stretching vibration frequencies of CH and CH$_2$ are in the range 2800-3000 cm^{-1}. The chemical environment of most of these groups does not alter with the transition, for example, from α- and β-D-glucoses to cellobiose and maltose. However, the spectra alter considerably in the frequency range of the stretching vibrations of CH and CH$_2$ groups, thereby indicating that there is a change in the interaction of the CH and CH$_2$ groups with adjacent structural elements when two D-glucose rings combine, and the interaction is affected by the location of the rings. In this connection, attention should be directed to the greater similarity in this region of the spectra of maltose and trehalose, than of the spectra of maltose and cellobiose. With the transition from α- and β-D-glucoses to maltose and cellobiose there are marked changes in the 1200-1500 cm^{-1} spectral region: The maximum of the band between 1400 and 1500 cm^{-1} is displaced by about 30 cm^{-1} toward longer wavelength, there is an increase in the number of bands with a simultaneous reduction in their

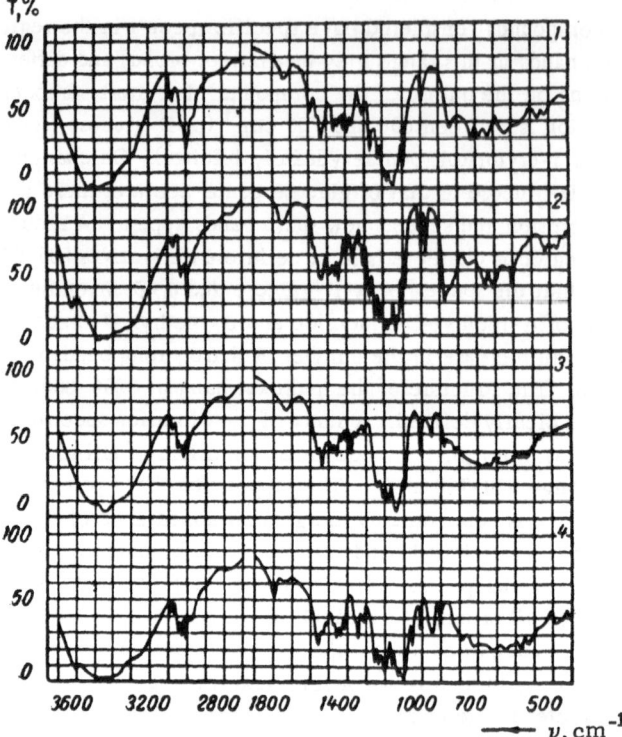

Fig. 28. Spectra of disaccharides. 1) Cellobiose; 2) lactose;
3) maltose; 4) trehalose.

relative intensity, etc. The considerable decrease in intensity of the sharp
bands at 1225 and 1205 cm^{-1} should be emphasized. It should also be noted
that the bands are more clearly separated in the 1400-1470 cm^{-1} and 2850-
2950 cm^{-1} regions (Fig. 28).

It has already been noted that the spectra of monosaccharides are
characterized by two strong bands between 1100 and 1200 cm^{-1}. With the
transition to disaccharides there is a reduction in band intensity in this region and, in the case of trehalose and lactose, there is an increase in the
number of bands. It was mentioned previously that the sharp bands between
1120 and 1170 cm^{-1} are missing from the spectra of straight-chain polyhydric alcohols. They can evidently be attributed to stretching vibrations of
C−O in rings. The reduced intensity of the bands in disaccharides can be
explained by a change in the environment of the C_1-O-C_5 bonds with the
transition from monosaccharides to disaccharides. The marked difference
between the spectra of cellobiose and maltose in this region indicates the effect of the mutual positioning of the rings on the stretching vibration frequencies of C−O.

It should be noted that the 1170 cm^{-1} band shows the greatest displacement, as compared with other bands in this region, with the transition from cellobiose to maltose. This supports the previous suggestion about the effect of the environment of the C_1-O-C_5 bond on the character of the spectrum in the region 1100-1170 cm^{-1}.

There are also specific differences between the spectra of di- and monosaccharides in the 400-1000 cm^{-1} range. Thus, the change from monosaccharides to disaccharides is accompanied by a considerable reduction in band intensity and changes in relative band intensities in the region 900-950 cm^{-1}. Broad weak bands appear in the interval 400-750 cm^{-1}; these can be explained as due to an increase in the number of vibrations, resulting from the more complex structure of disaccharides and the large number of isomers. It should be noted that the presence of several sharp bands in the range 700-950 cm^{-1} is accompanied by a considerable separation of bands in the region 2850-2950 cm^{-1}; however, when the bands between 700 and 900 cm^{-1} shift toward 900 cm^{-1}, the bands between 2850 and 2950 cm^{-1} amalgamate to give a band with its main maximum at 2900 cm^{-1}.

Polysaccharides

The spectra of polysaccharides (Fig. 29) are usually diffuse in character in all the regions suitable for analysis, so that the spectral differences are not so well defined (Appendix I, 36-42). This may be attributed to superposition of the spectra of the various isomeric forms of the molecules, resulting from the increased possibilities of conformational transitions, and the greater complexity of the various types of intra- and intermolecular interactions. In this connection, attention should be directed to the similarity between the spectra of amylose and glucose, obtained from aqueous solution (Fig. 30).

From a comparison of the spectra of cellobiose and cellulose, and of maltose and amylose, it is evident that the transition from disaccharides to polysaccharides is accompanied by the disappearance of most of the sharp bands and their amalgamation, and also by a redistribution of band intensity in the separate spectral regions. The greatest spectral differences, on changing from disaccharides to polysaccharides with the examples given above, are observable in the ranges 1200-1500 cm^{-1} and 400-900 cm^{-1}. It is, of course, accepted that spectra in the regions 1200-1500 cm^{-1} and 400-900 cm^{-1} are particularly sensitive to structural changes in the compounds under consideration.

Cellulose, of all the polysaccharides which have been analyzed, shows the least diffuse type of spectrum, and this indicates that it has the most

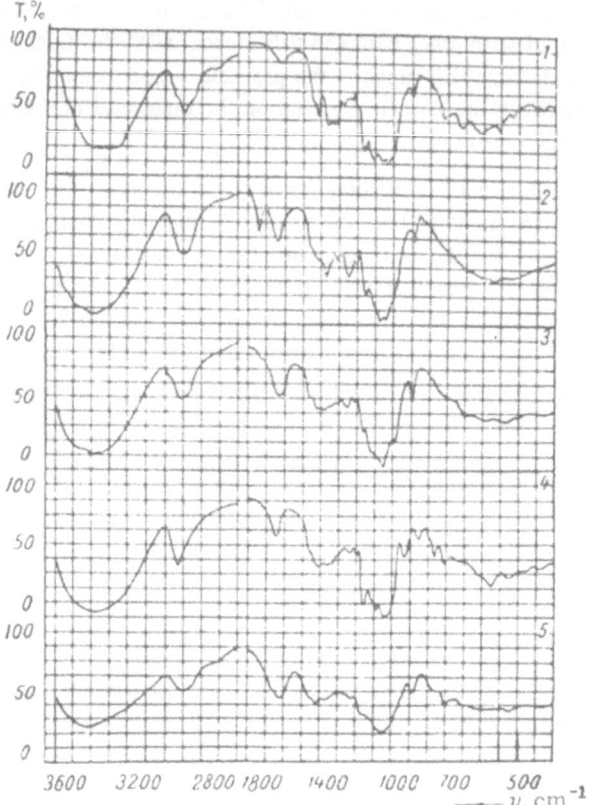

Fig. 29. Spectra of: 1) cellulose; 2) laminarin; 3) xylan;
4) amylose; 5) galactan.

ordered structure. A comparison shows that the spectrum of β -D-glucose is
more similar to that of hydrocellulose, and the spectrum of cellobiose is
closer to that of cellulose (Fig. 30a). This suggests that the elementary cell
of the hydrocellulose macromolecule differs from that of cellulose in that it
does not consist of cellobiose, but of its own characteristic cells.

Polysaccharide spectra in the 400-700 cm^{-1} range show a further
averaging out of absorption and are much less well defined. Apart from
cellulose, the spectra of the polysaccharides under consideration show only a
few inflections in this range, against a background of broad diffuse absorption.
It was accepted earlier that the OH nonplanar deformation vibration frequencies and the hydrogen bond overtones may lie in the 400-700 cm^{-1} region.
From this point of view the poorly defined structure of the spectra in this
range may be attributed to the complexity of the hydrogen bonding.

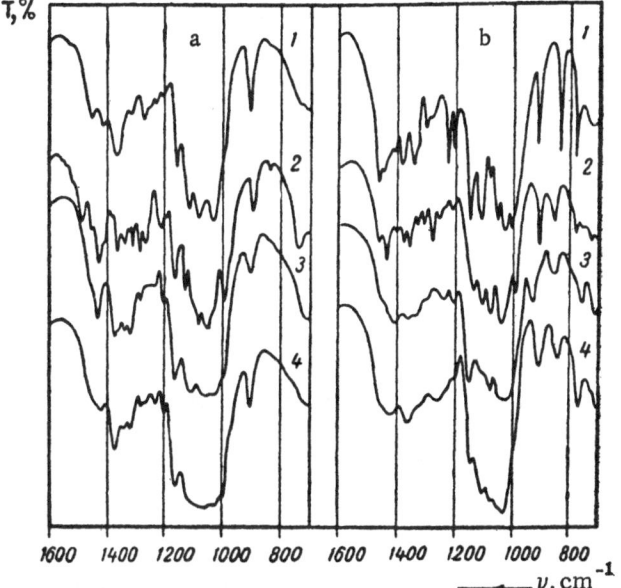

Fig. 30. Spectra of glucosides and their polymers. a. 1) β-
D-glucose; 2) cellobiose; 3) cellulose; 4) hydrocellulose. b.
1) α-D-glucose; 2) maltose; 3) amylose; 4) D-glucose obtained from aqueous solution.

Depending on the conditions, a redistribution of isomeric forms may
occur with the predominance of this or that concrete type of structure (cellulose or hydrocellulose). Great care must be taken in interpreting the disappearance or appearance of new bands in the spectra of carbohydrates after
various chemical reactions, since these may also be explained by structural
factors.

A fundamental problem, having both practical and scientific importance, is the development of methods for analyzing conformational transitions in the pyran rings of polysaccharides. There can be no doubt that this
factor has a substantial effect on the properties of monosaccharides and their
polymers.

The possibilities of infrared spectroscopy as a method for investigating
conformational transitions of the pyran rings in polysaccharides are not so
clear as with the simpler monosaccharide molecules. However, a comparison of the spectra of various polycarbohydrates with those of their monomers,
in the region 700-1000 cm^{-1}, which is the most suitable for carrying out
similar analyses, shows that the spectra of the polymers adequately reflect

the special features of the monomer rings (Fig. 30). It is natural to assume that a change in the conformation of the pyran rings should be reflected in the spectra of the polysaccharides. However, it must be appreciated that, in carrying out such an investigation, we first fix the conformational changes of the pyran rings of the predominant isomeric forms of the macromolecules, mainly responsible for the vibration spectrum of the polymer. This fact reduces the sensitivity of analysis.

The development of an effective spectroscopic method for investigating conformational transitions in the pyran rings of polysaccharides is retarded by the absence of enough reliable model compounds.

It is interesting to investigate cellulose derivatives containing a 3,6-anhydro ring. By analogy with pyranosides, it must be assumed that the presence of a 3,6-anhydro ring leads to a CI $\rightarrow$ IC conformational transition in the pyranose rings.

It has already been mentioned that Barker and his colleagues [125,127] showed, on the basis of the study of a large number of infrared spectra of carbohydrates and their polymers, that, for example, the presence of a band in the range 844 ± 8 cm^{-1} is a reliable criterion for the existence of a C_1H group in the equatorial position with respect to the plane of the pyranose ring. Thus, the appearance of a band in this region is a necessary condition for CI $\rightarrow$ IC conformational transitions in the case of β-D-glucose polymers. A band at about 840 cm^{-1} does appear in the spectra of cellulose derivatives containing a 3,6-anhydro ring. However, this necessary condition cannot be taken as an adequate proof of CI $\rightarrow$ IC transitions in the pyranose rings of cellulose, because the 3,6-anhydro rings themselves may absorb in this region. On the other hand, it should be recalled that we observed a new band at about 840 cm^{-1} in the spectrum of β-D-glucose, when the temperature was raised, and, on this basis, assumed that there were conformational transitions of the β-D-glucose pyranose rings when the temperature was increased. When certain reactions were carried out on the α-D-glucose polymer, dextran, we observed the disappearance of the 850 cm^{-1} band and the appearance of a new band at 870 cm^{-1}. This can be attributed to a conformational transition of the pyranose rings: the disappearance of the equatorial C_1H group and the appearance of equatorial C_2H and C_4H groups.

In conclusion, it should be noted that use of the infrared spectroscopic method for analyzing carbohydrate conformations is at present mainly based on Barker's work on bands of types 2a, 2b, and 2c [125, 127, 140]. However, attribution of bands to the CH group at individual carbon atoms is open to serious objections. It has already been pointed out that there are strong interactions between the individual structural elements in carbohydrates. It is difficult to distinguish between the presence of a considerable interaction between CH groups and the effect on this interaction of the stereochemical location of the CH groups.

We wrote out the structures of hydrocarbons (AAAAA for β-D-glucose, AEAAA for β-D-mannose, EAAEA for α-D-galactose, etc., where E and A denote equatorial and axial CH groups) whose spectra showed absorption bands in the ranges 820-855, 860-885, and 885-920 cm^{-1}. Comparison showed that compounds, showing absorption bands in these ranges, contained, respectively, the following single structural elements: EA, AE, AA.

Characterization of the frequencies of spatial combinations of equatorial and axial CH groups in hydrocarbons requires more detailed investigation, and is beyond the scope of this book.

We will now consider the spectra of various unmodified celluloses, in the range 400-3600 cm^{-1}, as obtained by the direct pressing method with subsequent impregnation of the fiber film by a liquid immersion medium. The use of this improved technique, as compared with pressing the ground-up fibers in a solid immersion medium, made it possible to show the special spectroscopic features of these materials more reliably (Figs. 31-37, and Appendix II, 54-83).

Cellulose from One-Year-Old Plants

Very little satisfactory work has been done on the infrared spectra of cellulose from one-year-old plants, with the exception of work on cotton cellulose. Consideration has been given [28] to changes in the spectra of celluloses from cotton, flax, corn, and straw, after alkaline treatment to isolate α-cellulose. The spectra of celluloses obtained from wood waste and from flax processing waste have been investigated [29]. The authors also noted, in particular, the similarities between the spectra of cotton and flax celluloses. Tsuboi used polarized infrared radiation to study fibers of linen, ramie, and cotton [143]. Liang and Marchessault [137] investigated the polarized spectra of doubly oriented films of Valonia ventricosa cellulose, ramie, and bacterial cellulose. O'Connor, Du Pre, and Mitcham [144] used the KBr disc technique to make a qualitative comparison of the spectra of native fibers of cotton, flax, ramie, hemp, and jute.

Most of the work listed above [28, 29, 143, 144] was carried out with instruments utilizing an NaCl prism dispersion system. This did not give sufficient resolution in the region of OH and CH group stretching vibrations, and it was not possible to investigate frequencies below 600 cm^{-1}. The techniques used [137, 143, 144] involved the grinding up of fibers (which may affect the structure) or specially prepared cellulose with the object of

Table 3. Chemical Composition Characteristics of the Cellulose from One-Year-Old Plants

Cellulose	Cellulose, %	Resins and waxes, %	Lignin by König method, %	Pentosans by Tollens method, %	COOH groups located at 6th carbon atom, %	CHO groups, %
Sulfate flax, unbleached	96.70	0.87	1.05	1.14	—	—
Sulfate flax, bleached	97.70	0.56	0.27	1.11	0.266	.043
Long-fiber bisulfite hemp (bast), unbleached	79.33	1.17	4.59	13.95	1.230	0.822
Short-fiber bisulfite hemp (tow), unbleached	71.26	2.04	5.39	—	1.175	1.533
Hemp tow waste	61.04	0.249	7.50	24.50	1.547	0.632
Sulfate corn, unbleached	86.50	2.80	—	26.76	1.623	0.290
Sulfate oats straw, unbleached	82.30	0.94	3.78	26.38	0.641	—
Ramie, bleached	—	trace	1.46	5.65	—	0.112

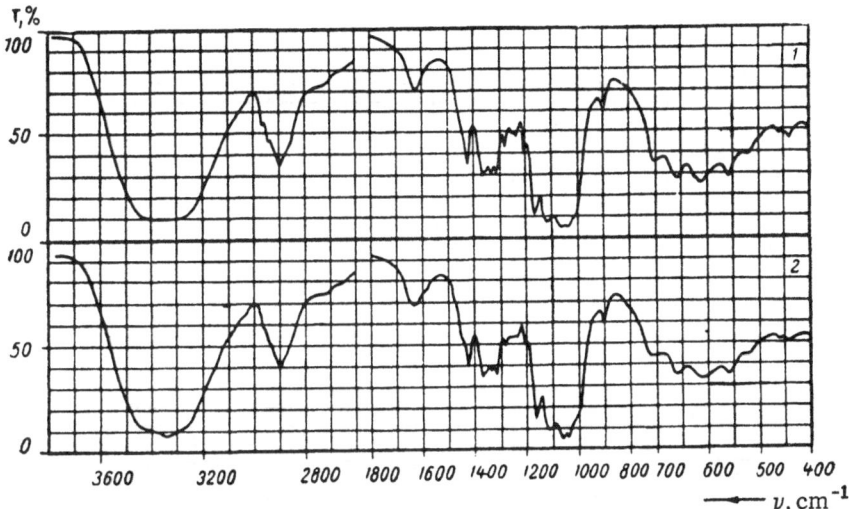

Fig. 31. Cellulose spectra. 1) Cotton fluff; 2) ramie.

obtaining oriented films. It was therefore of interest to study the spectra of undispersed fibers, using the direct pressing technique [6, 17, 18], which gives an optically more uniform medium so that special features of the spectra can be more clearly observed. The first studies of the spectra of fibers from various one-year-old plants were carried out [30] in the ranges 400-700 cm^{-1} (KBr prism), 700-2000 cm^{-1} (NaCl prism), and 2000-4000 cm^{-1} (LiF prism).

The celluloses investigated were obtained under laboratory conditions, in various ways. Flax cellulose was produced from tow fibrous waste from a flax processing plant by a method previously described [31], cellulose from flax tow by the "semi-kiering" method, long and short fiber cellulose from hemp stalks by the bisulfite method, cellulose isolated from corn stalks and oats straw by the sulfate method. Only the flax and ramie celluloses were bleached. All the celluloses investigated were characterized by chemical analysis for their contents of α-cellulose, resins, and waxes, pentosans, lignin, and CHO and uronic COOH groups (Table 3).

The spectrum of cotton fluff cellulose, which can be regarded as standard cellulose sample (Fig. 31) shows a broad diffuse band between 3000 and 3700 cm^{-1}, and bands or inflections at 2960, 2940, 2900, 2870, 1650, 1430, 1370, 1360, 1340, 1320, 1280, 1250, 1240, 1200, 1160, 1110, 1060, 1040, 1000, 970, 900, 720, 665, 615, 590, 560, 520, 490, 455, 435, and 400 cm^{-1}.

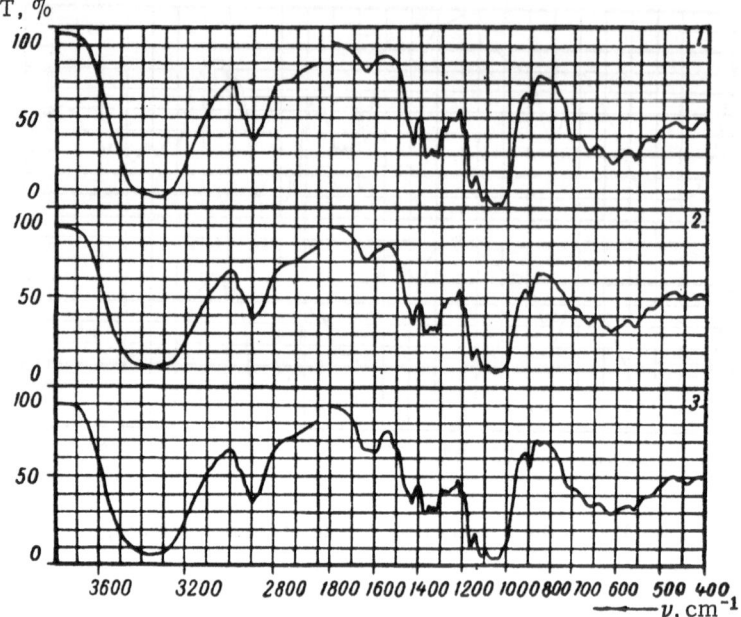

Fig. 32. Cellulose spectra. 1) Bleached flax; 2) unbleached flax; 3) hemp tow.

The attribution of the well separated absorption bands is clear and there are no difficulties in interpretation, but the attribution of the other bands requires more detailed consideration and comparison with model compounds.

A detailed interpretation of the cellulose spectrum will now be given below, based on a comparison of all the experimental material. Only a preliminary interpretation has been given previously.

The strong diffuse band between 3200 and 3600 cm^{-1} and the weaker band between 2800 and 3000 cm^{-1} can be attributed respectively to the stretching vibrations of hydroxyl groups, including hydrogen bonds, and of CH_2 and CH groups.

The relatively well defined band at 1430 cm^{-1} is associated with the internal deformation vibrations of CH_2 in CH_2OH groups. It has been shown previously that, in the spectra of sugars, the internal deformation vibration frequencies of these methylene groups are very sensitive to structural factors, so that the 1430 cm^{-1} band in the cellulose spectrum must be attributed to the CH_2 group in the force field of the surrounding structural elements, and not to the CH_2 group alone.

The deformation vibration frequencies of C—OH and CH groups are found in the 1300-1400 cm^{-1} region of the cellulose spectrum.

The strong absorption bands in the 1000-1200 cm^{-1} region can be attributed mainly to the stretching vibrations of C—O. Because of the strong interaction of the structural elements of cellulose, whose absorption bands fall in this spectral range, it is practically impossible to attribute frequencies in this range to definite groups or bonds.

The bands between 700 and 900 cm^{-1} can be attributed to the rocking vibrations of methylene groups, and to deformation vibrations of CH bonds and pyranose rings.

The diffuse absorption, with an ill-defined structure, in the 400-700 cm^{-1} region can be attributed to the deformation vibrations of hydroxyl groups and to hydrogen bond overtones.

The best-defined bands in this region can be seen in the spectra of cotton and bleached flax celluloses, and this is undoubtedly due to the higher degree of structural order in these types of celluloses.

The spectrum of bleached flax cellulose practically coincides with that of cotton cellulose, except, perhaps, for the band at 435 cm^{-1}, which is more clearly defined in the case of cotton. Differences in the intensity of this band might possibly be useful, in some cases, for identifying the type of cellulose.

The ramie spectrum (Fig. 31) shows some less well-defined absorption bands. The bands at 1360 and 1340 cm^{-1} are also less intense in the ramie spectrum than those at 1370 and 1320 cm^{-1} (the 1360 cm^{-1} band appears only as a hardly noticeable inflection on the 1370 cm^{-1} band), and there are some other relative intensity differences in the 600 to 700 cm^{-1} region.

The spectrum of unbleached flax cellulose, as compared with that of the bleached material, shows a new inflection at about 1600 cm^{-1}, and the whole spectrum is more diffuse in character (Fig. 32). The chemical analyses show that unbleached flax cellulose contains a considerable amount of lignin, pentosans, and other impurities. In the case of unbleached hemp cellulose, the band at 1600 cm^{-1} is stronger and the whole spectrum is still more diffuse in character (Fig. 33).

The spectra of short-fiber and long-fiber hemp celluloses differ from each other mainly in the absorption band between 1200 and 1500 cm^{-1}.

Celluloses from corn stalks and straw give the most diffuse spectra in all the regions discussed (Appendix II, 60, 61). Cellulose from rye straw shows two quite definite bands at 1600 and 1500 cm^{-1}, while cellulose from

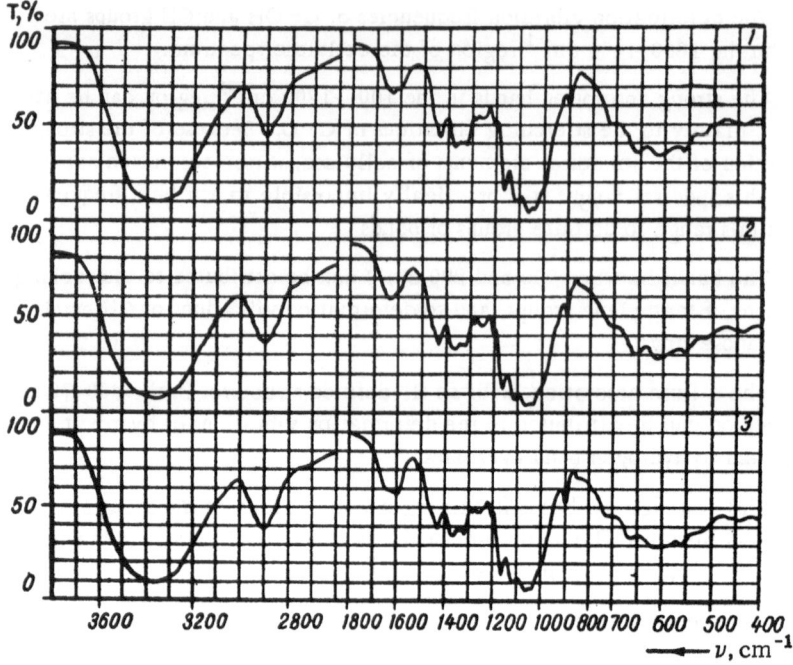

Fig. 33. Unbleached cellulose spectra. 1) Long-fiber hemp; 2) short-fiber hemp; 3) hemp tow.

corn stalks shows only one strong band at about 1600 cm^{-1}. The spectra of these celluloses are also characterized by some reduction in intensity of the bands at 1430, 1340, and 1320 cm^{-1}, as compared with that at 1370 cm^{-1}, and by an increased absorption in the 900 cm^{-1} region. It has been shown [7, 9, 28, 32, 145] that a reduction in intensity of the bands at about 1430, 1340, and 1320 cm^{-1}, together with an increased absorption in the 900 cm^{-1} region, accompanies the transition from a natural modification of cellulose to a modification of hydrocellulose and can be attributed to specific rotational isomerism of the CH$_2$OH groups. It is highly probable that the specific features of the celluloses discussed above may be explained by similar structural peculiarities.

It is interesting to consider the peculiarities of the spectra of celluloses obtained from flax and hemp tows (Figs. 32 and 33, curves 3). The spectra of these products differ from those of bast celluloses in that they show stronger absorption in the 1600 and 900 cm^{-1} regions, new weak bands at about 1500 and 850 cm^{-1}, and are generally more diffuse.

A comparison of the spectra of unbleached celluloses from flax, hemp, straw, and corn shows that, in this series, the spectra gradually become more diffuse, there is an increase in absorption in the 1600 cm^{-1} region, and new bands appear at about 1500 and 850 cm^{-1}. The more diffuse nature of the spectrum can be explained by a decrease in the degree of structural order of the cellulose, and/or by the presence of larger amounts of impurities.

The spectral bands of unmodified celluloses in the regions 1600, 1500, and 850 cm^{-1} may be attributed to accompanying impurities. The bands at about 1600 and 1500 cm^{-1} are probably associated with impurities having an aromatic structure (lignin). Indeed, the spectrum of lignin shows very strong bands at about 1600 and 1500 cm^{-1}. These bands have been observed in the spectra of thin wood sections [146] and have been attributed to absorption by lignin. The fact that an increase in absorption at 1600 cm^{-1} is not always accompanied by an absorption band at about 1500 cm^{-1} (cellulose from corn stalks) can evidently be explained by the specific nature of the aromatic impurities.

Attention should be directed to comparison of the results for short- and long-fiber hemp celluloses, since these samples were separated mechanically after digestion of the hemp stalks, i.e., the conditions for isolating the short- and long-fiber celluloses were identical. This shows that the differences observed between the spectra of these celluloses are due only to chemical and morphological differences in composition between the various parts of the hemp stalks.

It should be noted that the spectra of celluloses from one-year-old plants with high impurity contents (celluloses from rye straw and corn stalks) show reduced intensity of the bands at 1320 and 1340 cm^{-1}, as compared with the 1370 cm^{-1} band, and a simultaneous increase in absorption in the 900 cm^{-1} region. As stated above, this type of spectral change is characteristic of hydrocellulose. The composition of cellulose in plant tissues, and particularly its relation to accompanying materials and the nature of these materials, is probably one of the causes of the structural properties of cellulose.

Cellulose from Young Wood Shoots

The study of the infrared spectra of celluloses from young wood shoots is of considerable interest for elucidating the changes in cellulose structure during the process of formation and growth of wood.

The author, together with N. I. Garbuz, A. M. Shishko, A. I. Skrigan, and A. A. Bugaenkii, investigated the spectra of celluloses obtained under laboratory conditions from young shoots of pine, willow, poplar, alder, and

Table 4. Chemical Characteristics of the Celluloses Investigated

Cellulose	Ash, %	α-Cellulose, %	Resins and waxes,%	Lignin by König method,%	Pentosans by Tollens method, %
Mature pine wood June pine shoots	1.40	89.4	0.31	0.80	7.00
June shoots of Salix acutifolia (picked June 26)	4.15	93.30	1.40	15.19	17.47
June shoots of Salix acutifolia (picked July 26)	3.65	87.10	trace		19.96
June shoots of Salix acutifolia (picked September 26)	—	81.23	0.51	7.68	19.85
Year-old willow shoots (Salix acutifolia)	3.04	75.10	1.51	1.88	15.50
Year-old birch branches	1.19	80.14	1.41	6.9	
Year-old alder branches	1.50	82.34	1.02	6.8	
Mature Canadian poplar	1.04	82.34		2.01	18.83
Wood from July shoots of Canadian poplar	1.69	79.53		2.83	19.91
Bark from July shoots of Canadian poplar	0.99	72.11		4.97	9.75

birch. The willow, poplar, and pine celluloses were obtained from woods at different stages of vegetative development (Table 4).

Only a few papers have been published dealing with the spectra of celluloses obtained from pine wood samples at various stages of development. Comparisons have been made [33-35] of the infrared spectra of the products from alkaline treatment, deuteration, acetylation, and nitration of these celluloses. However, the authors did not detect any significant differences in the spectra of these celluloses, depending on age, over the frequency range 1400-3600 cm^{-1}.

A study of the spectra of celluloses from young wood, over a wider spectral range, using the direct pressing technique, enabled us to make a more thorough investigation of the spectroscopic special features of these products.

Figure 34 shows the spectra of celluloses obtained from May and June pine shoots, compared with that of cellulose from mature wood. It is

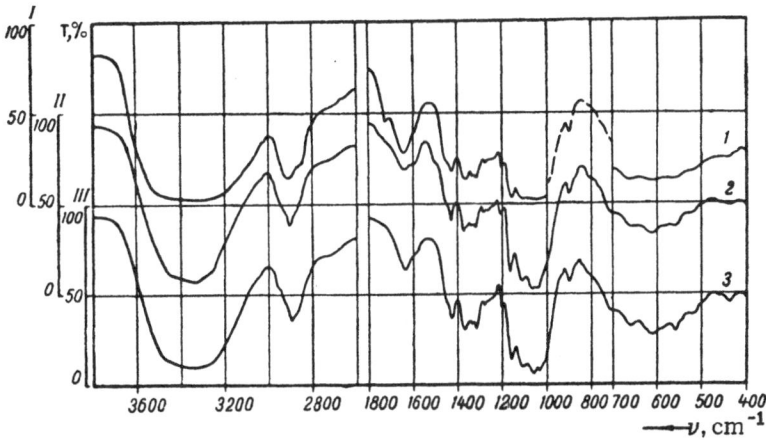

Fig. 34. Spectra of celluloses from pine wood. 1) May shoots; 2) June shoots; 3) mature wood.

particularly interesting to analyze the spectra of celluloses obtained from wood in the initial stages of its development. It should be noted that the spectrum of the product from May wood shoots shows all the characteristic features of the cellulose spectrum. This indicates that cellulose is present mainly in its finished state even in the early stages of development of the growing raw material. However, the spectrum of May cellulose has its peculiarities. For example, there is a well-defined band in the $C = O$ double bond region (1740 cm^{-1}), indicating a considerable carbonyl group content in the cellulose. Possible explanations are peculiarities of cellulose in the process of formation, or specific oxidation of this cellulose during the bleaching process. In any case, this characteristic distinguishes the cellulose from May wood shoots, as compared with cellulose from later shoots. Special properties of cellulose from May shoots were noted in a chemical investigation[36]. Attention should be given to the further application of spectroscopic methods in studies on the special structural features of celluloses in the various stages of its formation. We regret that, owing to experimental difficulties, it was not possible to obtain a film of May cellulose fiber of comparable thickness to compare with other celluloses, and this made it difficult to establish any other special features of the spectrum of this cellulose.

The spectrum of cellulose isolated from June pine shoots differed from that of normal mature cellulose in that it was rather more diffuse in the ranges 1200-1500 and 400-700 cm^{-1}, and exhibited more distinct aromatic impurity bands (1600 and 1500 cm^{-1}).

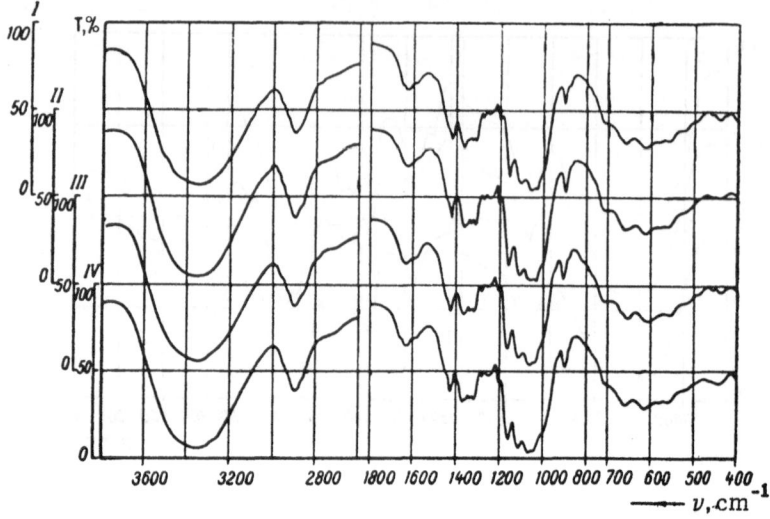

Fig. 35. Spectra of willow celluloses (<u>Salix</u> <u>acutifolia</u>). I) June; II) July;
III) September; IV) year-old shoots.

It was interesting to carry out spectroscopic investigations on celluloses
obtained from wood shoots at various periods of vegetative growth. For this
purpose, the spectra of willow celluloses (varieties <u>Salix</u> <u>vinimalis</u> and <u>Salix</u>
<u>acutifolia</u>) were studied, using samples taken from June, July, September, and
year-old shoots (Fig. 35). No significant differences were noted in the spectra of celluloses taken from August and later shoots, but the spectrum of the
youngest (June) sample had some distinguishing features; it showed a reduction in intensity of the bands at 1340 and 1320 cm^{-1}, as compared with the
band at 1370 cm^{-1}, and the band at 870 cm^{-1} was more pronounced. It has
been pointed out before [7, 9, 28, 32, 145] that a reduction in intensity of
the 1340 and 1320 cm^{-1} bands is characteristic of the hydrocellulose structural modification. Possibly, during the vegetative period, the formation of
cellulose is accompanied by a change in the ratio between various structural
modifications. The practical identity of the spectra of celluloses from July,
September, and later willow shoots, and their good agreement with the spectra of celluloses from mature wood of various species, indicate that the
period of willow cellulose formation is mainly completed within 3-5 months'
development of the wood. As might be expected, the spectra of unbleached
willow samples showed significant bands in the 1600 cm^{-1} region, attributable to the presence of the aromatic substance, lignin. There were no particular differences in the intensity of this band in the spectra of the various
willow celluloses investigated.

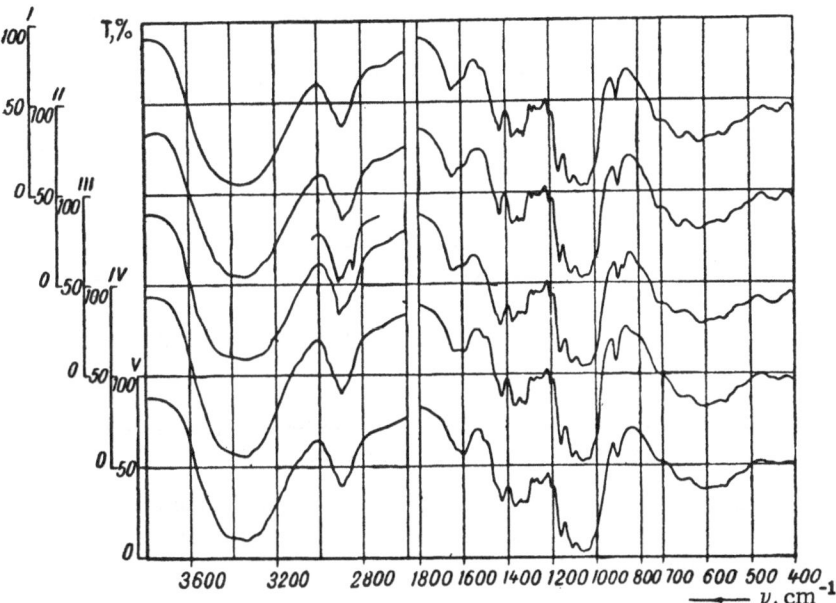

Fig. 36. Cellulose spectra. I) June shoots of Canadian poplar; II) mature wood from Canadian poplar; III) bark from June shoots of Canadian poplar; IV) year-old alder shoots; V) year-old birch shoots.

It is important to note that the spectroscopic method did not show any marked differences between the lignin contents of the various willow-shoot celluloses, whereas chemical analyses (by the König method) showed lignin contents ranging from 1.88 to 15.19% (Table 4). Such large differences in lignin content undoubtedly should have been reflected in the infrared spectra.

From this it follows that the so-called lignin determined by the König method cannot be identical in structure with lignin isolated from normal mature wood. The lignin found chemically must evidently include other compounds, particularly resinification products from hemicelluloses, since the products under consideration were especially rich in these. Hence, great care should be exercised in the determination of lignin in young celluloses by chemical methods.

An investigation was made of celluloses from poplar, isolated from mature wood bark, and wood from June shoots (Fig. 36). The spectrum of cellulose from the bark of the June shoots had some interesting features: an increase in intensity of the 1430 cm^{-1} band and reduced absorption at 1340 and 1320 cm^{-1}, as compared with the 1370 cm^{-1} band, and the appearance of a band at 870 cm^{-1}. It is unsatisfactory to suggest that these spectral changes are attributable only to low-molecular cogeners of cellulose. Indeed, celluloses with a high impurity content do not, as a rule, show these features in their spectra. Anticipating a little, we may state that similar spectroscopic features are characteristic of certain samples of bleached celluloses from fossil wood with low impurity contents, but are not observed in the spectra of samples with a high impurity content. It is possible that these spectral features of the bark from young poplar shoots may be due to structural peculiarities of this product. On the other hand, the spectrum of cellulose from the bark of June shoots from a Canadian poplar, which had not first been extracted with ether (Fig. 36, curve III, upper spectrum in the region 2800-3000 cm^{-1}), showed a band at 2830 cm^{-1} attributable to the presence of resinous impurities.

There were no significant differences between the spectra of poplar celluloses from June shoots and from mature wood.

Celluloses from alder and birch branches showed the most pronounced aromatic impurity (lignin) bands in the 1600-1500 cm^{-1} region, and this agreed well with results of chemical analysis (Table 4).

The spectrum of cellulose from birch branches was very diffuse in its general appearance, as compared with the spectra of analyzed cellulose samples from poplar, willow, alder, and pine branches. It has already been pointed out that this can be explained by the lower degree of structural order and/or the presence of significant amounts of impurities.

Celluloses from Woods of Various Ages

The study of celluloses, isolated from fossil woods of various ages, is of great scientific and practical importance, since it enables us to follow chemical transformations occurring in cellulose in the course of time, without complications by disturbing factors.

The only existent literature references [33-35] to the investigation of such celluloses by infrared spectroscopy have already been given. These describe studies on the cellulose and α-cellulose isolated from fossil wood 1000 years old. The spectra of these products were recorded over the comparatively narrow range of 1450 to 3500 cm^{-1}. The combination of chemical and spectral investigations made it possible to establish that pine wood stumps from marshland, by-products from peat extraction, should be considered as valuable cellulose-containing raw materials for the production of cellulose and α-cellulose.

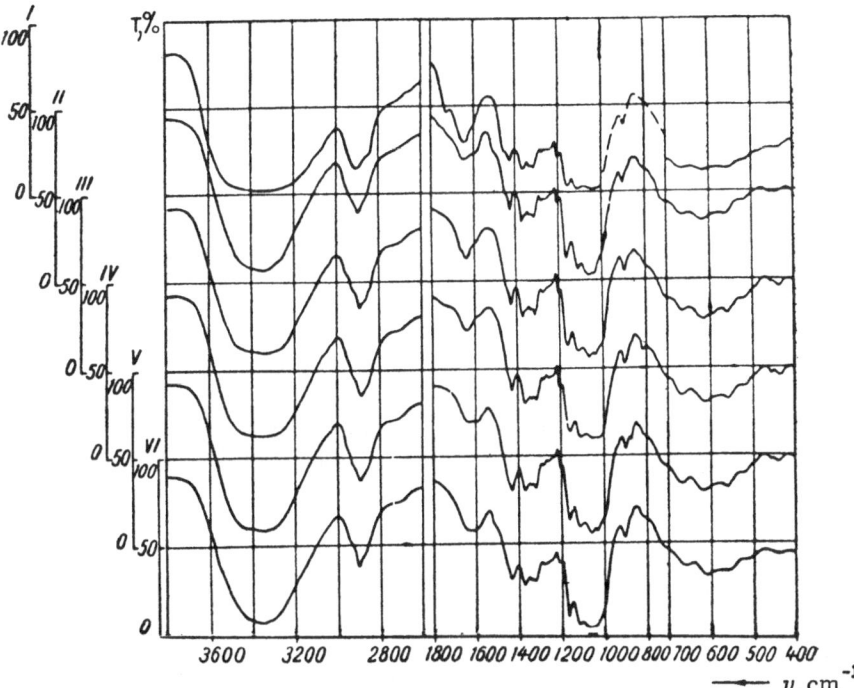

Fig. 37. Spectra of pine wood cellulose. I) May shoots; II) June shoots; III) mature wood; IV) 3000-year-old wood; V) 500-year-old wood; VI) interglacial wood 140,000 years old.

The author, together with N. I. Garbuz, A. M. Shishko, and A. I. Skrigan, carried out a more detailed investigation of celluloses isolated from woods of various ages, over the spectral range 400-3600 cm^{-1}. A comparison was made of the spectra of celluloses, isolated under laboratory conditions, from May and June pine shoots, mature wood, and woods 500, 3000, and 140,000 years old (Fig. 37). Table 5 shows the characteristics of these celluloses. The ages of the woods were determined by a known method [36].

As described above, the spectrum of the product isolated from the wood of May shoots shows all the specific features of the cellulose spectrum. The hydroxyl groups of this cellulose, as in normal cellulosic materials, are involved in hydrogen bonding. The spectrum of cellulose from 150,000-year-old wood also shows no new absorption bands, not attributable to compounds accompanying cellulose, and no significant frequency shifts. This demonstrates the high stability of the cellulose macromolecule, under normal conditions, to structural change.

Table 5. Chemical Characteristics of the Celluloses Investigated from Pine Woods of Various Ages

Cellulose	Ash, %	α-Cellu-lose, %	Resins, waxes, %	Lignin, König method, %	Pentosans, Tollens method, %	CHO groups, %
Mature pine wood (about 100 years old)	1.40	89.4	0.31	0.80	7.00	0.410
Roots of buried wood, 0.5 m depth (500 years old)	1.41	91.4	0.34	1.44	4.08	1.800
Roots of buried wood, 3 m depth (3000 years old)	1.91	89.9	0.75	3.00	4.79	0.693
Wood trunks from marshes, from inter-glacial period (140,000 years old)	6.39	88.1	0.66	–	6.95	0.731

Note: All the celluloses investigated were obtained by the sulfate method and were bleached with calcium hypochlorite solution.

Unlike normal mature wood, the June and interglacial celluloses show less well-defined spectra in the regions 1200-1500 and 400-700 cm^{-1}, and stronger impurity absorptions at 1600, 1500, and 800 cm^{-1}. The intensity of the impurity absorption bands is greater in the case of cellulose obtained from the interglacial wood. This cellulose has a more diffuse spectrum in the region 400-1800 cm^{-1} and also shows some increase in intensity of the band at 1430 cm^{-1}, as compared with the band at 1370 cm^{-1}. The latter can be attributed either to a specific impurity or to the occurrence with time of definite chemical processes or structural changes.

It has been shown [36] that cellulose isolated from interglacial wood has a higher density than normal cellulose. The experimental material presented above provides a basis for supposing that this is due to a higher packing density, made possible by a reduction in the structural order of the cellulose macromolecules (assuming that the possible effect of mineral impurities is eliminated).

It is interesting to note that the spectra of individual cellulose samples, obtained from fossil woods, show an increase in intensity of the band at 1430 cm^{-1} and a new band at about 870 cm^{-1}. Similar spectral features have been observed with some of the cellulose samples from young poplar bark. Consequently, these spectral features cannot be attributed to any specific impurities in fossil wood, particularly as these samples had relatively low impurity contents (Table 5).

It has already been pointed out that the increased intensity of the band at 1430 cm^{-1}, and the new band at 870 cm^{-1}, may be attributable to structural changes in the cellulose.

Samples of Technical Cellulose

There are only slight differences (mainly in the 1200-1500 and the 900 cm^{-1} regions) between the spectra of most of the analyzed samples of technical wood cellulose of various native and foreign brands. No special differences are observable in the spectra, depending on the method used for isolating the cellulose (sulfate or sulfite digestion), or on the nature of the wood, pine, spruce, or larch. None of the celluloses analyzed normally show any significant absorption bands characteristic of aromatic impurities.

However, some of the celluloses investigated show spectroscopic peculiarities. For example, the spectra of wood celluloses from the foreign firms "Bekai" and particularly "Lintra" are characterized by less intense bands at 1430, 1340, and 1320 cm^{-1}, increased absorption at 900 cm^{-1}, and leveling of band structure in the region 400-700 cm^{-1}. Certain other celluloses are characterized by similar features (see Appendix II). It will be shown below that these spectral features must be attributed to changes in the cellulose structure in the course of purification by alkaline treatment. The spectra of these celluloses also show differences in the relative intensities of individual bands in the 400-700 cm^{-1} region. These differences are undoubtedly structural in origin.

The spectra of the cotton celluloses investigated differ from those of the wood celluloses in that the absorption bands are more sharply defined. In the case of cotton celluloses, the bands at 1370, 1360, 1340, and 1320 cm^{-1} are of approximately the same intensity, while with wood celluloses the 1360 and 1340 cm^{-1} bands are somewhat weaker (in individual cases the 1360 cm^{-1} band practically disappears). Such spectral differences can be attributed to structural peculiarities of these celluloses.

CHAPTER III

HYDROCELLULOSE. PRODUCTS OF THE PARTIAL HYDROLYSIS OF CELLULOSE

Hydrocellulose has found wide application in the production of viscose silk, cord fibers, cellophane, and so forth.

Hydrocellulose has characteristic spectral features, as compared with natural structural modifications of cellulose. After treatment of natural cellulose fibers with a 17.5% solution of NaOH in water and subsequent regeneration, the spectrum of the product shows the following changes: broadening of the hydroxyl bands and the appearance of new bands at about 3480 and 3450 cm^{-1}; a change in the contour of the stretching vibration band of CH$_2$ and CH, with a more or less pronounced splitting of this band at the absorption maximum (2900 cm^{-1}); reduction in intensity of the bands at 1430, 1340, and 1320 cm^{-1}; a change in the intensity ratio of the bands at 1280 and 1240 cm^{-1}; the appearance of a new band at 1260 cm^{-1}; increased absorption in the regions 970 and 900 cm^{-1} (Appendix III, 84-109).

Other characteristic features of the hydrocellulose spectrum are: increased diffuseness of the 1000-1150 cm^{-1} band; more uniform general absorption in the region 400-700 cm^{-1}, with disappearance of the bands at 560 and 435 cm^{-1}.

The specific features of the hydrocellulose spectrum in the hydroxyl group region indicate that there are differences in hydrogen bonding, as compared with natural cellulose modifications. The characteristic features of the hydrogen bonds in hydrocellulose are very clearly reflected in the spectra of deuterated samples.

It is well known that, in the cellulose structure, there are parts which are more or less accessible to deuterium exchange, and this is usually explained by the presence of so-called "amorphous" and "crystalline" regions.

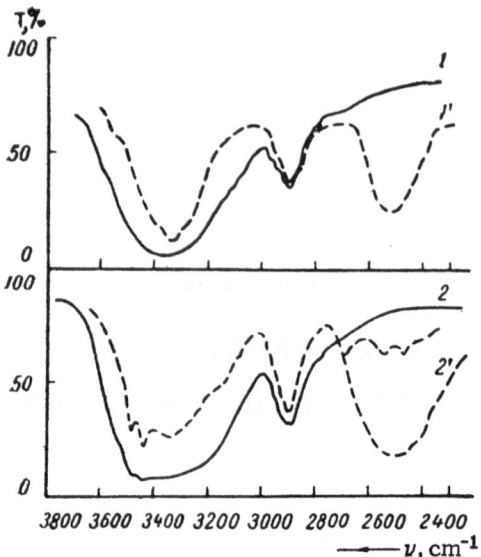

Fig. 38. Spectra of: 1) wood cellulose; 2) wood hydrocellulose, before and after (dotted line) deuteration.

In the spectra of deuterated products, the OH group bands have a sharper structure than the OH group bands before deuteration, and than the OD group bands after deuteration.

Figure 38 demonstrates that the spectrum of deuterated natural cellulose shows three bands at 3400, 3350, and 3300 cm^{-1}, while that of deuterated hydrocellulose shows a broad band with its main maximum at 3350 cm^{-1}, an inflection at about 3170 cm^{-1}, and two sharp bands at 3450 and 3480 cm^{-1}.

The spectra of deuterated cellulose fibers were obtained by the following method. The dried film of cellulose fiber, prepared by the pressing method without an immersion medium [6], was placed in a special vacuum cell with fluorite windows for optical contact, which was filled with liquid D_2O and allowed to stand for 30-40 min. The sample was then thoroughly dried under vacuum, the dry cell was filled with an immersion medium, and the spectrum of the sample was recorded.

The spectrum of hydrocellulose does not show any bands which can be attributed to free hydroxyl groups. There can be no doubt that the modifications of natural cellulose and hydrocellulose differ in the character of their hydrogen bonds, and not in the ratios of their free and combined hydroxyl groups, as has sometimes been supposed [37].

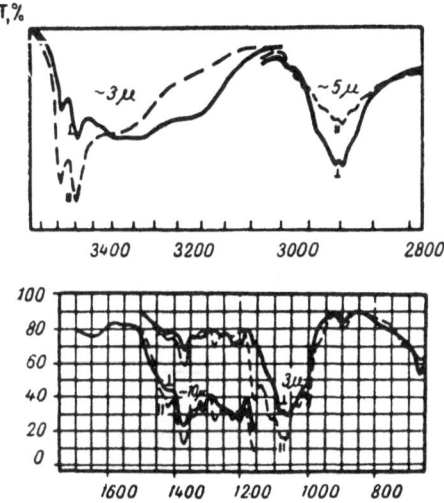

Fig. 39. Spectra of oriented samples of hydrocellulose from ramie [148]: ⊥ — the electric
vector perpendicular to the direction of the
macromolecular chain; ‖ — the electric vector
parallel to the direction of the macromolecular chain.

Interpretation of the special features of the hydrocellulose spectrum is
of considerable scientific and practical interest, since it should lead to definite views on the structure of this industrially important material.

Only a few papers have appeared on the infrared spectrum of hydrocellulose. Some of these have been concerned with the study of the hydrogen bonds in the structure. Nikitin [1] compared the spectra of fibers of
natural and mercerized cellulose in the hydroxyl group overtone region. He
showed that, in the spectra of mercerized, as compared with natural fibers,
there were band frequency shifts and some of the bands disappeared. Natural
cellulose showed bands at 1.49, 1.54, 1.58, and 2.11 μ, while mercerized
cellulose showed bands at 1.48, 1.58, and 2.09 μ. In the author's view, the
observed changes were associated with a weakening of the intermolecular interaction in mercerized cellulose (an increase in the OH . . . O spacing).
However, this opinion is not very convincing, since no frequency assignments
were made. The band shifts could also be attributed to a change in the nature of the intramolecular interaction, caused by structural rearrangement of
the cellulose macromolecules.

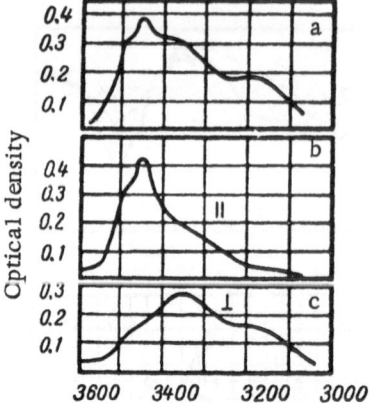

Fig. 40. Spectra of oriented samples of viscose fiber [147]: a) unpolarized radiation; b) electric vector parallel to direction of macromolecular chain; c) electric vector perpendicular to direction of macromolecular chain.

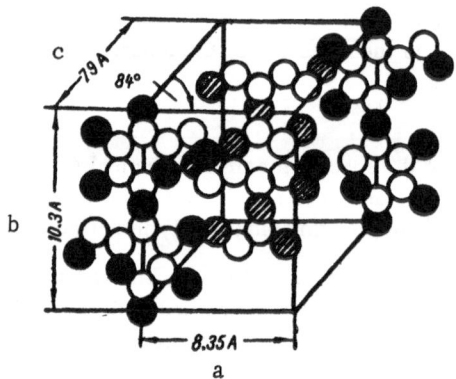

Fig. 41. Model of elementary cell of cellulose I, according to Meyer and Misch [197].

The most systematic investigations of the special features of hydrogen bonds in hydrocellulose were carried out by Marrinan and Mann [147], and by Marchessault and Liang [148], who used polarized infrared radiation to investigate dichroism in the region of the main hydroxyl frequencies, in the infrared spectra of oriented films of various cellulose structural modifications. The authors found that, in the spectra of mercerized cellulose, there was parallel dichroism of the bands at 3440 and 3480 cm^{-1}, and perpendicular

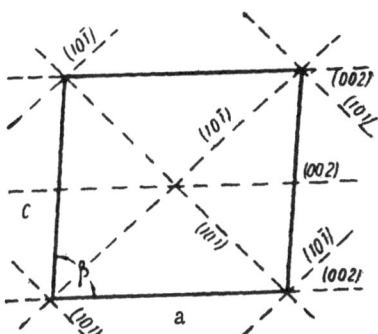

Fig. 42. Basic paratropic planes
in cellulose.

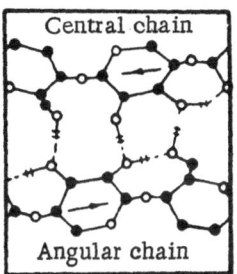

Fig. 43. Construction of hydrogen
bonds in hydrocellulose samples
having three-dimensional order
(Mann and Marrinan [195]).

dichroism of the broad bands at 3350 and 3200 cm^{-1} (Figs. 39 and 40). According to Marrinan and Mann, the angles formed by the hydrogen bonds giving rise to the 3440 and 3480 cm^{-1} bands were considerably less than 54°44',while the remaining hydrogen bonds formed angles greater than 54°44'.

The appropriate molecular models (Figs. 41 and 42) were used to show that only OH groups forming intramolecular hydrogen bonds between adjacent anhydroglucose rings in the cellulose macromolecule can form small angles with the direction of the chain. According to Marrinan and Mann, the possible bonds of this type are: $O_3-H \ldots O_5'$, $O_2-H \ldots O_6'$, and $O_2 \ldots$ H$-O_6'$ (the dashes denote atoms of the adjacent ring). These authors consider that the perpendicular dichroism of the double band at 3350 and of the band at 3200 cm^{-1} indicates the presence of three types of intermolecular hydrogen bond in hydrocellulose. Marrinan and Mann have produced a scheme for hydrogen bond formation in hydrocellulose, according to which all the hydroxyl groups participate in hydrogen bonding, but intramolecular hydrogen bonds are formed only between anhydroglucose rings in the $10\bar{1}$ plane (Fig. 43). Marchessault and Liang also attribute the 3440 and 3480 cm^{-1} bands to intramolecular hydrogen bonds, and the other longer-wavelength bands to intermolecular hydrogen bonds. However, allowing for the dichroism of the CH$_2$ group bands, Marchessault and Liang deny the possibility that intramolecular hydrogen bonds of the type $O_2 \ldots O_6'$ can exist, and conclude that the 3440 and 3480 cm^{-1} bands in the hydrocellulose spectrum can be attributed only to $O_3-H \ldots O_5'$ hydrogen bonds. In their opinion, the existence of two bands can be due to differences in the surroundings and in bond lengths. They propose two schemes for hydrogen bond formation in hydrocellulose (Fig. 44), which differ from those of Marrinan and Mann. Marchessault and Liang conclude that Scheme a of Fig. 44 agrees best

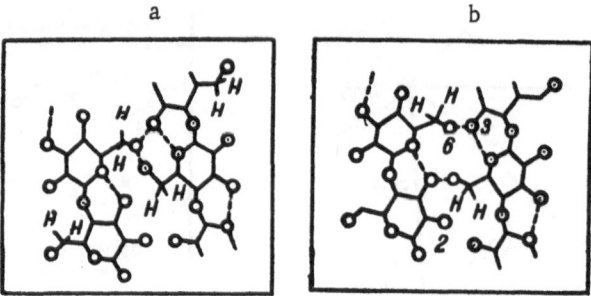

Fig. 44. Construction of hydrogen bonds in hydrocellulose samples having three-dimensional order
(Marchessault and Liang [148]). a) Displacement of
central with respect to angular chain for r = 0.30 V; b)
for r = 0.17 V and φ = 30°.

with the experimental data; this gives a closed chain $O_6' \rightarrow O_6 \rightarrow O_3 \rightarrow O_5'$ between adjacent chains in the 101 plane.

These schemes for hydrogen bonds in hydrocellulose are largely hypo-
thetical, in spite of their undoubted scientific interest. It is certain that the
nature of the hydrogen bonds in cellulose is more complex, and the number
of absorption bands appearing in the spectrum does not precisely reflect the
number of possible types of hydrogen bond. The hydrogen bond structures
suggested above are based on molecular models which do not fully reflect the
possible conformations of the cellulose macromolecule. Even a simple comparison of the spectra of cellulose and cellobiose (Fig. 30) shows that the
polymer spectrum has, as would be expected, a more diffuse character, which
can certainly be attributed to the large range of different conformations. It
would be difficult not to accept the existence, even in the crystalline parts,
of rotational isomers attributable to swinging or rotation of individual structural groups around single C—C bonds. It was shown above, by comparing
the spectra of various model compounds, that a change in the spatial location of even one of the hydroxyl groups can have a marked effect on the system of hydrogen bonds. In the present author's view, it is highly improbable
that the difference between cellulose and hydrocellulose can consist mainly
of peculiarities in the mutual location of the central and angular chains [147].

Changes in the mutual location of chains can result from changes in
the conformation of the whole macromolecule or, in particular, of separate
elementary units.

Until now, we have considered the possible types of hydrogen bonds in
the most ordered parts of cellulose, which are the most resistant to deuteration.

It is well known that there is a considerable part of cellulose — the so-called amorphous, easily deuterated fraction — in which there is a wider possible range of types of hydrogen bonds.

As stated above, the OD group band in the spectra of deuterated cellulose fibers is characterized by a broad diffuse contour without any significant structure.

In one of the first papers to compare the infrared spectra of various structural modifications of cellulose over a wide enough spectral range [111], the authors compared the spectra of native cotton and bacterial celluloses with those of regenerated, mercerized, and partially ground-up celluloses. The cellulose fiber samples for investigation were prepared by suspension in an organic oil, so that the cellulose spectrum could not be seen in the region where the oil absorbed. The authors did not interpret any changes they observed in the spectra and confined their discussion to general remarks.

Stepanov and his co-workers [14, 38] were the first to investigate the spectrum of cellulose which had been subjected to alkaline treatment in the process of viscose manufacture. The material investigated consisted of cellulose samples regenerated from a given batch at different stages in the technological process of viscose manufacture by the xanthate method. The authors showed that the special features of the spectrum of regenerated cellulose were not attributable to oxidation in the alkaline medium, in the course of mercerization or ripening. These features reflected the changes in the cellulose spectrum in the process of regeneration from chemical combination with alkali. It was shown spectroscopically that interaction of cellulose with alkali proceeded actively and was largely completed in the initial stages of the alkaline treatment of cellulose. The reduction in intensity of the CH_2 group band has been attributed to the characteristics of these groups in regenerated cellulose [14]. Since an increase in the time of interaction of cellulose with alkali (the ripening process) does not lead to any further reduction in intensity of the CH_2 group band, as compared with its intensity in the spectrum of mercerized cellulose, it may be concluded that, even in the time required for mercerization, practically all the primary hydroxyl groups have reacted chemically with alkali.

Analysis of the spectra of dried alkaline cellulose samples [38] has shown that a characteristic feature is a strong reduction in intensity of the OH group band. This makes it reasonable to assume that an alcoholate type of derivative is formed: cellulose−OH → cellulose−ONa.

Cellulose Regenerated After Treatment with Alkali of Various Concentrations

A detailed study has been published [28] of the spectra of natural cellulose fibers of diverse origins, the corresponding hydrocelluloses, and celluloses which had been regenerated after treatment with alkali of various concentrations. The authors showed that, with the change from natural cellulose to hydrocellulose (regardless of the origin of the material), there occurred a considerable reduction in intensity of the spectral bands at 1430, 1340, and 1320 cm^{-1}; an increase in absorption in the 900 cm^{-1} region; and the appearance of new bands at about 1260 and 970 cm^{-1}. These spectral changes were found to be sensitive, within certain limits, to the concentration of alkali with which the cellulose had been treated [145, 149].

The most obvious effects of the alkaline treatment on the character of the spectrum are the intensity changes of the bands at 1430 and 900 cm^{-1}. Figures 45 and 46 show that the intensity changes are slight after treatment with alkali of concentrations up to 10%. However, there is a marked reduction in intensity of the 1430 cm^{-1} band as the alkali concentration is further increased, to a steady value after treatment with 17.5% NaOH. An increase in alkali concentration to 40% does not have any further effect on the band intensity. It should be noted that the specific spectral features of mercerized cellulose are already noticeable after treatment with 12% NaOH. The alkali concentration has a similar effect on the increase in intensity of the 900 cm^{-1} band.

McKenzie and Higgins [145] also noted that conversion to hydrocellulose took place when cotton cellulose was treated with 9-13% alkali, or when eucalyptus cellulose was treated with 8-11% alkali. They investigated the ground-up fibers in alkali halide discs, a technique which does not show up the special features of the spectrum, as well as the direct pressing method, and thus were unable to establish more precisely the range of alkali concentration at which the conversion of cellulose to hydrocellulose first takes place.

Changes in the spectrum of cellulose, in the course of its treatment with alkali, provide a means for controlling the process of mercerization. Stepanov and others proposed methods for determining the extent of mercerization of cellulose [13], and for analyzing cellulose to determine with what concentration of alkali it has been treated [40].

The method for determining the degree of mercerization is as follows. A sample of the alkaline cellulose is centrifuged at 3 G and washed with 10% acetic acid, and the resulting hydrocellulose is washed, first with hot and then with cold water, until the washings are neutral, and dried under an infrared lamp for 5-10 min. A thin film of the air-dried hydrocellulose fiber is then

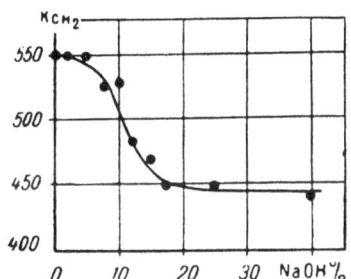

Fig. 45. The intensity change in the 1430 cm^{-1} band of the spectrum of cellulose, regenerated after alkaline treatment, as a function of the alkali concentration.

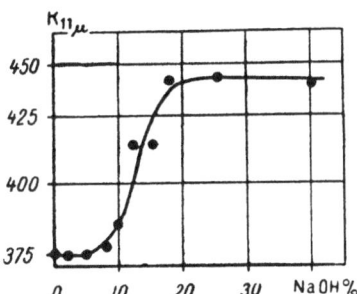

Fig. 46. The intensity change in the 900 cm^{-1} band of the spectrum of cellulose, regenerated after alkaline treatment, as a function of the alkali concentration.

prepared [6], and its infrared spectrum is recorded. The degree of mercerization (DM_1) can be obtained from the increase in absorption at 900 cm^{-1} by means of the following equation:

$$DM_1 = \left(\frac{k_{mer} - k_{init}}{k_{init}} \right)_{\nu=900 \text{ cm}^{-1}} = \left[\frac{\log\left(\frac{I_0}{I}\right)_{mer} - \log\left(\frac{I_0}{I}\right)_{init}}{\log\left(\frac{I_0}{I}\right)_{init}} \right]_{\nu=900 \text{ cm}^{-1}},$$

where k_{init} and k_{mer} are the extinction coefficients of the initial and mercerized cellulose. The analysis can be speeded up considerably if a calibration curve is constructed beforehand, showing the relation between the degree of mercerization and the absorption coefficient at 900 cm^{-1}. The bands at 1430, 1340, and 1320 cm^{-1} can be used as well in order to increase the analytical precision for control purposes. In this case it is convenient to use the intensity ratio of the bands at 1430 and 900 cm^{-1}:

$$DM_2 = \frac{\log\left(\frac{I_0}{I}\right)_{\nu=900 \text{ cm}^{-1}}}{\log\left(\frac{I_0}{I}\right)_{\nu=1430 \text{ cm}^{-1}}}.$$

Obviously, the degree of mercerization is only a ratio, characterizing the extent of structural change in the sample analyzed in the course of

alkaline treatment, as compared with a well mercerized cellulose. Determination of the degree of mercerization of cellulose by the infrared method
has been used industrially.

The intensity ratio of the bands at 900 and 1430 cm^{-1} has been recommended by O'Connor and his co-workers [144] as a means for determining
the so-called degree of crystallinity of cellulose, since, when cellulose is
ground up, there is a reduction in intensity of the band at 1430 cm^{-1} and an
increased absorption in the 900 cm^{-1} region. However, it is not in general
possible to determine the so-called crystallinity of cellulose in this way. The
existence of a crystalline state of cellulose is admitted, but there is no
reason to assume the existence of several crystalline structures, characteristic in particular of cellulose and hydrocellulose. As seen above, the spectrum of hydrocellulose is also characterized by a diminution of the band at
1430 cm^{-1} and increased absorption in the 900 cm^{-1} region. These spectral
changes cannot be attributed to a reduction in the degree of crystallinity of
cellulose. The same specific features in the spectrum of hydrocellulose are
also well developed in the case of crystallites [148]. It must be accepted
that these specific features of the spectrum of hydrocellulose correspond to
structural peculiarities of its macromolecules.

A study of the spectra of doubly oriented films of mercerized cellulose
[148] also showed a reduction in intensity of the 1430 cm^{-1} CH$_2$ group band,
as compared with native cellulose. The authors attributed this to a change
in the orientation of the circulatory momenta of CH$_2$ groups relative to the
fiber axis in the 10$\bar{1}$ plane. This explanation can only be applicable to
doubly oriented cellulose films when the planes of the glucose residues and
the macromolecular axes are oriented in a definite way relative to the film
surface. It has been shown [9, 10, 28, 38, 39] that similar spectral differences
between native and mercerized celluloses also occur when the fibers are
randomly located in the samples.

It is known that the chemical structures of natural cellulose and hydrocellulose are identical. For this reason, the diminution of the 1430 cm^{-1}
band, clearly associated with internal deformation vibrations of methylene
groups, cannot be attributed to a reduction in the number of CH$_2$ groups.
Stepanov, Zhbankov, and Marupov [17] set out to explain this phenomenon by
comparing hydrocellulose spectra with those of cellulose and the model compounds mono-, di-, and polysaccharides and polyhydric alcohols, in the region of the CH$_2$ group frequencies. These compounds, particularly the mono-
and disaccharides containing CH$_2$OH groups, showed sharp specific spectra in
this region, with a large number of absorption bands. For example, the spectra of all the mono- and disaccharides considered [17] showed a complex
band structure in the region of frequency corresponding to internal deformation

vibrations of CH_2 groups. The spectral bands observed were at 1460, 1450, 1430, and 1410 cm^{-1} for α-D-glucose; 1460, 1450, and 1430 for α-D-galactose; 1490, 1460, 1430, and 1410 for cellobiose; 1460 and 1435 cm^{-1} for maltose, etc. This sharp band structure diminished with the change to polysaccharides.

The CH_2OH groups can turn or rotate around the C_5-C_6 single bonds, so that the molecules under consideration, containing a CH_2OH group, can exist in various rotational isomeric forms. The structures of D-glucose, D-galactose, cellobiose, maltose, etc., differ in the location of atoms which can have different steric effects on the CH_2OH groups, and for this reason the properties of their rotational isomers, as revealed in the spectra, will also be different. The role of rotational isomerism is not so obvious in the case of stretching vibrations, but the complex structure of the bands in the region of the CH_2 group deformation vibrations would be difficult to interpret without allowing for the existence of rotational isomers. It should be noted that a complex band structure has been observed, in the region of CH_2 group deformation vibrations, in the Raman spectra of 1,2-dihalogenoethanes, and has been ascribed to rotational isomerism [41]. The number of possible molecular configurations increases with increasing degree of polymerization [42]. This can account for the disappearance of the sharp band structures, in the 2800-3000 and 1300-1500 cm^{-1} regions, with the transition from monosaccharides to disaccharides and polysaccharides. This disappearance of the sharp band structure has also been observed in the spectra of aqueous sugar solutions [132].

It is known that complete or partial rotation about a single bond is less difficult in the case of pure liquids or solutions. Thus, in the present case, as distinct from the case of aliphatic hydrocarbons [24], the conception that there is one characteristic CH_2 band loses any significance. For example, in the amylose spectrum there is not one sharp band in the 1400-1500 cm^{-1} region, but absorption by the CH_2 group is spread out over the entire region. The cellulose spectrum is complex in the ranges 1400-1500 and 2800-3000 cm^{-1}. It cannot be supposed that the frequency of the CH_2 internal deformation vibrations in the cellulose spectrum is only around 1430 cm^{-1}. The 1430 cm^{-1} band in the cellulose spectrum may correspond to one of the possible positions of the CH_2OH group in its rotation about the C_5-C_6 bond. Without allowing for rotational isomerism, it is impossible to construct a sound model of the elementary cell of the cellulose macromolecule, or to estimate the character of the participation of hydroxyl at C_6 in inter- and intramolecular interactions.

It has already been explained that the spectrum of hydrocellulose differs, in particular, from that of the natural cellulose modification, by the

reduced intensity of the band in the region of the internal deformation vibrations of the CH_2 groups. Stepanov, Zhbankov, and Marupov [17] suggested a
new hypothesis for the structure of hydrocellulose, according to which the
two structural modifications of cellulose differ with respect to the rotational
isomers of the elementary units, due to rotation of CH_2OH about the C_5-C_6
bond.* It is known that the character of the rotational isomers is very sensitive to temperature conditions. Thus, it is not surprising that there is a partial conversion of hydrocellulose to the natural modification on heating
under definite conditions [43, 150]. In this case, we have shown experimentally that there is an increase in intensity of the 1430 cm^{-1} band.

The author, together with K. M. Grushetskii, has recently made a
more detailed investigation of the problem of rotational isomerism of the
CH_2OH group. Calculations of the potential function for internal rotation
(bond orientation energy and steric interaction energy) showed that it was
quite possible for rotational isomerism of the CH_2OH groups to occur in carbohydrates. For example, with α-D-galactose, the most probable positions
for the CH_2OH group in its rotation about the C_5-C_6 bond are at φ equal to
about 65 and 170°, while with α-D-glucose the most probable positions are
at $\varphi = 80$, 177, and 300° (the angles are measured clockwise from the position of the O_6 and C_4 atoms). In order to establish the CH_2OH conformations
actually existing in hydrocarbons, it would be necessary to calculate the hydrogen bond energies.

Cellulose Subjected to Grinding

The transformation of cellulose to hydrocellulose can be achieved, not
only by chemical means (mercerization), but also by mechanical grinding
[19]. As in the case of mercerized cellulose, the spectrum of cellulose which
has been ground up mechanically (Appendix III, 92-95) is characterized by
diminution of the bands at 1430, 1340, and 1320 cm^{-1}, and by increased absorption in the 900 cm^{-1} region. However, in this case, the cellulose absorption bands become less sharp.

In a study of the change in the spectrum of cellulose on grinding [111,
144], it was noted that the sharp bands disappeared as grinding proceeded,
and that there was an accompanying increase in intensity of the band at 900
cm^{-1}. The authors attributed these changes to loss of crystallinity. It should
be noted, however, that the terms "crystalline" and "amorphous" bands were
widely used at the time these papers were published. In the present author's
view, these terms should be used very cautiously, particularly when the bands

---

* This conception has been confirmed by the work of Schneider and Vodňanský
[202].

have not been attributed to definite vibrations. For example, the band in the cellulose spectrum at about 900 cm^{-1} was called "amorphous" only because its intensity increased as the cellulose was ground up. However, as explained above, the intensity of this band also increases in the transition from the crystalline modification of cellulose to the crystalline modification of hydrocellulose. It would thus appear that the change in intensity of this band is due not to a breakdown in the three-dimensional order of the location of cellulose macromolecules, but rather to conformational changes in the macromolecules themselves. In support of our view, very strong bands have been observed in this region in the spectra of many crystalline mono- and disaccharides. *

The spectra of ground-up celluloses, which were subsequently heated to 170°C in an aqueous medium, differed from the spectra of the ground-up celluloses before heating in that there was some increase in intensity of the bands at 1430, 1340, and 1320 cm^{-1}, although full restoration of the spectrum of native cellulose was not achieved. Treatment with water at 0°C did not produce any significant change in the spectrum of ground-up cellulose. These observations are in agreement with the known results of x-ray analysis.

Specific features can be observed in the spectra of cellulose samples which have been subjected to grinding in HCl vapor, and which have a very low degree of polymerization (about 20). The spectrum of this product differs from those of natural cellulose and of hydrocellulose. Unlike the case of hydrocellulose, there is a leveling of the band at 1370 cm^{-1}, and new bands appear in the 950, 925, 850, and 770 cm^{-1} regions. The appearance of these new bands cannot be attributed to a change in the amorphous state, nor to a reduction in the degree of polymerization. The bands are not observed in the spectra of celluloses subject to intensive grinding, but do appear in the spectra of other polysaccharides of high molecular weight (dextran) [21].

It must be accepted that the process of grinding cellulose in HCl vapor leads to the appearance of a new structural modification, different from hydrocellulose. The new structure is responsible for its specific properties.

As stated above, a band in the 850 cm^{-1} region is characteristic of sugars with C$_1$H groups located equatorially relative to the plane of the

---

*Recent papers by O'Connor and co-workers [203] have also demonstrated the difficulty of using these bands for determining degree of crystallinity, in view of the complexity of the spectra of cellulose II and of amorphous cellulose.

pyran rings. The appearance of this band in the present case may be explained in terms of CI → IC conformational transitions of the pyran rings.

All the experimental results presented above show that the infrared spectroscopy method is very sensitive to any change in the cellulose structure, and that it can be used with advantage for both scientific and analytical purposes. Infrared spectroscopic methods appear very promising for the study of conformational changes in the cellulose macromolecule. Analysis of the spectra of model compounds shows that the infrared spectra of compounds of this class are very sensitive to changes in the spatial location of individual structural elements, thus indicating the strong interaction of the latter. The physicochemical properties of such compounds must be strongly influenced by conformational factors. This circumstance must always be kept in mind when the mechanisms of various chemical reactions of cellulosic materials are under consideration.

Viscose Fibers

The spectra of hydrocellulose viscose fibers are qualitatively similar to the spectrum of mercerized cellulose, but differ in being generally more diffuse. One difference is that in the spectrum of viscose fibers there is practically no resolution of the bands at 1340 and 1320 cm^{-1}, nor of those at 1280 and 1260 cm^{-1}.

There are also differences between the various types of viscose fibers. For example, in the case of high-strength cord fibers, the hydroxyl group band is usually less diffuse. This may be explained by a higher state of order in the hydrogen bonds.

No significant differences have been observed between the spectra of normal types of viscose fiber and of fibers made from celluloses reprecipitated from viscose solutions of various concentrations: 0.1, 0.5, and 8%. This circumstance gives no grounds for supposing that the structure of the initial cellulose affects the properties of the hydrocellulose fibers.

Effect of Hydrogen Bonds on the Structure of Cellulose

It is well known that the properties of polymeric materials are very much affected by the nature of the inter- and intramolecular interactions. This is particularly so with compounds containing numerous strongly polar groups. Cellulose is such a compound, since it contains three hydroxyl groups in each separate elementary unit. It may be taken as a well established fact that, in cellulose, practically all the hydroxyl groups are involved in hydrogen bonding. This, in its turn, should have a marked effect on the stabilization of definite conformations.

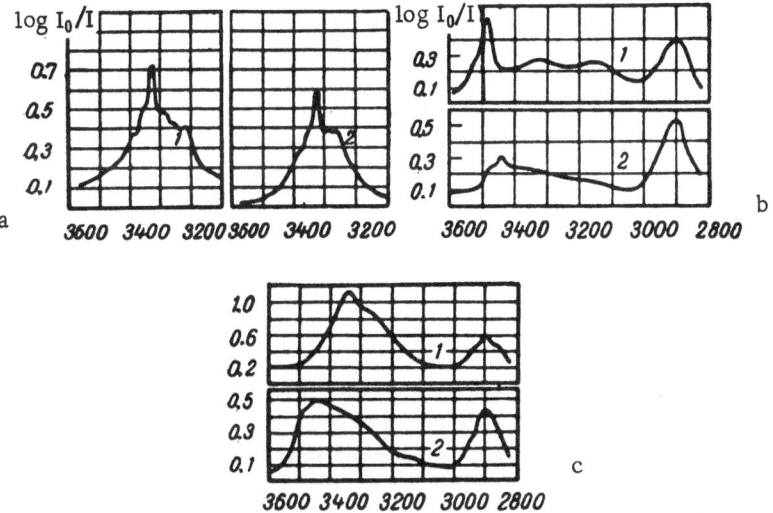

Fig. 47. Spectra of celluloses I (at liquid nitrogen temperature), III, and IV [147]. a. 1) <u>Valonia</u> <u>ventricosa</u> cellulose; 2) tunicin cellulose. b. 1) Cellulose III_I obtained from cellulose I; 2) cellulose III_{II} obtained from cellulose II. c. 1) Cellulose IV_I obtained from cellulose I; 2) cellulose IV_{II} obtained from cellulose II.

It has been seen above that, in the case of numerous crystalline mono- and disaccharides, the spectra show the existence of well-defined absorption bands over the entire frequency region for OH groups involved in hydrogen bonds: 3120, 3220, 3250, 3300, 3340, 3400, 3430, 3520, and 3540 cm^{-1} (Figs. 22 and 23). This provides a basis for supposing that, in compounds of this type, a wide range of energetically nonuniform hydrogen bonds can exist. However, in the spectra of cellulose fibers and other polysaccharides, there is a broad diffuse band between 3200 and 3600 cm^{-1} without any clearly defined structure. This indicates that there are many more different types of hydrogen bonds than in the simpler compounds. This is attributable to the larger number of possible conformations of the macromolecules.

If the spectrum of deuterated cellulose is considered in the regions of the undeuterated and deuterated hydroxyl groups (Fig. 38), it is clear that the former give discrete absorption bands, but that the OD group bands show no structure. It should be noted that deuteration is accompanied by a reduction in intensity over the whole contour of the hydroxyl band, and not only at selected parts of it. Comparison of the spectra of crystalline mono-, di-, and polysaccharides with those of deuterated cellulose fibers provides a basis

for believing that a definite part of fibrous cellulose is characterized by a
relatively high state of order in the hydrogen bonds. However, there are also
unordered parts in cellulose, and a wide range of possible types of hydrogen
bonds is specific to these. This latter circumstance also accounts for the diffuse contour of the hydroxyl group band in cellulose fibers [98].

As shown above, the hydroxyl group hydrogen bonds in compounds related to cellulose form a single hydrogen bond system, the individual parts
of which are related to each other, and which are very sensitive to the structural factor. It would therefore be expected that there would be marked differences in hydrogen bond structure in the various structural modifications of
cellulose. In fact, as already explained (Fig. 38), the spectra of fibers of
natural cellulose and hydrocellulose are specific in the region of the OH
group frequencies concerned with hydrogen bonding. Peculiarities in the hydrogen bond region also appear in the spectra of other structural forms of
cellulose (Fig. 47). If each specific cellulose structure corresponds to a particular arrangement of hydrogen bonds, then it also follows that, to obtain a
compound with a new structure, it is necessary to break down the previously
existing system of hydrogen bonds. In fact, the hydrocellulose structural
modification can be obtained either by hydrolyzing highly substituted cellulose esters, by grinding natural fibers, or by treating them with concentrated
alkalies. It will be shown below that, in the latter case, there is a strong reduction in intensity of the band of hydroxyl groups involved in hydrogen bonding.

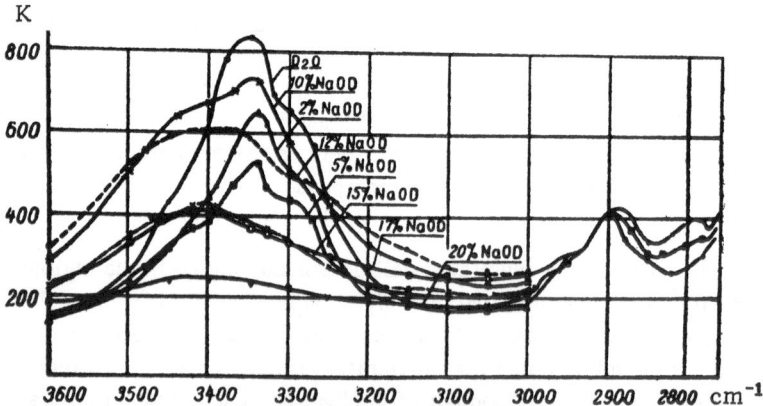

Fig. 48. Spectra of wood cellulose in alkali solutions of various concentrations.

Alkaline Cellulose

Infrared spectroscopy should provide a promising method for studying the structure of alkaline cellulose. However, it is very difficult to apply this method, owing to the strong absorption by water in the spectral regions to be investigated. The study of dried alkaline cellulose samples [14] gives an incomplete picture of their structure, since various additional processes may occur during drying. We used a previously described technique [16] to obtain infrared spectra of cellulose during the process of alkaline treatment. The spectra investigated were of wood cellulose in 2, 5, 8, 10, 12, 15, 17.5, and 25% solutions of NaOD in D_2O. They were recorded with a UR-10 double-beam infrared spectrometer, using an LiF prism for the region 2600-3800 cm^{-1}, and an NaCl prism for the region 1200-1500 cm^{-1}. It is known that alkali reacts chemically with cellulose by interacting with the hydroxyl groups of the elementary units. The nature of this interaction has not previously been elucidated. In principle, the reaction of alkali with cellulose could occur in two ways: to give cellulose$-$OH $\cdot$ NaOH (the molecular form) or cellulose$-$ONa (the alcoholate form) [19]. It is therefore of interest to analyze the spectra of alkaline celluloses precisely in the hydroxyl group region. Figure 48 shows the change in the character of the hydroxyl group band with the change in alkali concentration. The hydroxyl group band diminishes as the alkali concentration increases from 2 to 7%, then increases sharply in intensity as the alkali concentration increases to 10%, and subsequently diminishes again to a practically steady intensity when the alkali concentration reaches 18%. A further increase of alkali concentration to 25% has very little effect on the character of the spectrum.

The sharp increase in intensity of the OH group band in a solution of 10-12% alkali is accompanied by the maximum swelling of the cellulose. These two phenomena are probably related in that, if a large quantity of water penetrates the cellulose, then the concentration of alkali required to form alkaline cellulose cannot be maintained [44]. It should be noted

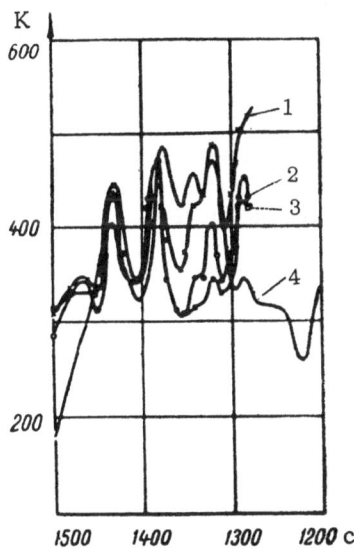

Fig. 49. Spectra of wood cellulose in alkali solutions of various concentrations. 1) Cellulose in D_2O; 2) in 2% NaOD; 3) in 5% NaOD; 4) initial cellulose.

that, simultaneously with the reduction in intensity of the band in the hydroxyl stretching vibration range, there is a disappearance of the bands at
1340 and 1320 cm^{-1} (Fig. 49). These bands are related to the deformation
vibrations of hydroxyl groups [9, 10]. The considerable diminution of the
hydroxyl bands of cellulose in solutions of NaOD in D_2O cannot be attributed to isotopic replacement of hydrogen by deuterium. It is known that an
increase in the extent of swelling of cellulose favors deuterium exchange
[19]. However, in the present case, the change in band intensity is least
when the swelling of the cellulose is at its maximum. Moreover, control
tests were carried out on the deuteration of cellulose which had previously
been swollen in alkali of various concentrations. The spectroscopically determined differences in the extent of deuterium exchange, for cellulose
treated with 5, 10, and 18% alkali solutions, were very slight. Consequently,
the decrease in absorption in the region of the cellulose hydroxyl groups
must be associated either with the formation of cellulose alcoholate or with
the effect of molecularly combined alkali. In order to estimate these factors, we compared the spectra of a saturated solution of alkali in methyl alcohol and of sodium methylate. In the first spectrum there was no significant reduction in intensity of the OH group band, whereas with the methylate the band was practically leveled off. The spectrum of cellulose placed
in concentrated alkali showed a new band at about 2800 cm^{-1}, whose appearance may be attributed to a change in the cellulose structure. It should be
pointed out that a band at this frequency also appears in the spectra of sodium methylate and ethylate, but not in the spectra of solutions of NaOD in
D_2O. Thus, the alcoholate form of interaction of cellulose with alkali appears very probable. However, it cannot be excluded that there may be a
simultaneous conversion to a molecular compound with alkali.

When cellulose is treated with a high concentration of alkali (>8%),
the contour of the hydroxyl band becomes more diffuse in character, thus indicating a rearrangement of the hydrogen bond structure.

Products of the Partial Hydrolysis of Cellulose

The spectra of cellulose which has been subjected to partial hydrolysis
show characteristic distinguishing features (Appendix IV, 110-117). The
spectra of these products are very clear and show sharp bands and inflections
in all the spectral regions investigated. For example, the spectrum of wood
cellulose, which has been hydrolyzed for 3 h with 10 or 20% H_2SO_4, shows
the gradual appearance of an asymmetric band at about 2900 cm^{-1}, inflections at about 2870, 2940, and 2960 cm^{-1}, and a band at 1360 cm^{-1}. The
spectrum of wood cellulose, which has been hydrolyzed for 3 h with 20%
H_2SO_4, shows bands at 1370, 1360, 1340, and 1320 cm^{-1} and, as in the case

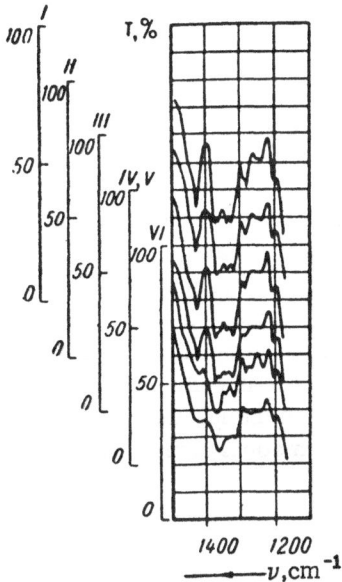

Fig. 50. Spectra of: 1) unhydrolyzed residue of cotton cellulose;
2) cotton cellulose; 3) unhydrolyzed residue of wood cellulose;
4) wood cellulose; 5) unhydrolyzed residue of viscose fiber; 6)
viscose fiber.

of cotton cellulose, these are all of approximately the same intensity. The
whole spectrum of this product is very
similar to that of cotton cellulose and,
in fact, it is very difficult to tell these
materials apart by purely spectroscopic
means.

The best-defined spectrum is obtained with partially hydrolyzed cotton
cellulose (Fig. 50 and Appendix IV,110
and 111). The spectrum of this product
differs from those of similar materials
in that the 2900 cm^{-1} band has a well-
defined structure (there are bands at
2960, 2940, and 2870 cm^{-1}), there is a
strong band at 1430 cm^{-1}, etc. It is
clear that this product is characterized
by a very high state of order in the
macromolecules.

It is interesting to compare the
spectra of hydrocellulose before and
after partial hydrolysis. If it is accepted that poorly hydrolyzed fractions
of cellulosic materials are characterized
by a definite family of structures, then
it would be anticipated that the spectra of the products of partial hydrolysis
of natural cellulose and hydrocellulose would be similar to each other. However, the product from the partial hydrolysis of hydrocellulose shows the specific features of its spectrum more sharply. For example, the bands at 1340,
1320, 1280, 1260, and 1230 cm^{-1} are practically unresolved in the spectrum
of viscose silk before hydrolysis, but are well defined in the spectrum of the
partially hydrolyzed product. More noticeable in this case is the increase in
intensity of the band at 900 cm^{-1}. A comparison of the spectra of the products of partial hydrolysis from natural cellulose and hydrocellulose shows
that the characteristic spectral features of these two structural modifications
become more clearly defined.

These characteristic features of the spectra of products from partial hydrolysis of cellulose provide a basis for supposing that the parts of cellulosic
materials which are difficult to hydrolyze are characterized by a higher degree of order of their macromolecules. However, the structural differences
between the macromolecules of natural cellulose modifications and

hydrocelluloses appear more sharply in these regions. Similar changes have been observed in the spectra of celluloses before and after partial methan-olysis (Appendix IV, 116 and 117). The spectra of products from partial methanolysis of cellulose are characterized by a clearer band structure in the regions investigated.

Generalizing from what has been said above, we may conclude that the spectrum of a cellulose is only an average indication of its structural features. It is also an established fact that cellulosic materials are hetero-geneous. Spectroscopic differences between separate fractions of cellulosic materials indicate that there must be differences in their physicochemical properties. A strictly scientific approach to the study, by infrared spectros-copy, of the various chemical reactions taking place in cellulose, and the disappearance or appearance of individual bands in the spectrum, certainly requires allowance for possible structural transformations in the materials under investigation.

CHAPTER IV

CELLULOSE ESTERS AND ETHERS

Cellulose esters and ethers are widely used in the production of films, plastics, artificial fibers, powder explosives, etc. It is not surprising that the first papers on the infrared spectrum of cellulose were concerned with these compounds. In a 1933 publication, Orlova and Fedorov [45] reported their work on the spectra of celluloid and benzyl- and ethylcelluloses directed to finding a substitute for normal industrial glasses. This investigation, like that of Wells [152], who considered the transparency to infrared radiation of ethylcellulose and acetylcellulose, are now only of historic interest. Nikitin [46] investigated the process of dissolution of dinitrocellulose and acetylcellulose in an ether−alcohol mixture of acetone, dioxan, pyridine, and acetic acid. He found that, in the dissolution process, the band of the unsubstituted hydroxyl groups in the esters broadened and shifted to a longer wavelength. On this basis he concluded that dissolution of the cellulose esters was molecular in nature. The spectrum of trimethylcellulose was investigated by Stelle and Pacsu [153]. Brown, Holiday, and Trotter [154] studied the infrared spectra of cyanoethylcellulose, ethylcellulose, acetylcellulose, acetobutyrylcellulose, etc., in the regions of the main frequencies of OH and CH groups. The spectra of a large number of cellulose esters and ethers between 2 and 15 μ were given by Kuhn, mainly for identification purposes [124]. Infrared spectroscopy has been used to investigate the characteristics of a number of cellulose esters, obtained from woods of various ages [33-35]. Gerbaux [155] used analysis of the symmetry of hydroxyl bands in the infrared absorption spectrum for determining specific features of hydrogen bonding in cellulose acetates, obtained by homogeneous and heterogeneous methods. Spectra of a large number of cellulose esters were obtained in the work of O'Connor and his co-workers [112, 114]. Bouriot [156] investigated the polarization spectra of cellulose acetates with various degrees of acetylation. The infrared spectra of cellulose fibers, with different degrees of acetylation− from the triacetate to the almost completely hydrolyzed product − have been studied in the region 400-3600 cm^{-1} [25]. A spectroscopic investigation of ethylcellulose, with an emphasis on infrared spectroscopy, was carried out by Katibnikov, Ermolenko, and others [47]. Hürtubise [157] attempted the quantitative determination of acetyl groups in acetylcellulose, and interpreted some of the main frequencies. Infrared spectroscopy was used by Kurlyankina, Polyak, and Koz'mina [48] to investigate the thermal oxidation of cellulose esters.

Kozlov, Zhbankov, Zueva, Ivanova, and Podgorodetskii [49-50] applied infrared spectroscopy in their chemical investigations on the special features of hydrogen bonds in homogeneous and heterogeneous cellulose acetates, and arrived at some conclusions on the effects of these bonds on the specific properties of these products.

Elina, Gusev, and Ermolenko [104] investigated the possibility of determining the contents of both carboxyl and acetyl groups in partially acetylated samples of cellulose.

The large amount of experimental material now available provides a basis for believing that infrared spectroscopy may be effective for elucidating the following problems, concerned with the structure of cellulose esters and ethers: 1) identification of compounds; 2) analysis of hydrogen bonding of the free hydroxyl groups; 3) investigation of the extent of substitution of hydroxyl groups; 4) elucidation of the structural features of a product, depending on the conditions of esterification or hydrolysis; 5) studies on the kinetics of esterification and hydrolysis; 6) establishment of the existence of ester bonds.

Identification of a cellulose ester, by means of its infrared spectrum, is a relatively simple experimental task, provided that the compound is not too unusual, and that an adequate catalog of spectra is available. Cellulose esters and ethers usually have spectra which are specific for each compound, reflecting the presence of new functional groups and bonds and any changes in the polymer structure (Appendix V, 118-142). Thus, the spectrum of ethylcellulose is characterized by the presence of sharp bands at 2970, 1450, and 1380 cm^{-1} (C_2H_5); that of acetylcellulose by bands at 1740, 1380, and 600 cm^{-1} (CH_3CO); that of nitrocellulose by strong bands at about 1650, 1280, and 840 cm^{-1} (ONO_2); and so on. The spectra of cellulose ethers containing aromatic groups show various bands characteristic of the substituent benzene rings: at 3090, 3060, 3030, 1610, 1500, 740, and 700 cm^{-1} for benzylcellulose; 3090, 3060, 3030, 1600, 1490, 765, 750, and 700 cm^{-1} for tritylcellulose. In these cases, the intensity ratios of the bands, corresponding in frequency to the stretching vibrations of CH in benzene rings (2900-3100 cm^{-1}), also have special features.

The spectra of cellulose esters and ethers also show other absorption bands, which have hitherto been difficult to interpret. For example, acetylcellulose shows a sharp spectral band at 900 cm^{-1}, with an inflection at 880 cm^{-1}, and weaker bands at 950 and 840 cm^{-1}; ethylcellulose shows bands at 920, 880, and 820 cm^{-1}; tritylcellulose shows a sharp band at 900 cm^{-1}, with an inflection at 920 cm^{-1}, and a weak band at 850 cm^{-1}; benzylcellulose shows bands at 920, 850, and 820 cm^{-1}; cellulose benzoate shows sharp bands at 940, 850, and 805 cm^{-1}, and so on. It is possible that some of the bands in this region reflect structural features of the cellulose derivatives, caused by the introduction of new functional groups.

Important information on the structure of cellulose esters can be obtained from the analysis of OH group absorption bands. Even a qualitative inspection in this region makes it possible to estimate the completeness of the esterification process and its approximate value. The spectra of many samples of triacetylcellulose show a weak band at 3480 cm^{-1}; this cannot be attributed to an overtone of the $C=O$ (1750 cm^{-1}) band, since it is absent from the spectra of other cellulose esters of organic acids, e.g., cellulose tristearate. We must therefore accept the presence of a few unesterified hydroxyl groups in triacetylcellulose. The changes in contour intensity and frequency of the main maximum of the OH group band, in the 3200-3600 cm^{-1} region, can be used to investigate special features in the reactions of esterification and hydrolysis.

It is now an accepted fact that practically all the hydroxyl groups in cellulose are involved in hydrogen bonding. However, an opposite view has been advanced with respect to cellulose esters. Thus, Nikitin considered that the introduction of bulky ester groups into the cellulose structure should lead to rupture of the hydrogen bonds. It was his view that the hydroxyl groups in nitrocellulose and acetylcellulose are free from hydrogen bonding [46]. It should be noted, however, that this view was based on analysis of the spectra in the hydrogen bond overtone regions, and also in the fundamental frequency region, where the dispersive power of the NaCl prism system used was relatively poor. Analysis of spectra in the first overtone region can give rise to confusion between the frequencies of OH groups involved in hydrogen bonding of the type OH . . . O and of free hydroxyl groups [22]. It is not possible to obtain sufficiently reliable results, when analyzing the hydroxyl group region, if a spectrometer with an NaCl prism is used. For example, Nikitin was unable to distinguish the difference between the OH group frequencies of nitrocellulose and acetylcellulose.

No systematic investigation has yet been carried out on the hydrogen bonding of hydroxyl groups in cellulose esters of various structures. Our own work on the analysis of highly substituted ($\gamma \geq 200$) cellulose esters and

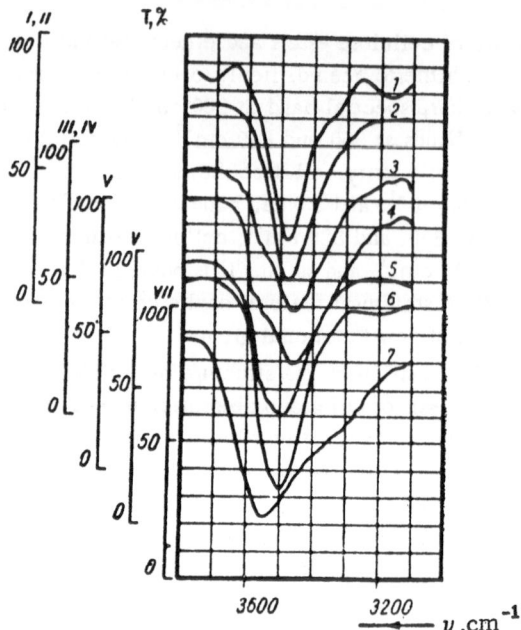

Fig. 51. Spectra of highly substituted (≥2) cellulose esters and ethers. 1) Ethylcellulose; 2) nonylcellulose; 3) benzylcellulose; 4) tritylcellulose; 5) cellulose acetoacetate; 6) cellulose 2-dimethylpropionate; 7) nitrocellulose. The film thicknesses are arbitrary.

ethers (apart from nitrocellulose) has shown that the maxima of the main hydroxyl group bands are located within the fairly narrow frequency range of $3470-3520$ cm^{-1} (Fig. 51). These maxima frequencies are as follows: tosylcellulose ($\gamma = 200$), 3520 cm^{-1}; cellulose 2-dimethylpropionate ($\gamma = 290$), 3490 cm^{-1}; acetylcellulose ($\gamma = 230-290$), nonylcellulose ($\gamma = 190$), and ethylcellulose ($\gamma = 230$), 3480 cm^{-1}; benzylcellulose ($\gamma = 200$), 3470 cm^{-1}, and so on. The range $3470-3520$ cm^{-1} lies in the frequency region of hydroxyl group hydrogen bonds of the dimer type OH . . . OH [24]. It should be noted that, in the spectra of highly substituted cellulose esters and ethers, there are no absorption bands which can with certainty be attributed to free hydroxyl groups.

Nitrocellulose is an interesting exception to the above. The hydroxyl group band in its spectrum is considerably displaced to a longer wavelength and has its main maximum in the $3550-3580$ cm^{-1} region. The very diffuse character and complex structure of the hydroxyl group bands of nitrocellulose may indicate the presence of several types of hydrogen bonds.

It is known that there is a correlation between the frequency of a hydrogen bond and its energy [23]. It must therefore be assumed that in cellulose nitrates there are several types of energetically nonuniform hydrogen bonds, with a predominance of considerably weaker bonds than in other cellulose esters and ethers.

The specific character of the hydroxyl group hydrogen bonds in nitrocellulose is of considerable interest and requires special consideration. There is no doubt that the fundamental nature of these bonds is to some extent determined by a number of the specific properties of this ester.

In the spectra of highly substituted cellulose esters the relatively small variation in frequency of the hydroxyl group band with the nature of the esterifying group gives grounds for supposing that the variation is mainly due to intramolecular hydrogen bonding.

It is well known that the esterification of cellulose can be carried out in a homogeneous or heterogeneous medium. Although such reactions, going to completion in a homogeneous medium, are relatively seldom used under industrial conditions, it is nevertheless of considerable interest to compare the special features of esters obtained by the two methods. For example, a cellulose esterification, beginning in a heterogeneous medium and finishing in a homogeneous medium, has been widely used [19].

Cellulose esters, obtained under various esterification conditions, can differ widely in their properties: solubility, viscosity, etc. It has been pointed out [19] that the degree of esterification does not in itself provide a basis for predicting the solubility of a cellulose ester; a knowledge of the previous history of the product is also necessary.

The possibilities of infrared spectroscopy in the study of the structural features of cellulose esters, obtained under various conditions of esterification and hydrolysis, can be illustrated by the example of highly substituted cellulose acetates, products which are widely used in industry.

The spectra of highly substituted cellulose acetates, obtained by homogeneous and heterogeneous methods, have been studied and compared with viscosity data [49, 50]. It is well known that cellulose acetate can be produced industrially by two different methods, homogeneous and heterogeneous. In the homogeneous process, the acetylation of cellulose and partial hydrolysis of the resulting cellulose triacetate are carried out under homogeneous conditions, i.e., with dissolution of the primary product in the acetylating mixture, and with subsequent partial hydrolysis also in solution. On the other hand, the heterogeneous process is carried out under heterogeneous conditions, i.e., without any change in the fibrous structure of the

Table 6. Cellulose Acetates, Produced under Homogeneous Conditions of Acetylation and Partial Hydrolysis

Properties	Samples of cellulose acetate produced in the USSR			Cellulose acetate made by the Hercules Co.
	No. 451	No. 132	No. 1257	
Combined acetic acid content, %	60.22	60.3	60.52	60.12
Ball viscosity of 12.5% solution in 9:1 mixture methylene chloride and methanol, sec	46.4	40.0	24.5	10.8
Specific viscosity of 0.25% solution in 9:1 mixture methylene chloride and methanol	0.74	0.78	0.89	0.52
Acidity, %	0.002	0.003	0.006	0.001
Ash, %	0.05	0.09	0.05	0.04
Moisture content, %	1.5	0.6	0.9	1.0
Content of low-molecular fractions, %	5.2	3.4	1.85	8.26
Thermal stability, °C	190.0	195.0	195.0	235.0
Filtration modulus	164.0	101.0	79.0	79.0

initial cellulose. It is natural to suppose that the special features of the processes of acetylation and hydrolysis must affect the properties of the final products.

Although the differences in properties of these products have been studied by numerous investigators [19, 51, 52, 158], their work has not completely elucidated the differences in solubility.

We investigated [49, 50] industrial samples of partially hydrolyzed cellulose triacetates, produced by homogeneous and heterogeneous acetylation and partial hydrolysis. The infrared absorption spectra were obtained with a UR-10 double-beam recording spectrophotometer, using an LiF prism. In order to eliminate the effects of the morphological macrostructure, the spectra were all obtained with 20- to 25-μ films, produced by dissolving the samples in methylene chloride and drying carefully at 100°C. Special vacuum cells were used for recording the spectra. The humidity of the samples was checked by means of the absorption band at 3630 cm^{-1}. This band was very strong in undried samples, and it could therefore be attributed to water adsorbed by the cellulose acetates. The methylene chloride, used as solvent,

Table 7. Cellulose Acetates, Produced under Heterogeneous Conditions of Acetylation and Partial Hydrolysis

Properties	Industrial samples			Laboratory samples	
	No. 187	No. 25	No. 162	No. 1	No. 2
Combined acetic acid content,%	60.0	60.5	62.5	60.57	61.08
Ball viscosity of 12.5% solution in 9:1 mixture of methylene chloride and methanol, sec	37.0	36.0	58.0	–	–
Specific viscosity of 0.25% solution in 9:1 mixture of methylene chloride and methanol	0.70	0.79	0.78	0.79	0.79
Acidity, %	0.005	0.005	0.005	–	–
Ash, %	0.026	0.02	0.02	–	–
Moisture content, %	0.65	1.45	2.40	–	–
Content of low molecular fraction, %	9.26	18.40	2.00	–	–
Thermal stability, °C	220.0	220.0	220.0	–	–

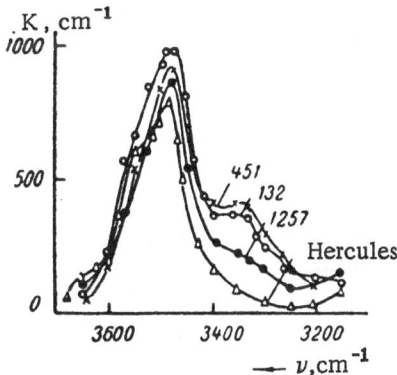

Fig. 52. Spectra of cellulose acetates produced under homogeneous conditions. The numbers on the curves refer to the sample numbers in Table 6.

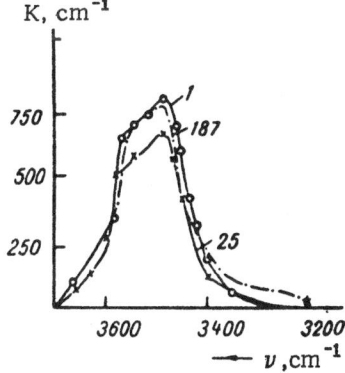

Fig. 53. Spectra of cellulose acetates produced under heterogeneous conditions. Numbers on the curves refer to the sample numbers in Table 7.

was previously neutralized, dried over ignited potash, and redistilled. The fraction boiling between 40 and 41°C at 760 mm Hg was collected; this had d^{20} = 1.3275. The solutions of cellulose acetates in methylene chloride, used for the viscosity measurements, were prepared by stirring for 5-6 h and were freed from mechanical impurities by filtration through Nos. 1 and 2 glass filters under pressure. Each primary solution was divided into several parts, to which increasing quantities of methanol were added, together with enough methylene chloride to give the same total volume, and the resulting solutions were then further stirred for 3-4 h. The cellulose acetate concentrations were expressed in grams per 100 ml of solution. The amount of methyl alcohol in the mixture was expressed as $\log(10^2 \omega)$, where ω was the ratio of the amount of methyl alcohol to the amount of hydroxyl groups in the cellulose acetate, expressed as g mole of alcohol and of OH group. The viscosities of the solutions were measured with suspended-level Ubbelohde viscometers, of capillary length not less than 80 mm and radius 0.015-0.049 cm. The calibration constants of the viscometers were determined with liquids of known absolute viscosity (water and m-cresol). The observed viscosities of the cellulose acetate solutions were expressed in absolute units in centipoises. The results shown were the means of several measurements. Tables 6 and 7 show the characteristics of the products investigated, and the spectra are shown in Figs. 52 and 53.

Cellulose Acetates Produced under Homogeneous Conditions of Acetylation and Partial Hydrolysis

In the spectra of the homogeneous cellulose acetates investigated there was a very strong band at about 3480 cm^{-1} in the hydroxyl group region, and with some of the samples there was a complex band in the region 3250-3400 cm^{-1} (Fig. 52). There was a less distinct band, which did not appear in all the samples, at 3550 cm^{-1}. The 3480 cm^{-1} band also appeared in the spectrum of cellulose triacetate. However, this band cannot be attributed to an overtone of the $C = O$ carbonyl group frequency. It has been shown [154] that the first overtone of the stretching frequency of the carbonyl groups in cellulose esters makes no significant contribution to the hydroxyl group band. For example, there is no absorption at all here in the spectrum of cellulose tristearate. Thus, the 3480 cm^{-1} band in the spectrum of acetylcellulose must be wholly attributed to unesterified hydroxyl groups. Bouriot [156], on the basis of the difference in frequency between the maxima of the OH group in the spectra of acetylcellulose (3480 cm^{-1}) and cellulose (3400 cm^{-1}), concluded that the hydrogen bonds in the two compounds were of different types. He attributed the 3480 cm^{-1} band to hydroxyl groups involved in hydrogen bonding of the type OH . . . O $=$ C, and considered that the presence of this type of hydrogen bond was demonstrated by a shift in the $C = O$ frequency of

11 cm^{-1}, toward longer wavelength, on changing from cellulose triacetate to acetyl cellulose of γ = 74. Bouriot supported his interpretation by reference to known literature data on the values of the OH frequencies in some compounds where hydrogen bonding of this type is possible. However, there is really insufficient evidence to confirm all this. The frequency differences between the OH band maxima in the spectra of acetylcellulose and cellulose do not establish the existence of different types of hydrogen bonds in the two compounds. The broad diffuse hydroxyl band in the cellulose spectrum is undoubtedly complex in character because of the existence of many types of hydrogen bonds, whereas only one of these types may predominate in acetylcellulose. The 10-11 cm^{-1} shift of the C$=$O frequency, on changing from cellulose triacetate to a partially hydrolyzed product, is not in itself a proof that the C$=$O group is associated, although the possibility of this is not excluded. It is known that the C$=$O frequency is sensitive to a change in electron density, caused by adjacent polar groups [53], and it is difficult to accept Bouriot's view that each unesterified hydroxyl group in acetylcellulose is associated to form a C$=$O . . . H$-$O type of hydrogen bond. The presence of only one band at 3480 cm^{-1}, with the acetylcelluloses he investigated, does not provide a basis for his opinion. The 3480 cm^{-1} band is not specific for hydrogen bonds of the type C$=$O . . . H$-$O. As noted above, this band has been observed in the spectra of many highly substituted cellulose ethers. In particular, a band in this position has been observed in the spectrum of tritylcellulose. In this case, the large volume of the trityl group suggests that the bond is attributable to intramolecular hydrogen bonding. The trityl ether of cellulose is known to be characterized by the presence of an ether group at the sixth carbon atom. Therefore, the band under consideration should be attributed, in the case of trityl cellulose, to intramolecular hydrogen bonds involving secondary hydroxyl groups. It is highly probable that the 3480 cm^{-1} band is of similar origin in the case of acetylcellulose.

In order to confirm our suggestion, that this band is attributable to intramolecular hydrogen bonding in the case of acetylcellulose, we obtained the spectra of dilute solutions of acetylcellulose in chloroform. No reduction in the intensity of the 3480 cm^{-1} band was detected with the change in solvent. The hypothesis that this band is attributable to intramolecular hydrogen bonding has also been put forward [156] on the grounds of the parallel dichroism of the band (Fig. 54). However, as already explained, in this case the 3480 cm^{-1} band was attributed only to hydrogen bonds of the type C$=$O . . . H$-$O. It should also be noted that the 3480 cm^{-1} band has been observed in the spectra of cellulose films, regenerated from viscose [151] and alkaline cellulose [148], and has been attributed to intramolecular hydrogen bonding [148]. With some samples, the most clearly evident bands are in the region 3300-3400 cm^{-1}, which is known to include the stretching vibration

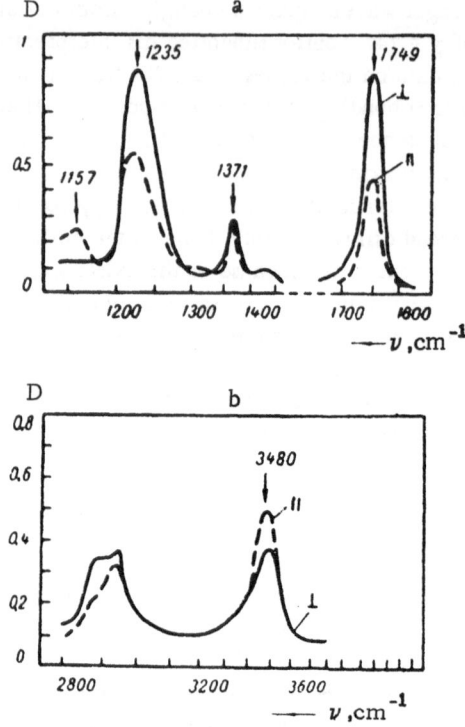

Fig. 54. Spectra of oriented films of secondary cellulose acetates. ⊥ — electric vector
perpendicular to direction of macromolecular
chain; ‖ — electric vector parallel to direction of macromolecular chain [156].

frequencies of OH groups involved in relatively strong hydrogen bonds [22,
24, 53]. The existence of strong hydrogen bonds in cellulose acetates must
affect the solubilities of the products and structure formation in solutions. A
comparison of the above results with data on the filtration moduli of the
cellulose acetates investigated should lead to some conclusions on the filtration quality of their solutions. The existence of a 3300-3400 cm^{-1} band and
an increase in its intensity corresponded to a deterioration in filtrability.
However, it should be noted that repeated checks did not show that there was
a precise relation between filtrability and the intensity of this absorption
band. On the other hand, in all cases where there were bands in the 3300-
3400 cm^{-1} region, the solutions were of poor filtrability. Cellulose acetate
samples, which did not show a band in the 3300-3400 cm^{-1} region, or showed
only a weak band, gave solutions of high filtrability (samples No. 1257 and
the Hercules Co. sample).

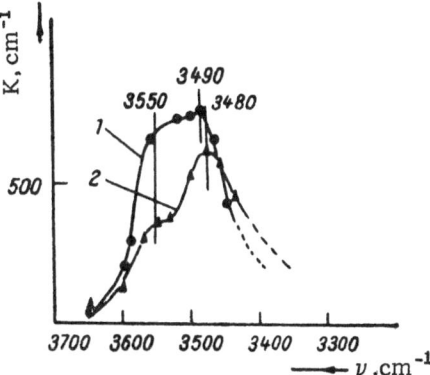

Fig. 55. Spectra of cellulose acetates,
produced under heterogeneous conditions
of acetylation and partial hydrolysis: 1)
before; 2) after introduction of plasticizer.

The conclusion is that infrared spectroscopic methods, for the analysis
of cellulose acetates produced by homogeneous means, can undoubtedly
supply useful information for normalization of the technological process and
improvement of the quality of cellulose acetates.

Cellulose Acetates Produced under Heterogeneous Conditions of Acetylation and Partial Hydrolysis

The spectra of partially hydrolyzed heterogeneous cellulose acetates
show the band at about 3480 cm^{-1}, but are also characterized by a distinct
band in the 3550 cm^{-1} region (Fig. 53). The latter is hardly detectable in
the spectrum of cellulose triacetate, and greatly increases in intensity at the
start of the hydrolysis process [25]. The interpretation of this band, which appears most distinctly in the spectra of partially hydrolyzed samples of heterogeneous acetylcellulose, is undoubtedly of considerable interest.

This band was first observed by Brown and his co-workers [154], when
analyzing the spectra of films of acetylcelluloses, produced by acetylation
in a heterogeneous medium. These authors attributed it to unassociated hydroxyl groups at the sixth carbon atom, which they believed could not take
part in hydrogen bonding because of steric factors. However, in our opinion,
attribution of the 3550 cm^{-1} band to free hydroxyl groups is inadequately established. The band lies in the frequency range of hydroxyl groups involved
in hydrogen bonding [22, 53]. For example, the spectrum of tritylcellulose
shows a band of still higher frequency — 3570 cm^{-1}, superimposed on the
strong background band at about 3480 cm^{-1}.

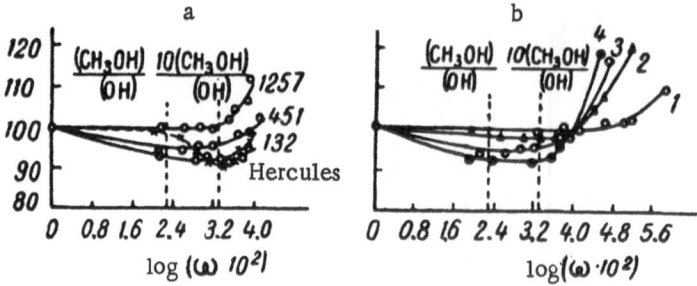

Fig. 56. Changes in the absolute viscosity of cellulose acetates,
produced under homogeneous conditions. a) Solutions of cellulose acetates in methylene chloride with addition of methyl alcohol (the numbers on the curves are the sample numbers); b) cellulose acetate solutions (sample 451), of various concentrations, in
methylene chloride with addition of methyl alcohol: 1) 0.16; 2)
0.92; 3) 2.32; 4) 3.84 g per 100 ml of solution.

It is known that, in the initial stages of hydrolysis, there is preferential
restoration of free hydroxyl groups at the sixth carbon atom [19]. It is
reasonable to suppose that one of the most probable types of intermolecular
hydrogen bonds in cellulose is one involving participation of primary hydroxyl groups. The work of Purves [159, 160] has established that the hydroxyl groups are distributed more locally in the structure of cellulose acetates produced heterogeneously, than with materials produced homogeneously. Gerbaux [161] pointed out correctly that, with a localized distribution of
a small number of hydroxyl groups (in our case, 12-15 per 100 rings), there
is a greater probability of formation of OH . . . OH intermolecular hydrogen
bonds than in the case of a uniform distribution. This argument provides a
basis for attributing the 3550 cm^{-1} band, which appears particularly in the
initial stages of hydrolysis of heterogeneous cellulose acetates, to intermolecular hydrogen bonds. To confirm this hypothesis, we investigated [49, 50]
the spectra of films of partially hydrolyzed heterogeneous cellulose acetate,
which had been plasticized with the intermicellular plasticizer trimono-
chloroethyl phosphate. Figure 55 shows the very great reduction in intensity
of the 3550 cm^{-1} band when the plasticizer was introduced.

It is known that the frequency of hydroxyl groups involved in hydrogen
bonding is a definite measure of the bond energy [23]. Thus, the 3350 cm^{-1}
band must be attributed to very weak hydrogen bonds. The existence of such
weak intermolecular hydrogen bonds is in all probability due to steric factors,
associated with the introduction of a large number of acetyl groups.

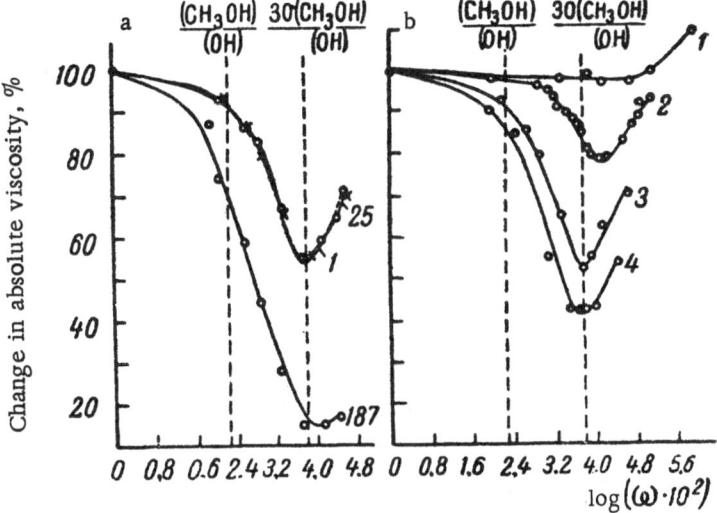

Fig. 57. Changes in the absolute viscosity of cellulose acetates, produced under heterogeneous conditions. a) As a function of the quantity of methyl alcohol added (the numbers on the curves are the sample numbers); b) as a function of the quantity of methyl alcohol added and of the polymer concentration: 1) 0.15; 2) 0.98; 3) 2.31; 4) 3.82 g per 100 ml of solution.

Comparison of Spectral Data with the Results of Viscometric Analysis of Cellulose Acetates

Figures 56 and 57 show the results of investigations on the absolute viscosities of cellulose acetate solutions in methylene chloride, with increasing contents of methyl alcohol [49, 50]. It follows from Fig. 56 that, as a rule, highly substituted cellulose acetates, produced by the homogeneous method, are characterized by a characteristic change in viscosity of a solution in methylene chloride, with increasing methanol content. The viscosity of the system initially decreases, as small amounts of methyl alcohol are added, up to a mole ratio of methyl alcohol to OH groups in cellulose acetate of 1:1, but subsequently remains practically constant.

On the other hand, Fig. 57 shows that highly substituted cellulose acetates, produced by the heterogeneous method, are characterized by a different characteristic pattern for the change in viscosity of a solution in methylene chloride, with increasing methanol content. Initially, with addition of small amounts of methyl alcohol, sufficient for complete blocking of the OH groups in the cellulose acetate, i.e., with 1 g-mole of methanol per g-mole

of OH group (CH_3OH/OH), there is a slow drop of 8-12% of the initial viscosity; there is then a further sharp decrease in viscosity as more methanol
was added; finally, the viscosity begins to increase again because of the reduction in the solvent power of the mixture.

Analysis of the effects of solution concentration, and of the degree of
hydrolysis of the polymer, established that the sharp decrease in viscosity
could be fully explained by rupture of the intermolecular bonds of the OH
groups in cellulose acetates, and not be depolymerization or conformational
changes of the molecules.

As shown above, the special features of the infrared spectra of cellulose acetates, produced by homogeneous and heterogeneous acetylation and
partial hydrolysis, are dependent on the characteristics of the hydroxyl group
hydrogen bonds.

If we exclude the possibility of the existence in partially hydrolyzed
cellulose triacetates of relatively strong hydrogen bonds (the 3300-3400 cm^{-1}
band), which in our view is due to defects in the technology of their production, then a characteristic of these products is the preferential existence of
intramolecular hydrogen bonds (the 3480 cm^{-1} band).

With these products the possibility of intermolecular interaction is
small, and the addition to their solutions in methylene chloride of small
amounts of methanol, sufficient to block the OH groups, completely eliminates the structural component of the viscosity, associated with the interaction of OH groups in the cellulose acetate macromolecules. For this reason,
the absolute viscosities of such solutions are relatively small, and the structural component does not exceed a few percent of the total value of the viscosity.

The matter becomes more complicated with the establishment of a
relation between the viscosity of a solution and the character of the intermolecular interaction, in the case of highly substituted cellulose acetates
produced by heterogeneous means. The observed anomalous viscosity with
solutions of these products can be explained by the existence of weak intermolecular hydrogen bonds, which give rise to the 3550 cm^{-1} band in the
spectra (Fig. 53). As explained above, this band is typical for such products
and can be attributed to weak intermolecular hydrogen bonds, in which it is
mainly primary hydroxyl groups that participate. However, it should be
understood that, to break down these bonds it is necessary to use a large
amount of alcohol, greatly exceeding the amount calculated on the assumption that each OH group can be blocked by one molecule of alcohol. The
absolute viscosities of solutions of these products in methylene chloride, but
with the structural component eliminated by addition of the optimum quantity
of methanol, amount to only about ten percent of the original total viscosities.

If we assume that the cause of the anomalous viscosity of these solutions lies in the possibility of forming weak intermolecular hydrogen bonds, as the result of localized distribution of hydroxyl groups (as suggested by Gerbaux [155]), then we are probably concerned with hydrogen bonds formed by the primary hydroxyl groups of the polymer. The experimental results indicate that these weak intermolecular hydrogen bonds are formed to a much greater extent with highly substituted cellulose acetates produced by heterogeneous means, than in the case of cellulose acetates produced homogeneously. Thus, the absolute viscosity of the solutions and its structural component should be much greater in the case of heterogeneous than of homogeneous cellulose acetates, and this should give rise to viscosity anomalies.

However, this hypothesis introduces some doubts as to whether the explanation is clearly established, because it is difficult to understand how a limited number of primary hydroxyl groups, quite separate from each other in localized positions, can so strongly affect the reduction in viscosity of a cellulose acetate solution resulting from blockage of the OH groups by large amounts of alcohol.

It would be interesting to explore the possibility of formation, in such products, of chemical bonds giving a three-dimensional structure [19]. It also remains to be considered whether the intermolecular interaction between the OH groups of cellulose acetate and alcohol is as strong or less strong than the interaction between the OH groups of cellulose acetate macromolecules. In the latter case it would require more alcohol to block these groups and produce a marked reduction in viscosity of the solution by the value of the structural component. In this connection, it would be valuable to find effective low-molecular substances for blocking the hydroxyl groups of the analyzed products in solution [50]. Infrared spectroscopy should be very useful in solving these practically important problems.

Acetylcellulose Fibers with Different Acetyl Group Contents

It has already been mentioned that the change in character of the hydroxyl group band can be used to investigate special features of the reactions of esterification and hydrolysis of cellulose.

We made a study [25] of the spectra of cellulose acetate fibers, ranging from the triacetate to almost completely hydrolyzed material (Table 8).

It is clear from Fig. 58 that the initial stages of hydrolysis were accompanied by a restoration of the longwave part of the band contour, corresponding to hydroxyl groups with the lowest hydrogen bond energy. In the later stages of hydrolysis there was preferentially increased absorption in the longer wavelength part of the band.

Table 8. Characteristics of the Cellulose Acetates Investigated,
with Various Acetyl Group Contents

Sample	Hydrolysis time, h	Combined acetic acid, %	Acetal groups,%	No. of substituted hydroxyl groups in 100 rings
1	0	62.57	44.80	300
2	24	61.08	43.77	288
3	35	60.30	43.24	282
4	46	58.24	41.79	266
5	72	57.45	41.22	260
6	80	55.71	39.97	247
7	97	53.73	38.45	232
8	146	49.73	35.64	206
9	384	37.90	27.12	139
10	696	24.63	17.59	80
11	1850	1.47	1.05	4

Note. The cellulose triacetate was hydrolyzed at room temperature, which varied from 21 to 26°C.

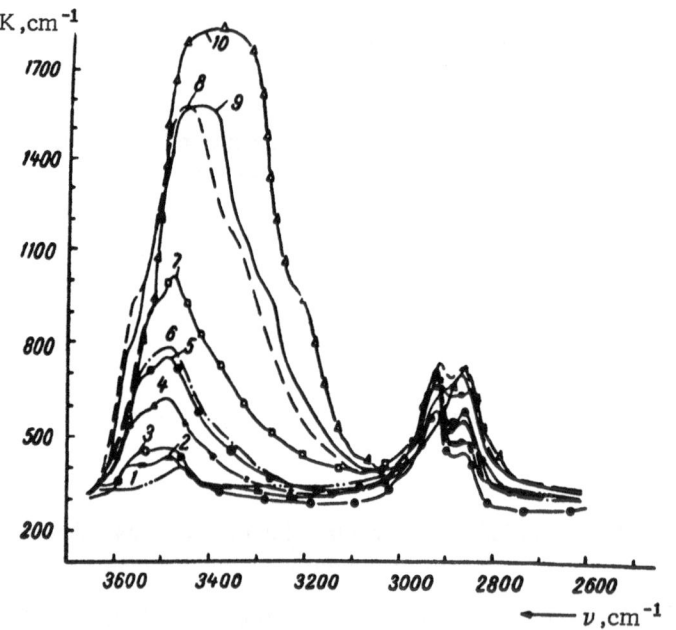

Fig. 58. Spectra of: 1) cellulose triacetate, of the same product hydrolyzed for 2) 24; 3) 35; 4) 46; 5) 86; 6) 97; 7) 147; 8)
384; 9) 696 h; 10) wood cellulose (see Table 8).

For example, the spectrum of cellulose acetate containing 37.9% of combined acetic acid shows bands at about 3550 and 3480 cm^{-1}, but also a distinct band in the 3370 cm^{-1} region.

The appearance of the latter band indicates that a new type of hydrogen bond was being formed. In order to characterize the nature of the restoration of the OH group band contour, we plotted the relations between the extinction coefficients at 3550, 3480, 3370, and 3000 cm^{-1} and the logarithm of the hydrolysis time (Fig. 59), and the relations between the extinction coefficients and the quantity of combined acetic acid (Fig. 60). It is evident from these figures that, in the course of the first 100-150 h of hydrolysis, there was a preferential increase in intensity of the 3550 and 3480 cm^{-1} bands; absorption in the 3370 and 3300 cm^{-1} regions increased only slightly. As hydrolysis continued, the band at 3550 cm^{-1} was the first to reach its maximum intensity, followed by the band at 3480 cm^{-1}, while the 3370 and 3350 cm^{-1} bands rapidly increased in intensity. A similar rule was then observed with these latter bands: The 3370 cm^{-1} band increased most rapidly at first, but eventually attained its steady maximum value, while the 3300 cm^{-1} band continued to increase in intensity. These results were compared with viscosity data on dilute solutions of the same materials. It has been shown [25] that the viscosity of cellulose acetate in m-cresol increases with the splitting off of combined acetic acid. Since practically all the hydroxyl groups in cellulose acetates are involved in hydrogen bonding, this increase in viscosity can be attributed to an increase in the degree of association, resulting from intermolecular hydrogen bonds. Samples of cellulose triacetate, which have been hydrolyzed until they contain only 49.7% of combined acetic acid, are no longer soluble in m-cresol. It can be seen from Fig. 59 that it is precisely these samples which showed a marked increase in absorption in the 3370 and 3300 cm^{-1} regions. These frequencies must therefore correspond to relatively strong intermolecular hydrogen bonds of the type OH . . . OH . . . OH.

It is natural that the form of the curves, shown in Figs. 59 and 60, should be determined by the nature of the ester group, the location of the unsubstituted hydroxyl groups in the hydrolysis product, the conditions under which the reaction was carried out, etc. A comparison of the character of the restoration of the various parts of the hydroxyl group band contour can give definite information as to the properties of the cellulose under investigation, the conditions under which it has been treated, its solubility, etc. In order to achieve this, it is useful first to construct standard curves, based on the spectral analysis of model compounds.

In the spectrum of some cellulose esters — the acetate and nitrate — there is a reduction in intensity of the band at about 2900 cm^{-1}, as compared

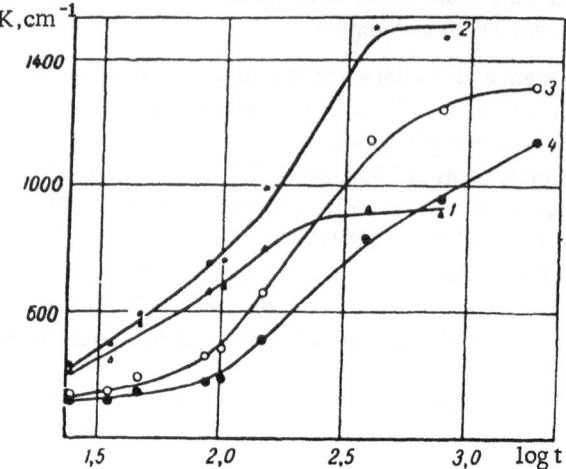

Fig. 59. Variation of the extinction coefficient in the spectrum of cellulose acetates at: 1) 3550; 2) 3480; 3) 3370; 4) 3300 cm^{-1}, with the logarithm of the hydrolysis time.

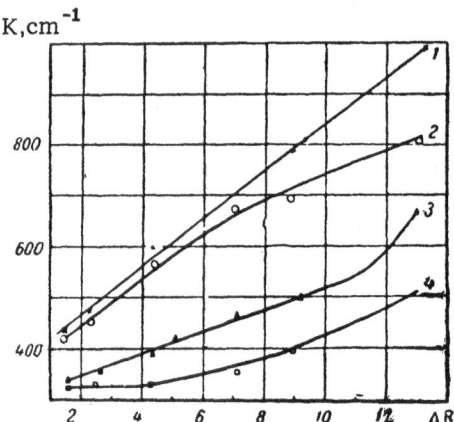

Fig. 60. Variation of the extinction coefficient in the spectrum of cellulose acetates at: 1) 3550; 2) 3480; 3) 3370; 4) 3300 cm^{-1}, with the difference between the combined acetic acid contents of the sample and of cellulose triacetate.

with its intensity in the cellulose spectrum. The stretching vibrations of CH_2 and CH lie in this region. It is reasonable to suppose that the CH_2 groups play an important part in determining the integral absorption here [17]. A reduction in intensity of this band can be interpreted as due to the effect of double bonds in the ester groups. A similar effect has been noted with hydrocarbons [24].

Consequently, the change in absorption at 2900 cm^{-1} can be used to follow the kinetics of accumulation of acetyl and nitrate groups at the sixth carbon atom. It is convenient to carry out a similar type of analysis by means of the 2900 cm^{-1} band, in all cases when it is impossible to use the band corresponding to CH_2 group deformation vibrations. As shown above, the 1430 cm^{-1} band should be used very cautiously in analytical work, since it is extremely sensitive to structural transformations.

Consideration has also been given [25] to the kinetics of the increase in intensity of the 2900 cm^{-1} band during the hydrolysis of acetylcellulose. Investigations showed that the intensity of this band increased most sharply in the initial stages of the process. These conclusions agreed well with the results of chemical analyses, showing that there was preferential restoration of primary hydroxyl groups in the initial stages.

Special Features of the Spectra of Ethers and Esters of Cellulose

Differences in the spectra of cellulose ethers and esters are clearly visible in the 1000-1250 cm^{-1} region (Appendix V). Cellulose ethers show a broad diffuse absorption band here, without any well-defined structure. On the other hand, the spectra of all the cellulose esters investigated show not less than two quite well-defined absorption bands in the regions 1150-1250 cm^{-1} (the stronger band) and 1000-1100 cm^{-1}. The character of these bands is specific for each compound. Thus, with acetylcellulose, the bands are located at 1230 and 1050, with cellulose methylxanthate at 1220 and 1060, with cellulose 2-dimethylpropionate at 1150 and 1060 cm^{-1}, and so on.

We can trace the changes in the nature of these bands with the transition from acetylcellulose to cellulose esters containing more bulky groups — propionyl $OOCCH_2CH_3$, butyryl $OOC(CH_2)_2CH_3$, stearyl $OOC(CH_2)_{16}CH_3$, etc. Analysis of the spectra shows that changes of a specific type in the character of both bands accompany changes in the nature of the organic acid residue (Fig. 61). Thus, with the transition from triacetylcellulose to cellulose tripropionate, there is a considerable change in the frequency of the shorter wavelength band — from 1230 to 1170 cm^{-1}. However, a further increase in the volume of the acid residue does not greatly affect the frequency of this

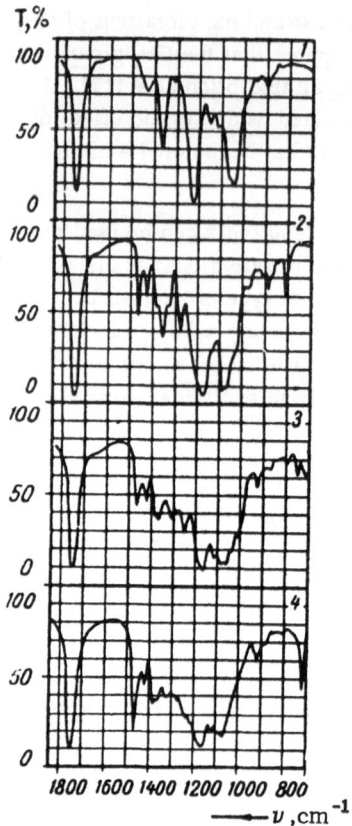

T,%

1800 1600 1400 1200 1000 800

$\longrightarrow \nu$, cm^{-1}

Fig. 61. Spectra of: 1) acetylcellulose; 2) cellulose propionate; 3) cellulose butyrate; 4)
cellulose stearate.

band. The change in frequency of the
longer wave band is less marked and proceeds in the opposite direction; thus, the
frequency is about 1050 cm^{-1} for triacetylcellulose and about 1070 cm^{-1} for cellulose tripropionate and tristearate.

The band in the 1000-1100 cm^{-1}
region differs from that in the 1150-1250
cm^{-1} region in having a more complex
structure. It has inflections at 1070 and
1050 cm^{-1} in the case of triacetylcellulose, at 1030, 1070, and 1085 cm^{-1} in the
case of cellulose tripropionate, and at
1010, 1040, 1060, 1080, and 1110 cm^{-1}
in the case of cellulose tributyrate, etc.
In our view this should be attributed
mainly to structural factors.

With increasing size of the organic
acid residue in the above series of cellulose esters (acetate, propionate, butyrate,
stearate), the resolution of the two main
bands between 1000 and 1250 cm^{-1} becomes less distinct. This circumstance
clearly arises from the appearance of the
new sub-bands (Fig. 61).

In order to elucidate the effects of
this "subdivision" by ester groups, we investigated the spectrum of cellulose 2-
dimethylpropionate. In the case of this
ester, the frequency of the band between 1150 and 1250 cm^{-1} was the lowest
observed, about 1150 cm^{-1}, while the band between 1000 and 1100 cm^{-1} was
a well-defined doublet with peaks at 1040 and 1070 cm^{-1}.

Thompson and Torkington [162], in their studies of a number of the
simplest esters (formates, acetates, butyrates, etc.), revealed the presence of
two strong spectral bands. The strongest was in the 1050-1250 cm^{-1} region,
while the weakest, in the range 1000-1200 cm^{-1}, was shifted to a varying extent, depending on the nature of the alcohol residue. For this reason, the
first band was attributed to the C—O bond in the carbonyl group, and the
second to the C—O bond in the alcohol residue. On the other hand, the alcohol group remained the same in the cellulose esters which we investigated,

and only the character of the R group varied in cellulose $-O-C{\displaystyle {\nearrow \atop \searrow }}_R$. It should be emphasized that we observed a considerable change in band frequency in the region 1100-1250 cm^{-1} with the change in R from CH_3 to CH_2CH_3 and $C(CH_3)_3$, but there was little change in frequency up through the series CH_2CH_3, $(CH_2)_2CH_3$, and $(CH_2)_{16}CH_3$. It might be supposed that the band in the 1100-1250 cm^{-1} region, in the spectra of cellulose esters, was due to vibration of the bond $-O-C{\displaystyle {\nearrow \atop \searrow }}$, and that the band in the 1000-1100 cm^{-1} region was due to alcoholic $C-O$. In fact, the spectrum of cellulose nitrate (cellulose$-O-NO_2$) shows no band in the 1100-1250 cm^{-1} region; there is only a band at about 1070 cm^{-1}.

We attach special importance to the above-noted band splitting and, in many cases, to the more complex band structure in the region 1000-1150 cm^{-1}. It will be remembered that the structure of this band becomes more complex with the transition from acetylcellulose to cellulose esters with more bulky ester groupings. This cannot be attributed to absorption by the $C-C$ bands of alkyl radicals. For example, the bands at 1040 and 1010 cm^{-1}, observed in the spectrum of cellulose tributyrate, do not appear in the spectrum of cellulose tristearate. It must be accepted that the specific features of the band in the 1000-1100 cm^{-1} region are due to conformational peculiarities of the cellulose esters under consideration, together with possible differences in the steric location of the ester groups.

It should be noted that, in individual cases, the band in the 1100-1250 cm^{-1} region was also complex in character. For example, the cellulose esters of chlorinated fatty acids showed spectral bands at 1170 and 1130 cm^{-1}. A further elucidation of the reasons for the splitting of the bands in the regions 1000-1100 and 1100-1250 cm^{-1}, in individual cases, would be of undoubted interest from the viewpoint of interpretation of the structural features of cellulose esters.

The spectra of some steroid acetates show splitting of the 1240 cm^{-1} acetate band, and this has usually been attributed to stereochemical factors. The band remains single when the acetate group occupies an equatorial position, and splits when the acetate group has an axial conformation. It is possible that the band splitting may be due to rotational isomerism [53].

Not much work has been published on unsaturated esters of cellulose. According to Colthup [163], all the esters which he investigated – the acrylate, malelate, benzoate, etc. – showed two strong spectral bands in the intervals 1250-1310 and 1000-1200 cm^{-1}. The short-wavelength ester group band in the spectrum of cellulose tribenzoate was at 1270 cm^{-1}. It should be noted that this is a higher frequency than for any of the esters which we investigated, and the band may be useful for identifying the nature of the $\overset{\diagdown}{\underset{\diagup}{C}}-O-C\overset{\diagup}{\diagdown}$ group.

The stretching vibration frequency of the $C=O$ group, in the spectra of cellulose esters containing the $\overset{\diagdown}{\underset{\diagup}{C}}-O-C\overset{\diagup}{\diagdown}$ grouping, is at about 1750 cm^{-1}. There is no significant displacement of this band with the change from acetylcellulose to cellulose stearate. This demonstrates that the $C=O$ frequency in such ester groupings is little affected by the chain length of the organic acid residue. The relative constancy of the carbonyl frequency in the spectra of the simplest esters, acetates, and butyrates, has been noted [24]. However, in the case of cellulose 2-dimethylpropionate, the $C=O$ frequency is somewhat displaced (about 10 cm^{-1}) toward longer wavelength, as compared with acetylcellulose, and this may indicate a specific effect of branching. A similar displacement of the $C=O$ frequency has been observed in the cellulose tribenzoate spectrum, and has been attributed to conjugation of the $C=O$ group with the benzene ring [24].

It follows, from what has been said above, that the presence of a band in the 1100-1250 cm^{-1} region demonstrates the presence in a cellulose ester of a $\overset{\diagdown}{\underset{\diagup}{C}}-O-C\overset{\diagup}{\diagdown}$ type ester grouping, while the character of the bands in the 1100-1250 and 1000-1100 cm^{-1} regions gives a definite measure of the nature of the ester group and its structural features.

The 1150-1250 cm^{-1} band is the most convenient to use for identification purposes, because it overlaps least with the absorption of cellulose itself. The intensity of the band in this region can be used directly to investigate the existence of ester bonds. It is desirable to make sure at once that there is no absorption in this region by any other groupings. As will be shown below, a band at about 1200 cm^{-1} in the spectra of cellulose esters of phosphorus-containing acids can be attributed, not to ester bonds, but to stretching vibrations of $P=O$.

The spectral region 700-1000 cm^{-1} must be considered separately. It has been mentioned that new bands appear in the spectra of cellulose esters and ethers in this region, which are difficult to interpret. Thus, the spectra of tritylcellulose, benzylcellulose, ethylcellulose, and cellulose tristearate show a band at 920 cm^{-1}. A sharp band at about 900 cm^{-1} is characteristic of acetylcellulose and tritylcellulose. The spectra of tritylcellulose, cyanoethylcellulose, cellulose benzoate, benzylcellulose, and acetylcellulose show bands of variable intensity at about 840-850 cm^{-1}, while those of ethylcellulose and tertiary butylcellulose show bands in the region 800-820 cm^{-1}.

Cellulose tribenzoate has a very characteristic spectrum in this region. There are sharp bands of approximately equal intensity at 940, 850, and 805 cm^{-1}.

In any consideration of the spectra of cellulose esters and ethers, it must be appreciated that absorption in this region is sensitive to structural changes in the cellulose macromolecule. As noted earlier, the transition from natural cellulose to the hydrocellulose structural modification is accompanied by an increase in intensity of the spectral band at 900 cm^{-1}. Marked differences in the 700-900 cm^{-1} region have been observed between the spectra of the α- and β-anomers of monosaccharides, and these have been attributed [125], in particular, to a change in the spatial location of the CH groups. It is reported [125] that an equatorial location of the C_1H group in the pyranoside structure is always accompanied by the appearance of a band at about 840 cm^{-1}.

Considerable changes in this spectral region are observed, for example, with the transition from pyranoside to 3,6-anhydropyranoside, corresponding to a different conformation of the pyranose rings [140].

It is not impossible that the series of bands in the 700-950 cm^{-1} region, in the spectra of cellulose esters and ethers, reflects unique conformational transformations in their structure. Indeed, it is a reasonable assumption that the introduction of ether or ester groupings, depending on their nature and positioning, must have a definite influence on the conformation of the pyran rings and the location of the CH_2OH groups.

The study of conformational transformations in cellulose esters and ethers, and the development of methods for their analysis, are of considerable scientific and practical interest. There are favorable prospects for the use of infrared spectroscopic methods in this direction, since these methods are known to be very sensitive to structural factors.

This region is of particular interest for studying the process of crystallization of cellulose esters and ethers. Investigations by the author, in cooperation with V.P. Komarom, P. V. Kozlov, and M.I. Rodionova, have shown that the process of crystallization is accompanied by the following main changes in the spectra of these compounds: 1) an increase in the intensity of all or most of the absorption bands; 2) a selective increase in the intensity of individual absorption bands; 3) the appearance of new absorption bands.

An improvement in the definition of the spectrum and an increase in the intensity of the absorption bands on crystallization are inherent with the cellulose esters and ethers under consideration. Selective increase in intensity of certain bands and the appearance of new bands are determined by the specific properties of the ester or ether. For example, on crystallization, the bands at 495, 585, and 850 cm^{-1} show a selective increase in intensity with cellulose tribenzoate, and the bands at 850 and 955 cm^{-1} with cellulose tributyrate, while new bands appear at about 625 cm^{-1} with triethylcellulose, and at 495 and 525 cm^{-1} with triacetylcellulose.

The general improvement in definition of the spectra of cellulose esters and ethers, which indicates an increase in the state of order of the macromolecules, is also more noticeable in the region 400-1000 cm^{-1}.

As noted earlier, the cellulose spectrum in the 400-700 cm^{-1} region consists of a diffuse broad absorption band as background, with several separate indistinct bands superimposed on it. The spectra of the cellulose esters and ethers, which have been investigated, show decreased absorption in this region as esterification proceeds. In fact, with highly substituted cellulose esters and ethers, absorption in this region is practically leveled out, except when a vibration frequency of the substituent group lies in this spectral range (e.g., the 600 cm^{-1} band of the CH_3COO group in acetylcellulose, the 650 cm^{-1} band of the C—Cl group in cellulose esters of chlorofatty acids, etc.). It must be accepted that absorption in the 400-700 cm^{-1} region of the cellulose spectrum is mainly associated with the hydroxyl groups of this polymer. This spectral region corresponds to the out-of-planar deformation vibrations of hydroxyl groups [138] as well as to the overtones of the hydrogen bonds. The use of this region for analytical purposes would, in all probability, give valuable information as to the nature of the hydroxyl groups in the structure of cellulose derivatives.

CHAPTER V

OXIDATION PRODUCTS OF CELLULOSE.
SALTS OF OXIDATION PRODUCTS OF CELLULOSE

The first spectra of a number of oxidized celluloses were obtained by Rowen, Hunt, and Plyler [164]. They used films of regenerated cellulose, oxidized by nitrogen dioxide or periodic acid. These authors erroneously attributed bands at 1640 and 1280 cm^{-1}, in the spectrum of cellulose oxidized by nitrogen dioxide, to the carboxyl group.

In a later paper [165], Rowen and his co-workers compared the spectra of cellulose oxidized with nitrogen dioxide with those of alginic acid, sodium alginate, and cellulose nitrate, and correctly concluded that the bands at 1640 and 1280 cm^{-1}, in the spectrum of cellulose oxidized with nitrogen dioxide, were due to the nitrate group.

The infrared spectra of cellulose oxidized with periodic acid were studied in a third paper [166]. The authors used strongly oxidized samples of cotton cellulose (50, 66, and 79% conversion of the rings), prepared for analysis by the Nujol emulsion method. It was found that the oxidation with periodate led to an unexpectedly small absorption in the $C = O$ frequency region. There was no absorption band at about 1610 cm^{-1}, which some authors had observed [124] in the spectra of D-glucose derivatives oxidized by periodic acid. On the basis of calculations, and analysis of the spectra of methyl-4,6-benzylidene-β-D-glucopyranoside and of the dihydrate of the product from its oxidation with periodic acid, Rowen and his co-workers concluded that the aldehyde groups in cellulose are either hydrates, or form hemiacetal bonds according to the scheme shown on page 118.

Forziati, Rowen, and Plyler [167] attempted a quantitative determination of the carboxyl groups in the cellulose structure. They constructed an

analytical curve, using 1 : 1 mixtures of a cellulose sample with a determined carboxyl group concentration and polystyrene. The 1610 cm^{-1} band
of polystyrene made it possible to use this material as an internal standard.
The components were ground up in a ball mill (to 100 mesh) and were placed
in a mineral oil medium. Cellulose samples with various carboxyl contents
were obtained by mixing unoxidized cellulose powder in various proportions
with the oxidation product obtained from it. The calibration curve was constructed with

$$k = \frac{\log\left(\dfrac{I_0}{I}\right) \; 1740 \; cm^{-1}}{\log\left(\dfrac{I_0}{I}\right) \; 1601 \; cm^{-1}}$$

as ordinate, and the carboxyl concentration as abscissa. The authors concluded that they could detect 0.1 mmole of carboxyl per gram of material.
However, they were wrong. The observed linear relation between the intensity of 1740 cm^{-1} band and the COOH content of the sample can be explained on the grounds that the calibration curve was constructed on the basis
of chemical analysis of only one sample of oxidized cellulose, diluted with
unoxidized cellulose powder. However, since other $C{=}O$ groups, as well as
carboxyl groups, produce absorption at 1740 cm^{-1}, the proposed method really
only gave the total number of $C{=}O$ bonds.

   Higgins and McKenzie [168] investigated the infrared spectra of
eucalyptus cellulose, oxidized by periodate under various conditions, and
showed that moisture affected the intensity of the 1740 cm^{-1} band. The fact
that this band increased in intensity during drying provided an explanation of
the discrepancy between the chemical evidence that these samples contained
aldehyde groups and the absence of the $C{=}O$ band from their spectra.
Higgins and McKenzie suggested that this was due to hydration of the aldehyde groups. Their results indicated that, under the experimental conditions,
oxidation of cellulose at C_2 and C_3 resulted in a more diffuse spectrum in the
region 900-1450 cm^{-1}. Carboxyl groups were only detectable in the spectra
of well-oxidized samples.

The present author was the first to investigate the spectra of a large number of selectively oxidized celluloses over the whole spectral region 650-4000 cm^{-1} [10]. Consideration was given to distinguishing features of the spectra of celluloses, oxidized to various extents by nitric acid (monocarboxycellulose*), periodic acid (dialdehydocellulose*), periodic acid and sodium chlorite (dicarboxycellulose*), and of oxidation products of cellulose nitrate. Ermolenko, Zhbankov, Ivanova, Lenshin, and Ivanov [54] considered the course of some of these oxidation reactions. Ermolenko and Zhbankov [55] used spectroscopic methods to investigate the kinetics of cellulose oxidation by nitrogen dioxide. Gusev and Ermolenko [56] considered the possibility of determining carboxyl groups quantitatively in cellulose by means of infrared spectroscopy. The spectra of celluloses oxidized by periodate were also investigated by Spedding [169], but he did not publish any recordings of his spectra. Kurlyankin, Polyak, and Koz'mina [48, 57] studied processes of thermal oxidation of cellulose esters by atmospheric oxygen.

The possibilities of spectroscopic methods for investigating chemical processes occurring during the oxidation of cellulose have been described in detail by Ermolenko [12]. Our work has been confined to the spectroscopic features of oxidized celluloses and their possible interpretation. The most interesting in this respect are selectively oxidized celluloses — monocarboxycellulose (the C_6 hydroxyl grouping is preferentially oxidized to carboxyl), dialdehydocellulose (the C_2 and C_3 hydroxyl groupings are preferentially oxidized to aldehyde groups, with rupture of the pyranose ring), dicarboxycellulose (the aldehyde groups of dialdehydocellulose are oxidized to carboxyl), samples of dialdehydocellulose oxidized with nitrogen dioxide, and dialcoholcellulose (dialdehydocellulose with restored hydroxyl groups at C_2 and C_3). The spectra of these and other oxidized celluloses are given in Figs. 62 and 63 (Appendix VI, 143-160).

Monocarboxycellulose

In the spectrum of monocarboxycellulose, as in the spectra of other oxidized celluloses, an increase in the degree of oxidation is accompanied by a reduction in the intensity of the hydroxyl group band and the appearance of a strong C=O band at about 1740 cm^{-1}. Dialdehydocellulose behaves differently, and the special features of its spectrum will be considered later.

A feature which is more pronounced in the monocarboxycellulose spectrum than in the spectra of other cellulose oxidation products is the reduction in intensity of the bands in the 2900 and 1300-1450 cm^{-1} regions.

---

* The terminology is taken from a monograph [19].

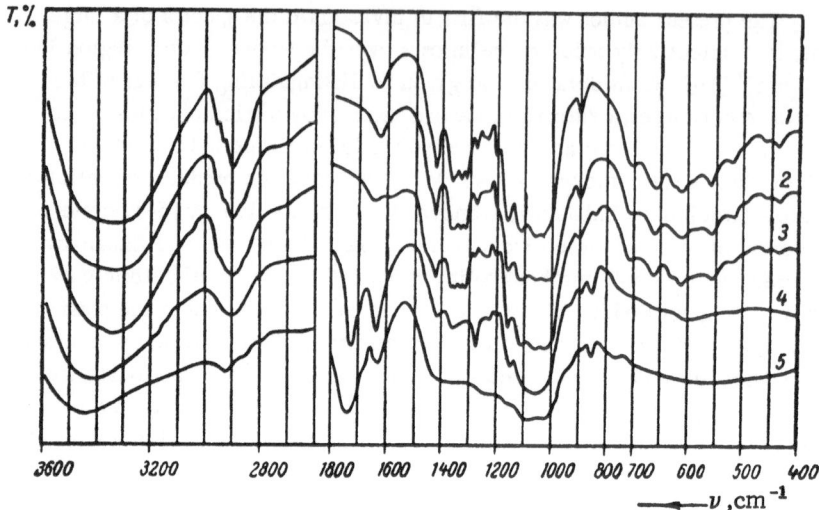

Fig. 62. Spectra of: 1) cotton cellulose and of the same material; 2) oxidized with periodic acid (14% CHO groups); 3) oxidized with periodic acid
(25.8% CHO groups) and with the aldehyde groups restored (dialcoholic cellulose); 4) oxidized with nitrogen oxides (12.3% COOH groups); 5) oxidized
with periodic acid (7.2% CHO groups) and then with nitrogen oxides (23.2%
COOH groups).

For example, the spectrum of oxidized cotton cellulose, containing 12.3% of
COOH (Fig. 62), shows only two weak bands at 1420 and 1370 cm^{-1} in the
1300-1450 cm^{-1} region, while the bands at 1360, 1340, and 1320 cm^{-1} have
practically leveled out. The band structure in this region generally disappears with further oxidation.

Similar features characterize the spectrum of hydrocellulose oxidized
by nitrogen oxides (Appendix VI, 141).

The spectrum of monocarboxycellulose shows a band of variable intensity at about 1650 cm^{-1} in addition to the $C = O$ stretching vibration band at
1740 cm^{-1}. The appearance of this 1650 cm^{-1} band, and of the bands at
1280 and 850 cm^{-1}, must be attributed to combined nitrogen in the form of
nitrate groups. Indeed, intense bands in these regions occur in the spectrum
of cellulose nitrate, but are not present in the spectra of other cellulose
oxidation products.

Bands in the 1200-1300 cm^{-1} region are characteristic of the spectra
of carboxylic acids [24]. Several authors have attributed these either to deformation vibrations of carboxylic OH groups or to stretching vibrations of

C$-$O [24]. An increase in absorption in the 1200-1300 cm^{-1} region occurs also in the case of monocarboxycellulose. It will be seen below that this absorption practically disappears on salt formation, and this provides a basis for attributing the increased absorption here to planar deformation vibrations of the bond angles $-C\overset{\displaystyle O}{\underset{\displaystyle O^-}{\diagup}}$ H [7].

The spectra of highly oxidized samples of monocarboxycellulose show a weak band at about 940 cm^{-1}, which may be attributed to nonplanar deformation vibrations of OH in carboxyl groups [7, 10, 24].

From a consideration of the spectra of monocarboxycelluloses oxidized to various extents, it appears that the sharp band structure disappears, over the whole spectral range investigated, as the number of carboxyl groups increases. This may indicate a reduction in the degree of structural order of the cellulose.

It is known that, in the cellulose spectrum, a CH$_2$ group is present only at the sixth carbon atom of the elementary unit. Consequently, it should in principle be possible to follow the kinetic oxidation of primary hydroxyl groups in the course of reaction by means of the intensity of this band. However, when carrying out such analyses, it must always be appreciated that the intensity of this band is sensitive to structural factors, so that the leveling of this band in the course of the oxidation reaction may be attributable to changes in the cellulose structure. Moreover, as explained above, the 1430 cm^{-1} band can be attributed to only one of the possible isomeric forms of CH$_2$OH, in its rotation or turning about the C$_5$$-C_6$ bond. The behavior of this band must therefore be used with considerable reserve as a basis for theories of the mechanism of oxidation of primary hydroxyl groups.

The leveling of bands in the 1200-1450 cm^{-1} region, which accompanies oxidation of cellulose by nitrogen dioxide, gives grounds for supposing that the character of these bands in the cellulose spectrum is largely determined by the nature of the CH$_2$OH groups.

The spectrum of monocarboxycellulose, obtained with an NaCl prism instrument, shows only one absorption band in the C$=$O bond region (about 1750 cm^{-1}). Since there are a number of keto groups, as well as carboxyl groups, in the structure of this product, it is impossible to determine the carboxyl content from the spectrum in this region. Even the use of a prism with a higher dispersion in this region (CaF$_2$) [48, 57] does not provide an adequate resolution of the carboxyl and carbonyl bands. However, it has been found [7, 58] that it is possible to determine carboxyl and carbonyl separately by spectroscopic means if the monocarboxycellulose is first treated with a solution containing metal cations. There is then an ionic exchange

COOH → COOMetal. The frequency of the COOMetal group is considerably displaced toward longer wavelength and depends on the nature of the cation.

Analysis of the monocarboxycellulose spectrum in the hydroxyl group region shows that, as oxidation proceeds, there is a shift of the main maximum of the hydroxyl band to shorter wavelength. For example, if with cellulose the main maximum is at 3350 cm^{-1}, then, with monocarboxycellulose containing 12% COOH, it will be displaced to 3400 cm^{-1}, thus indicating that there is a change in the ratio of the various types of hydrogen bonds.

A band appears at about 2500 cm^{-1} [7, 11] in the spectra of highly oxidized samples of monocarboxycellulose, and this must be attributed to stretching vibrations of carboxylic OH groups, which are strongly associated by hydrogen bonding. A band in this region has also been observed, for example, in the spectra of dicarboxylic acids, and it is therefore attributed to the stretching vibrations of hydroxyl groups in the dimerized forms:

$$R-C \overset{O \cdots\cdots HO}{\underset{OH \cdots\cdots O}{}} C-R .$$

Dialdehydocellulose

The spectra of dialdehydocelluloses (even of highly oxidized samples) are characterized by the absence of strong C=O double bond bands, less well-defined resolution of the bands at 1370, 1360, 1340, and 1320 cm^{-1}, a leveling of the band at 1200 cm^{-1}, and an increased absorption in the 900 cm^{-1} region with a shift of the main band maximum to longer wavelength, but the absorption here is more diffuse in character than in the case of hydocellulose. The dialdehydocellulose spectrum also shows a less sharp band structure over the whole range investigated, as compared with the original cellulose spectrum.

A linear relation has been established between the buildup of aldehyde groups and the increase in absorption in the 900 cm^{-1} region [7, 10].

As mentioned above, the absence of a 1740 cm^{-1} band gives grounds for supposing that the aldehyde groups are chemically bound, either hydrated or involved in hemiacetal bonds. The increased absorption in the 900 cm^{-1} region may therefore be a result of this bonding.

According to our measurements, there is no band in the 900 cm^{-1} region in the spectrum of chloral hydrate $CCl_3-CH(OH)_2$, which suggests that the absorption here in the case of dialdehydocellulose should be attributed to hemiacetal bonds [7, 10, 54]. However, in view of the sensitivity of the 900 cm^{-1} band to structural factors, this argument cannot be regarded as conclusive, although it is supported by the fact that the increased absorption in the 900 cm^{-1} region, associated with the transformation of cellulose to another structural modification, is always accompanied by a reduction in intensity of the 1430, 1340, and 1320 cm^{-1} bands, whereas this does not happen in the case of dialdehydocellulose. Higgins and McKenzie [168] and Spedding [169] established the sensitivity of the free aldehyde group band at 1740 cm^{-1} to the moisture content of the product, but even after thorough drying they did not obtain really strong free aldehyde group bands. His studies of carefully dried diacetaldehydocellulose samples enabled Spedding to conclude that, in spite of the presence of several forms of combined aldehyde groupings — hydrated aldehyde, hemiacetal, and hemialdol — in dialdehydocellulose, they exist mainly in the hemialdol form.

It should be noted that dialdehydocellulose differs from monocarboxy-cellulose in that there is no marked leveling of the spectral structure in the region 1200-1450 cm^{-1}. As pointed out above, the spectra of sugars and related compounds are very sensitive, not only to substitution of even one of the hydroxyl groups in the pyran ring, but also to a change in the spatial location of a hydroxyl group. This indicates that there is a relatively strong interaction between the individual structural elements in such compounds. It is therefore difficult not to accept that rupture of the pyran rings and the appearance of aldehyde groups in the C_2 and C_3 positions must lead to substantial changes in the spectrum.

The relative constancy of the spectrum of cellulose, in the course of its oxidation by periodate under the given conditions, must clearly be attributed to preferential attack of the least ordered parts of the cellulose, which have the more diffuse spectra.

Dialdehydocellulose Oxidized by Nitrogen Oxides

The oxidation of dialdehydocellulose with nitrogen oxides can yield products containing carboxyl groups at C_2, C_3, and C_6 [59].

The highly converted product shows a very diffuse spectrum, with leveling of the sharp absorption bands. It is characterized by a very diffuse $C=O$ band in the region 1650-1850 cm^{-1}, with a poorly defined structure, and the practical fusion of the absorption bands in the regions 2800-3700 and 950-1480 cm^{-1}. The spectrum of a sample containing 42.4% of COOH (Appendix VI, 147) showed the following ill-defined bands: 3440, 3260, 2940, 2850, 1750, 1640, 1400, 1360, 1220, 1160, 1100, 1050, 960, 810, 750, 720, 660, and 625 cm^{-1}.

It should be noted that, in spite of the high carboxyl group content, the spectrum of this compound does not show the free hydroxyl group absorption bands.

The broad diffuse hydroxyl-group band, observed in the spectrum of this compound, is characteristic of carboxylic acids, and implies the formation of a chelate structure [24]. It is necessary to suppose, in this case, that the carboxylic OH groups enter into a similar association.

A feature of the spectra of dialdehydocelluloses, which have been oxidized with nitrogen dioxide, is a displacement of the main maximum of the CH-group stretching-vibration band from 2900 to between 2920 and 2940 cm^{-1}. This cannot be explained without further investigation.

The very diffuse nature of the spectrum of dialdehydocellulose, oxidized with nitrogen oxides, indicates a very low state of structural order.

Dicarboxycellulose

The spectrum of dicarboxycellulose differs from that of monocarboxycellulose in that there are no bands of variable intensity in the regions 1650, 1280, and 850 cm^{-1} (attributed to the presence of combined nitrogen as nitrate groups in the monocarboxycellulose structure); there are several bands showing high absorption in the region 1200-1300 cm^{-1}; and there is less pronounced leveling of the bands in the 1300-1400 cm^{-1} region. In general, the spectrum of dicarboxycellulose is less diffuse than that of monocarboxycellulose over the whole spectral region investigated of 400-3600 cm^{-1}. However, there is a considerable reduction in intensity of the bands at 1430, 1370, 1360, 1340, and 1320 cm^{-1}, as compared with the spectrum of dialdehydocellulose. By analogy with the leveling of these bands in the monocarboxycellulose spectrum, this gives reason for believing that the primary hydroxyl groups are partially oxidized [7, 10, 54]. The spectrum of dicarboxycellulose differs from that of dialdehydocellulose in that there is some reduction of absorption in the 900 cm^{-1} region [7], attributable, in all probability, to a reduction in the aldehyde group content as the result of oxidation by sodium chlorite.

Dialcoholcellulose

The spectrum of dialcoholcellulose differs from that of cellulose in that the 2900 cm^{-1} band is more diffuse, new bands appear at 1600, 1400, 940, and 840 cm^{-1}, and there is some increase in absorption in the 1240 and 900 cm^{-1} regions.

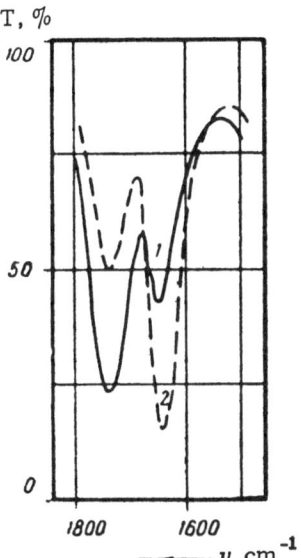

T, %

Fig. 63. Spectra of celluloses oxidized by nitrogen dioxide. 1) Cotton (3.5% COOH); 2) dialcoholcellulose (7.2% of reduced CHO, 1.5% of COOH).

The band at 1400 cm^{-1} and the increased absorption in the 1240 cm^{-1} region can be attributed to deformation vibrations of CH$_2$ groups at the second and third carbon atoms of the cellulose macromolecule. Finding reasons for the 940 and 840 cm^{-1} bands is more difficult. At present we can make no definite suggestions, and there is no evidence that these bands are also due to CH$_2$OH groupings at C$_2$ and C$_3$, because absorption in this region is very sensitive to structural changes.

It should be noted that dialcoholcellulose oxidized by nitrogen dioxide differs from similarly oxidized samples of cellulose or dialdehydocellulose in that the nitrate group band shows anomalously high intensity, even exceeding that of the C=O group band (Fig. 63). This experimental fact is in good agreement with the results of chemical investigation [63].

Salts of Oxidation Products of Cellulose

Salts of cellulose oxidation products are used as cation exchangers, detergents, brightening agents, etc. Cellulose ion-exchange adsorbents are very cheap and have a high adsorptive surface. Even a small number of cations has a marked effect on the properties of cellulosic materials: ability to take up dyes, viscosity, strength, thermal stability, electrical insulation, and other properties [19]. Cellulosic materials are always oxidized to some extent and interact with solutions containing cations in the course of various technological processes, or in use.

Little work has been published on the infrared spectra of salts of selectively oxidized cellulose products.

The first papers dealing with this topic [58, 64] were concerned with changes in the spectra of various selectively oxidized celluloses (mono- and dicarboxycellulose, dialdehydocellulose) on treatment with the cations of various metals: Li, Na, Mg, Al, Ca, Mn, Fe, Ni, Cu, Ag, Cd, Ba, Hg, Pb, and UO$_2$. The authors showed that treatment of these cellulose oxidation products with metal cations produced a marked change in the spectrum in the region of carboxyl group absorption, and that various cations had specific

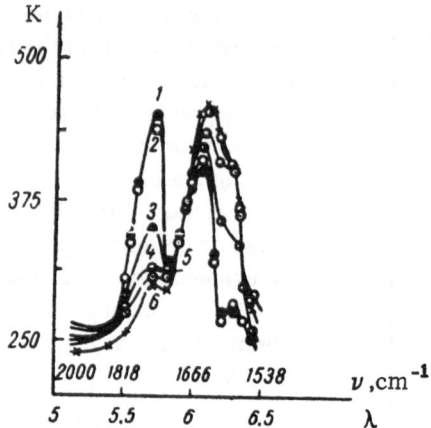

Fig. 64. Spectra of dicarboxycellulose
(COOH,5.5%) treated with 0.1 M FeSO₄ at
a pH of: 1) 0.80; 2) 1.28; 3) 1.32; 4) 2.05;
5) 2.45; 6) 2.90.

effects. A study has also been made of the infrared spectra of dicarboxycellulose, treated with ferrous ions at various pH values [58]. It was found
that the process of salt formation from oxidized cellulose greatly depends on
the pH of the medium. The sorption is small at a pH of 0.8-1.2, but greatly
increases and practically achieves saturation with a small increase in pH to
the region 1.2-1.35 (Figs. 64 and 65). The effect of washing the samples on
the quantities of combined ferrous ions in the products obtained at various
pH values was also investigated; this emphasized the advantages of the spectral method, which made it possible to investigate the process of salt formation without resorting to washing. Ermolenko, Zhbankov, and Rozenberg [65]
considered changes in the spectra of oxidized cellulose, when ferrous cations
were adsorbed at various pH values, in connection with the technological
process of viscose production.

Gusev and Ermolenko [66] used infrared spectroscopic methods for investigating the sorption of UO₂ cations by oxidized celluloses. The same
techniques have been applied to the study of cation exchange on carboxymethylcellulose [67]. The infrared spectra of oxidized celluloses, treated
with various basic dyes, have been investigated [68].

We will consider the main spectroscopic features of the salts of selectively oxidized celluloses. * Spectra of monocarboxycellulose salts, over the

* This matter is considered in more detail in the monograph by I.N. Ermolenko,
The Spectroscopy and Chemistry of Oxidized Celluloses (Minsk, 1959).

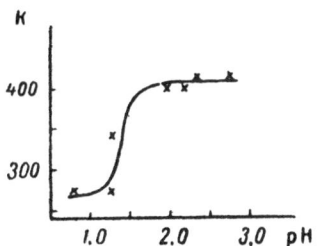

Fig. 65. Variation of the sorption of ferrous cations by the carboxyl groups of dicarboxy-cellulose with pH, as measured by the intensity of the carboxyl absorption band (at about 6.3 μ).

entire range, 400-3600 cm^{-1}, are shown in Appendix VII, 161-166.

Salts of Cellulose Oxidation Products with Inorganic Cations

Carboxyl groups, whose hydroxylic hydrogen atoms are capable of ion exchange, play an important role in the sorption of metal cations by oxidized celluloses [69, 171]. It is therefore interesting to follow the changes in the spectra of oxidized celluloses containing carboxyl groups. It has been shown [58, 64, 65] that when carboxyl-containing celluloses are treated with solutions of metal salts, depending on the pH of the medium there is a more or less sharp reduction in intensity of the 1740 cm^{-1} band, and new bands appear in the regions 1580-1650 cm^{-1} and 1350-1450 cm^{-1} (Figs. 64 and 65). These spectral changes can be attributed to the process COOH → COOMetal. A similar phenomenon has been observed by Lecomte and co-workers [172] in their studies of the spectra of the salts of organic acids. On dissociation with formation of the COO− grouping, instead of the C=O carbonyl band there appear two bands in the regions 1550-1610 and 1350-1450 cm^{-1}, which can be attributed to

$$C\!\!<^{O-}_{O}$$
[24]. The same cause is responsible for the appearance of bands in the regions 1550-1650 and 1350-1450 cm^{-1} in the spectra of oxidized celluloses treated with metal cations (Appendix VII, 159-163). It is characteristic of such products that their spectra show a leveling of the increased absorption in the regions 1200-1300 and 940 cm^{-1}, which is specific for carboxyl groups. This leveling confirms the previously expressed view which attributed absorption in these regions to angular deformation vibrations of

$$-C\!\!<^{O}_{\hat{O}}\!\!-H.$$
As would be expected, displacement of the cations by acid leads to restoration of the spectrum of the initial oxidized cellulose.

The process of cation exchange by carboxyl-containing celluloses depends on the pH of the medium and the character of the cation. When there is a mixture of cations, these exchange concurrently.

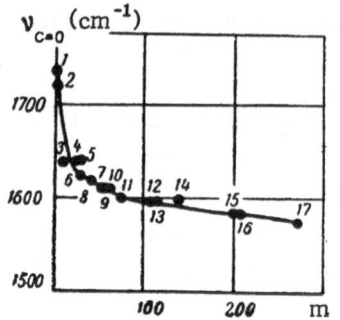

Fig. 66. Variation of the frequency

of C = O in the group $-C\overset{\displaystyle O}{\underset{\displaystyle OK}{\diagup}}$ with

the atomic weight of the cation in
the spectrum of dicarboxycellulose
(14.7% COOH). 1,2,3,4,5,6,7,8,9,
10,11,12,13,14,15,16, and 17 refer
respectively to the cations H, D, Li,
Na, Mg, Al, Ca, Mn, Fe, Ni, Cu,
Ag, Cd, Ba, Hg, Pb, and UO$_2$.

The frequency of the 1550-1650
cm^{-1} band in the spectra of salts of
oxidized celluloses (Fig. 66) depends
on the mass of the metal cation and is
displaced to a longer wavelength as
this mass increases (Appendix VII, 161-
166). A similar phenomenon has been
noted, for example with the salts of
fatty acids [173]. The authors of this
work have shown that the difference in
displacement for the salts of Li and Pb
amounts to 77 cm^{-1}. This is not a
random effect: Calculations, carried
out with the simpler M−C = O model,
lead to a similar result [70]. In some
cases there is a splitting of the 1550-
1650 cm^{-1} band, and this splitting also increases with the mass of the metal
cation. The intensity ratio of the two
bands depends on the nature of the cation, and may differ greatly for cations
with similar masses. For example,
with lead, the longer wavelength band is the more intense, while with uranyl
the reverse is true; the shorter wavelength band is stronger. This circumstance prevents resolution of the bands for cations of high mass. In many
cases the bands are not resolved at all. Band splitting in this region may be
attributed either to differences in the bonding of the cations (to one or several
carboxyls simultaneously), or to the existence of ionized and unionized metal
carboxylate groups.

The effect of the nature of the cation on the spectrum of a salt of a
carboxyl-containing cellulose has made it possible to develop a spectroscopic means for analyzing metal cations in solution [64]. The technique is
as follows. A phial, with a ground stopper, is charged with 5-20 ml of the
neutral solution to be investigated and about 200 mg of the oxidized cellulose. When equilibrium has been attained (the time for every type of solution has been investigated and amounts to 5-30 min), the cellulose sample is
squeezed out and dried. If the oxidized cellulose is in fiber form, then a
semi-transparent film is prepared by pressing the sprayed fiber [6]. The spectrum is recorded in the normal way. The position of the carboxyl C = O
band is determined, and the mass of the metal cation is obtained from this
frequency, using a previously prepared calibration curve. Most impurities do
not affect the analysis, and this eliminates the need for prolonged washing of
the sample after sorption.

For the concentration of cations it is most convenient to use an oxidized cellulose with a small carboxyl group content, about 4-6%. Such materials are stable in solution, and the intensity of their $C=O$ band is quite
adequate for analysis. Regeneration of these cation exchangers is carried out
by treatment with dilute acid and washing.

The spectra of oxidized celluloses, which have been treated with various salt solutions, show a band of variable intensity in the 1740 cm^{-1} region.
In the absence of other groups than carboxyl-containing $C=O$, this indicates
the existence of a number of carboxyl groups which do not participate in ion
exchange. It should be noted that the intensity of the residual $C=O$ band
depends on the nature of the cation, and this means that purely spectroscopic
means can be used to determine the extent of ion exchange at equilibrium,
in the case of various cations. A comparison of the spectra of oxidized celluloses, treated with salts of a given metal with various anions, has shown that
the nature of the anion does not affect the frequencies of the new absorption
bands appearing after ion exchange.

The spectra of salts of oxidized celluloses do not show any reduction
or significant increase in intensity of the alcoholic hydroxyl group absorption
bands of the polymer, which would indicate the participation of these groups
in the ion exchange reactions. However, under our experimental conditions,
there was some broadening of the OH group band contour in the case of monocarboxycellulose, particularly when this was treated with Al^{3+} cations. This
observation requires further detailed confirmation, to increase the precision,
by tests on more samples, and to elucidate its mechanism.

The process of salt formation by oxidized celluloses is accompanied
also by a spectral change in the frequency region of the CH bond stretching
vibrations (2800-3000 cm^{-1}), involving a displacement of the main maximum
from 2900 to 2920 cm^{-1}. Interpretation of this fact is still difficult. It may
be associated with the effect of the cationic charge on the nature of the force
field.

Salts of Cellulose Oxidation Products with Organic Cations

The spectra of oxidized celluloses, treated with organic cations, show
changes similar to those observed after treatment with metal cations, with
the difference that the absorption bands of the organic cations themselves also appear. For example, the spectrum of dicarboxycellulose treated with
streptomycin shows reduced intensity of the 1740 cm^{-1} band, a new band appears at about 1600 cm^{-1}, and bands also appear at 1725 and 1675 cm^{-1}; the
latter are probably due to absorption by streptomycin itself.

Great practical importance attaches to studies on the spectra of oxidized celluloses treated with dye cations, because the ability of the material
to become colored is determined by the composition and properties of the
fibers.

It has been established [68] that, when mono- and dicarboxycelluloses
have been treated with the cations of various types of dyestuffs* — Auramine,
Methylene Blue, Rhodamine 6J — their spectra also show a reduction in intensity of the 1740 cm^{-1} band and a new band in the region 1610 cm^{-1}. It is
significant that the residual 1740 cm^{-1} band varies in intensity. This indicates that there are differences in the extent of chemical interaction of oxidized cellulose with the various types of dyes. The 1740 cm^{-1} band is particularly suitable for analyzing the process of salt formation between oxidized celluloses and organic cations, when the cation does not contain a
$C = O$ bond. In the contrary case, it is necessary to allow for superposition of
the band due to this bond.

It is important to note that, in the spectra of all the salts investigated
[68] of oxidized celluloses with organic cations, the frequency of the band in
the 1550-1650 cm^{-1} region did not depend on the nature of the cation, and
was approximately 1600 cm^{-1}. Analysis indicated that oxidized celluloses,
treated with the above-mentioned organic cations, had a characteristic spectral feature which distinguished them from oxidized celluloses treated with
metal cations. In the case of organic cations, the 1550-1650 cm^{-1} band had
a sharp nondiffuse contour, in none of the examples considered [68] was
there any splitting and, most important, the band frequency did not depend
on the mass of the sorbed cation. All this is presumably determined by the
predominance of ionized carboxyl groups. The infrared spectra make it possible to investigate both the number of chemically bound organic cations
(from the reduction in intensity of the 1740 cm^{-1} band and the new band in
the region 1550-1650 cm^{-1}) and the total number of these cations (from the
intensity of their own bands). The latter possibility is particularly useful in
the case of dyes, since absorption of dyes by fibers is usually determined indirectly. It is well known that the capacity of a material to become colored
and the mechanism of the process are determined by the composition and
properties of the fibers. An important role in dyeing is played by the carboxyl groups, which are present to some extent in all celluloses. Thus, the
study of the process of dye bonding by selectively oxidized celluloses has considerable practical importance. It has been established [68], in particular, that
the carboxyl groups at positions C_2 and C_3 are more active in reacting with dye
cations than carboxyl at C_6.

---

* The dyeing was carried out under static conditions, using 0.0001 M aqueous
solutions, at $20°C$, and the bath modulus was $1000:1$ before equilibrium.

CHAPTER VI

NEW TYPES OF CELLULOSE DERIVATIVES

Cellulosic materials have various disadvantages — they crease, are relatively unstable to heat and light, and are inflammable. These disadvantages can be reduced by modifying the properties of cellulose and its derivatives.

"For many years to come, in spite of all the progress in the field of synthetic fibers, cellulose fibers of viscose and cotton will provide the bulk of textile fibers. The chemical improvement of these natural fibers is therefore a very important problem, since, with the consumption of a small amount of chemicals, it is possible to give them properties which approximate to those of the best synthetic fibers" (from the speech of Academician V. A. Kargin to the December Plenum of the Central Committee of the Communist Party of the Soviet Union, 1963).

In the last few years, various new classes of cellulose derivatives have been synthesized, and these show many specific valuable properties.

Modification of the properties of cellulose can now be achieved in the following main directions: the synthesis of new cellulose derivatives; the introduction of new functional groups into the cellulose macromolecule and the carrying out of additional chemical conversions based on these; the synthesis of copolymers of cellulose with carbon chain or heterochain polymers.

For our investigations we used compounds first synthesized in the Problems Scientific Laboratory of the Department of Chemical Fibers at the Moscow Textile Institute.

These products showed a number of specific valuable properties: noninflammability, ion exchange, bactericidal properties, high stability, resistance to the action of chemical reagents, etc.

The infrared spectra of these compounds are our main concern here, so that it will be convenient to describe these first in detail and carry on with the interpretation of the main frequencies.

Esters of Cellulose with Phosphorus-Containing Acids

The production of noninflammable cellulose materials is a very important part of modification of the properties of cellulose, with the object of giving it valuable new qualities.

One of the methods for obtaining fire-resistant cellulose fabrics is to treat these with solutions of phosphorus-containing compounds, leading to partial esterification.

Recent investigations have resulted in the synthesis of new types of cellulose esters with phosphorus-containing acids [72-74], having the following structures:

$$(\text{cell.}-\text{O})_2\text{P(O)CH}_3(\text{I}); \text{ cell.} \underset{\text{O}}{\overset{\text{O}}{\diamond}}\text{P(O)OCH}_3(\text{II}),$$

$$\text{cell.} -\text{O}-\text{P(O)(OC}_6\text{H}_5)_2(\text{III}),$$

$$\text{cell.} -\text{O}-\text{P(O)(OH)OC}_6\text{H}_5, \quad (\text{cell.}-\text{O})_2\text{P(O)C}_6\text{H}_5(\text{IV}),$$

$$\text{cell.} -\text{O}-\text{P(O)(OC}_6\text{H}_5)\text{ONH}_4(\text{V}); \quad \text{cell.} -\text{OP(O)(OC}_2\text{H}_4\text{Cl})_2(\text{VI}).$$

The infrared spectra of these compounds [71] were obtained from KBr-pressed discs under standard conditions. They were compared with the spectra of the model compounds: $\text{CH}_3\text{P(O)(OC}_2\text{H}_5)_2$, $\text{P(O)(OC}_2\text{H}_4\text{Cl})_3$, $\text{P(O)Cl} \cdot (\text{OC}_6\text{H}_5)_2$, $\text{P(O)Cl}_2(\text{OC}_6\text{H}_5)$, and H_3PO_4. The model compound spectra were obtained using cells without spacers, which gave layers less than 10 μ thick.

The spectra of all these compounds investigated showed a broad diffuse band in the region 1200-1220 cm^{-1}. This band must be attributed to stretching vibrations of the P=O group. It has been established that, for most phosphorus-containing compounds, the frequency range for this band is 1250-1300 cm^{-1} [24, 53,75-77, 174-176]. Absorption in the 1200-1220 cm^{-1} region cannot be attributed to stretching vibrations of the ester bond by analogy with the $\text{C}-\text{O}-\text{C} \overset{\text{O}}{\diagup}$ grouping [24-27]. The fundamental work of Bellamy and Beecher [177], and of others, has shown that the spectra of aliphatic esters of phosphoric and alkylphosphonic acids have strong $\text{C}-\text{O}-\text{P} \overset{\text{O}}{\diagup}$ bands in the region 990-1030 cm^{-1}. In our case, attribution of the 1200-1220 cm^{-1} band to the P=O bond is quite unambiguous only for esters containing the grouping P-O-C(aryl). However, in the spectra of the corresponding model compounds, such as diphenyl phosphochloridate,

$$Cl-P\begin{matrix} O \\ \| \\ \\ \end{matrix}\begin{matrix} O-\bigcirc \\ O-\bigcirc \end{matrix},$$

the P−O−C band appears as a doublet at 1160-1185 cm^{-1}.

The P=O frequency is very sensitive to any form of association. It has been shown [174, 177] that the P=O band is displaced by 40-80 cm^{-1} in the case of hydrogen bond formation with OH or NH groups. The frequency of the unassociated P=O group normally lies in the region 1250-1300 cm^{-1} [24, 53,76].

In the spectra of the model compounds for the cellulose esters under consideration, the vibrations of the P=O group appear in the form of sharp bands at 1250 cm^{-1} for $CH_3P(O)(OC_2H_5)_2$, at 1260 and 1310 cm^{-1} for $P(O) \cdot (OC_2H_4Cl)_3$, and at 1310 cm^{-1} for $ClP(O)(OC_6H_5)_2$, while in the case of phosphoric acid the P=O band is diffuse in character and is considerably displaced toward longer wavelength.

It has already been explained that the frequency of the P=O bond in the spectra of the cellulose esters investigated lies at about 1200-1220 cm^{-1}, and that the band itself is broad and diffuse. We must assume that there is hydrogen bonding between the cellulose hydroxyl and P=O groups. A low frequency for the P=O bond (1230-1232 cm^{-1}) has also been observed, for example, in the spectra of dialkyl α-hydroxyalkylphosphonates and has been attributed to hydrogen bonds of the type [178]:

$$\begin{matrix} > \!\!-P-C\!\!< \\ \quad\| \quad | \\ \quad O...HO \end{matrix}.$$

It is important to note that the spectra of some of the esters investigated show, in addition to the band at 1200-1220 cm^{-1}, another band at shorter wavelengths. For example, there is a sharp band at 1290 cm^{-1} in the spectrum of

$$cell.-O-P\begin{matrix} O \\ \| \\ \\ \end{matrix}\begin{matrix} O-\bigcirc \\ O-\bigcirc \end{matrix}.$$

This band should be attributed to unassociated P=O groups. In fact, it is within the range of the free P=O group frequency, while the spectrum of the compound in the range 1200-1300 cm^{-1} shows the single sharp band of unassociated P=O groups at 1300 cm^{-1}.

The special feature noted above for the ester

can evidently be explained by steric factors, associated with the presence of two benzene rings around the P=O group, these rings being linked through oxygen atoms to the same phosphorus atom.

Investigations have shown that, in the case of phosphorus-containing compounds which show the property of noncombustibility, the P=O groups are involved in hydrogen bonding.

From this point of view, it is interesting to compare the characteristics of the P=O bonds of esters I and II. These compounds differ only in the means of attachment of the methyl group to the phosphorus atom, but they differ considerably as regards noncombustibility. The spectrum of the ester

P(O)OCH$_3$ in the region 1200-1300 cm^{-1} shows bands or steps at 1280, 1230, and 1200 cm^{-1}, while the spectrum of (cell. −O)$_2$P(O)CH$_3$ shows only one broad band at 1200-1220 cm^{-1}.

It follows from the above that, in the ester II structure, a definite proportion of the P=O groups are in an unassociated or weakly associated state. However, the increase in absorption at 1200 cm^{-1} in the spectrum of this ester cannot be wholly attributed to P=O involved in hydrogen bonding. The vibration frequencies of the CH$_3$−O− group also lie in this region. For example, the (P)O−CH$_3$ group has a characteristic band at 1190 cm^{-1} [174].

The spectra of these compounds show new bands in the regions 950-980 and 700-800 cm^{-1}, and increased absorption at about 900 cm^{-1}. Bands in the regions 950-1050 and 700-800 cm^{-1} appear in the spectra of many phosphoric esters [24, 53, 76, 174-176]. However, there are conflicting views as to the interpretation of these bands. Thus, Bergmann, Litauer, and Pinchas [175] attribute the 980 cm^{-1} band to symmetrical and the 1000-1050 cm^{-1} band to asymmetrical vibrations of ternary P−O−C groups, whereas Bellamy and

Beecher [177] attribute the 1000-1050 cm^{-1} band to vibrations of C$-$O$-$(P). The attribution of bands in the region 960-1050 cm^{-1} to asymmetrical and symmetrical vibrations of ternary P$-$O$-$C groups is not very convincing. Sharp bands at 1030 and 970 cm^{-1} appear in the spectrum of

$$CH_3-P \overset{\overset{\displaystyle O}{\|}}{\underset{\displaystyle}{<}}\overset{OC_2H_5}{\underset{OC_2H_5}{}}$$

(Fig. 67, curve 4).

Absorption in the region 800-900 cm^{-1} is not specific for organic esters of phosphoric acids. There is no absorption in this region in the spectra of such compounds as

while cellulose esters containing similar phosphoric groupings show increased absorption between 800-900 cm^{-1}.

Thus, it is necessary to attribute bands in this region either to structural factors or to the bonds of cell.$-$O$-$P.

The spectra of the esters investigated have their individual characteristics, reflecting the specific features of the new functional groups introduced into the cellulose. For example, the bands at 2920 and 3000 cm^{-1} in the spectrum of (cell.$-$O)$_2$P(O)CH$_3$ can be associated with the stretching vibrations of the methyl group. The absorption at about 3000 and 2920 cm^{-1}, observed in the spectrum of CH$_3$P(O)Cl$_2$, has been attributed to the asymmetrical and symmetrical stretching vibrations of the methyl group [178]. The spectra of esters with a high phosphorus content show a reduction in the intensity of the cellulose band at about 2900 cm^{-1}. Similar effects have been observed in the spectra of many cellulose esters and have been attributed to the influence of double bonds on the stretching vibrations of the CH$_2$ groups [25-27]. The band at 1310 cm^{-1} can be attributed to deformation vibrations of a methyl group attached to a phosphorus atom. For example, a sharp band in this region has been observed in the spectra of phosphonate esters [179] and other phosphorus derivatives containing a P$-$CH$_3$ group, and has been interpreted in a similar way.

The spectra of the esters investigated, which contain benzene rings, show sharp bands at 1600, 1500, 770, and 690 cm^{-1}. The presence of such

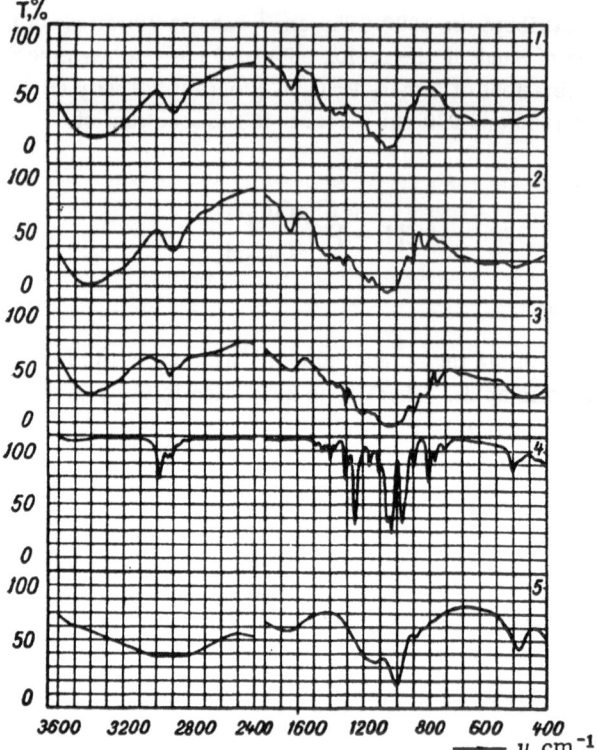

Fig. 67. Spectra of phosphorus-substituted esters of cellu-

lose: 1) cell. $\overset{O}{\underset{O}{\diagdown}}$ P(O)OCH$_3$; 2) (cell.$-$O)$_2$P(O)CH$_3$; with

2% P; 3) (cell.$-$O)$_2$P(O)CH$_3$ with 8% P; 4) CH$_3$P(O)(OC$_2$H$_5$)$_2$; $_1$.
5) H$_3$PO$_4$.

bands unambiguously demonstrates the presence in the structure of a monosubstituted benzene ring [24].

The ester IV, containing a phenyl group attached to phosphorus, gives
a sharp band at 1440 cm^{-1}. According to Daasch and Smith [174], who investigated a large number of compounds containing phenylphosphorus groups,
a band at about 1440 cm^{-1} appears in the spectra of all these compounds and

is convenient for determination of the group P.

The bands at 3340 and 1620 cm^{-1} in the spectra of the compound V
can be attributed to the stretching and deformation vibrations of NH.

Specific features in the region 400-700 cm^{-1} appear in the spectra of esters containing a large number of ester groups. It was suggested above that the broad diffuse absorption between 400 and 700 cm^{-1} in the spectrum of cellulose can be attributed to the superimposed overtones of the vibrations of the hydrogen bonds themselves or of the nonplanar deformation vibrations of hydroxyl groups. The basis for this view is that a reduction in the number of OH groups in the structure is accompanied by a reduction in absorption between 400 and 700 cm^{-1}, except, of course, when the absorption of the introduced functional group itself is in this region. A reduction in absorption in individual parts of the 400-700 cm^{-1} region, with increasing degree of esterification, has also been observed with the esters under consideration. Attribution of spectral differences between these esters, in the range 400-700 cm^{-1}, is not excluded, but cannot be unambiguous owing to the possibilities of absorption in this region by the new functional groups introduced into the cellulose. Analysis of the spectrum between 400 and 700 cm^{-1} can be useful for the identification of a product.

Esters of Cellulose with Chlorinated Aliphatic Acids

A study has been made [26] of the esters of cellulose with chlorinated aliphatic acids, cell.$-O-CO(CH_2)_4Cl$, and of mixed esters of acetic and chlorinated aliphatic acids, synthesized by known methods [80] (see Table 9). These compounds are of interest for the production of films and plastics without use of plasticizers (the so-called internal plasticization process).

The spectra of esters with a high degree of substitution ($\gamma \sim 280$) do not show the strong diffuse hydroxyl group band at about 3100-3600 cm^{-1}. The disappearance of this hydroxyl group band is indeed a direct confirmation of the high degree of esterification of these compounds.

The spectra of esters with chlorovaleric acid (Fig. 68) are also characterized by the appearance of bands in the region of the stretching and deformation vibrations of methylene groups (2960, 2870, 1460, and 1420 cm^{-1}), and of C=O (1750 cm^{-1}) and C-Cl (650 cm^{-1}) bonds.

The existence of two bands (at 1420 and 1460 cm^{-1}) for the internal deformation vibrations of methylene groups can probably be explained by the specific location of these groups. Indeed, the chlorovaleric acid residue

$-C\overset{\displaystyle \nearrow O}{\underset{\displaystyle \searrow (CH_2)_4Cl}{}}$ contains CH_2 groups characteristic of aliphatic compounds and

of structures of the type $-C\overset{\displaystyle \nearrow O}{\underset{\displaystyle \searrow CH_2}{}}$ and $-CH_2Cl$.

Table 9. Composition and Degree of Substitution of Esters of Cellulose with Chlorovaleric Acid

Sample No.	Composition of elementary unit of macromolecule	Degree of esterification (value of γ)	
		n	n_1
1		289	—
2	$C_6H_7O_2(OH)_{3-n}[OCO(CH_2)_4Cl]_n$	280	—
3		270	—
4		30	270
5	$C_6H_7O_2(OCOCH_3)_{n_1}[OCO(CH_2)_4Cl]_n$	30	270—260
6		27	272

Note. A study of the spectra of samples, which differed little in degree of substitution, made it possible to estimate the reproducibility of the chemical analyses and of the spectral data obtained.

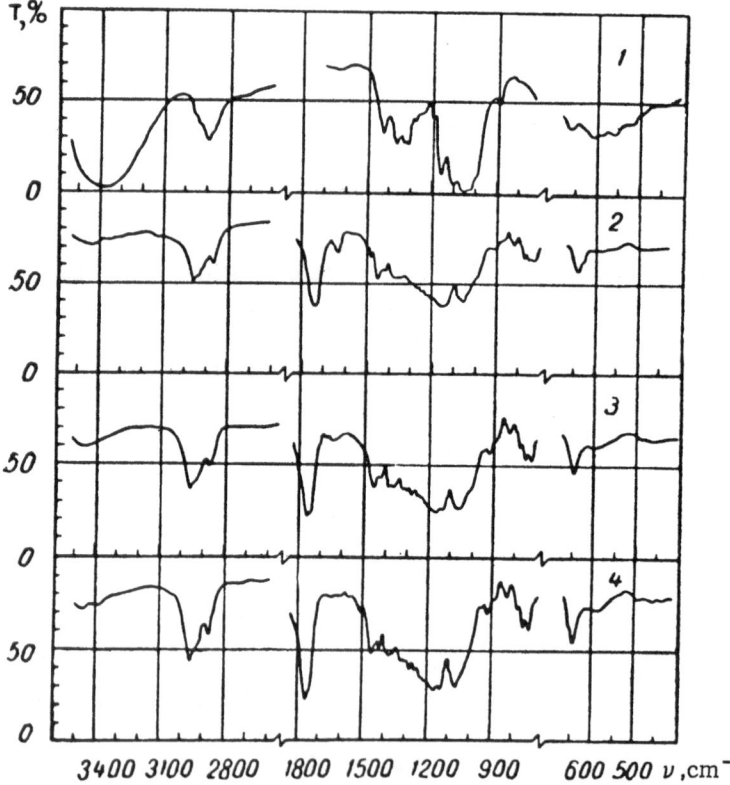

Fig. 68. Spectra of: 1) cellulose; and of esters of cellulose with chloro-valeric acid with γ equal to: 2) 289; 3) 280; 4) 270.

It is well known that the frequency of the CH_2 group for aliphatic compounds is highly characteristic and lies in the region 1460 cm^{-1} [24, 53]. It may be noted in passing that a band at 1460 cm^{-1} appears in the spectrum of cellulose acetobutyrate. On the other hand, it is known that when a $C=O$ double bond is present, and when the carbon atom of the methylene group is attached to chlorine, the CH_2 deformation vibration band is displaced to a lower frequency. For example, the band is at 1429 cm^{-1} in the spectrum of methylene chloride [24, 53].

The cellulose esters of chlorinated aliphatic acids are characterized by the existence of spectral bands at 1350, 1360, 1320, and 1300 cm^{-1}, which can be attributed to the external deformation vibrations of the CH_2 in the ester groupings.

The appearance of two intense bands at 1130-1170 and 1060 cm^{-1} can also be attributed to ester bonds. It has been seen above that the spectra of cellulose esters show two bands in the regions 1150-1250 and 1000-1100 cm^{-1}. For example, there is a band at 1170 cm^{-1} in the spectrum of cellulose propionate.

The spectra of samples 1-3 of cellulose esters of chlorinated aliphatic acids show weak bands at 1280, 1260, 1230, 1215, and 1200 cm^{-1} against a background of intense absorption in the region 1200-1300 cm^{-1}. The interpretation of this is difficult. This frequency region can include the stretching vibrations of the ester groupings and the external deformation vibrations of the CH_2 groups in the ester groupings.

The spectra of these esters show bands at 920 and 820 cm^{-1}, which can also be observed in the spectra of other cellulose esters and ethers. For example, a band at 920 cm^{-1} is present in the spectra of cellulose ethers (the trityl ether, ethylcellulose, etc.) and of certain polysaccharides (dextran, amylose). Thus, a band in this position cannot be attributed unambiguously to ester bonds. It has already been suggested that absorption in the region 800-900 cm^{-1} may reflect a change in the structure of the cellulose macromolecule, caused by the introduction of ester groupings.

A system of bands in the region 700-800 cm^{-1} (bands at 780, 750, and 720 cm^{-1}) is characteristic of the spectra of cellulose esters of chlorinated aliphatic acids. These bands, or any any rate some of them, can be attributed to twisting vibrations of the CH_2 groups in the acid residues. It is known that, in the case of aliphatic hydrocarbons containing four or more methylene groups, a characteristic band appears at about 720 cm^{-1}. The twisting vibrations of the methylene groups are more affected by adjacent groups than are the deformation vibrations of other groupings. Accordingly, as correctly noted in the literature [53], it is to be expected that the twisting vibration

bands will be characteristic, for example, for the group $-C\overset{\displaystyle O}{\underset{\displaystyle CH_2-}{\big<}}$. The
appearance, in the compounds which we investigated, of several bands in the
frequency region of the CH_2 group twisting vibrations is in accordance with
this prediction. The possibility of the existence of rotational isomers is not
excluded. It was shown earlier that the spectra in the 700-1000 cm^{-1} region
are highly specific for many cellulose derivatives. The data presented in
this book confirm these views.

A sharp band at 650 cm^{-1} is characteristic of the esters of cellulose
with chlorinated aliphatic acids. This band can be attributed to the stretching vibrations of C$-$Cl bonds. It is known that, with compounds which contain two or more unbranched carbon atoms, the frequency of the stretching
vibrations of C$-$Cl bonds lies approximately in the region 650 cm^{-1} [81]. It
should be noted that a band at 658 cm^{-1} has been observed, for example, in
the spectrum of ethyl chloride, CH_3-CH_2Cl, and has been attributed to this bond
[81]. The spectra of all the cellulose esters investigated show a reduction in absorption in the region 400-550 cm^{-1}. This reduction in absorption is indeed characteristic of all cellulose derivatives in which the OH groups have been replaced.

Stable Xanthate Derivatives of Cellulose

Cellulose xanthates are very important in the viscose industry. This
accounts for the large number of papers dealing with the conditions for formation of these esters and their properties [19].

It is well known that cellulose xanthates are an intermediate product
in the formation of viscose fiber. In view of their instability, it is not possible to make use of fibers obtained directly from cellulose xanthates. Accordingly, in the process of producing viscose fibers, the cellulose xanthate
fibers are subjected to decomposition and regeneration as hydrocellulose.
Hence, there is a need to investigate the possibilities of stabilizing cellulose
xanthates, and to use products based on these.

Infrared spectroscopy has not been used, with the exception of one
paper [27], to study the structure of cellulose xanthates, largely because
these compounds are so unstable.

However, in recent years we have developed conditions for obtaining
a series of stable cellulose xanthate derivatives [82, 83] (Table 10).

Our analyzed samples made it possible to compare the spectra of
stable cellulose xanthate derivatives with the same chemical structure but
with different degrees of substitution, and also of such derivatives with different chemical structures but with approximately the same degree of substitution.

Table 10. Compositions of Stable Cellulose Xanthate Derivatives Investigated

Sample	Structural formula	Degree of esterification (value of γ)
Methylxanthate	Cell. $-$OC$\diagup^S_{\diagdown SCH_3}$	$\left\{\begin{array}{l}8\\30\\160\end{array}\right.$
p-Nitrophenylxanthate	Cell. $-$OC$\diagup^S_{\diagdown SC_6H_4NO_2}$	20
Diethylacetamidoxanthate	Cell. $-$OC$-\diagup^S_{\diagdown SCH_2CON(C_2H_5)_2}$	30
Phenylthiourethane	Cell. $-$OC$\diagup^S_{\diagdown SNHC_3H_5}$	22

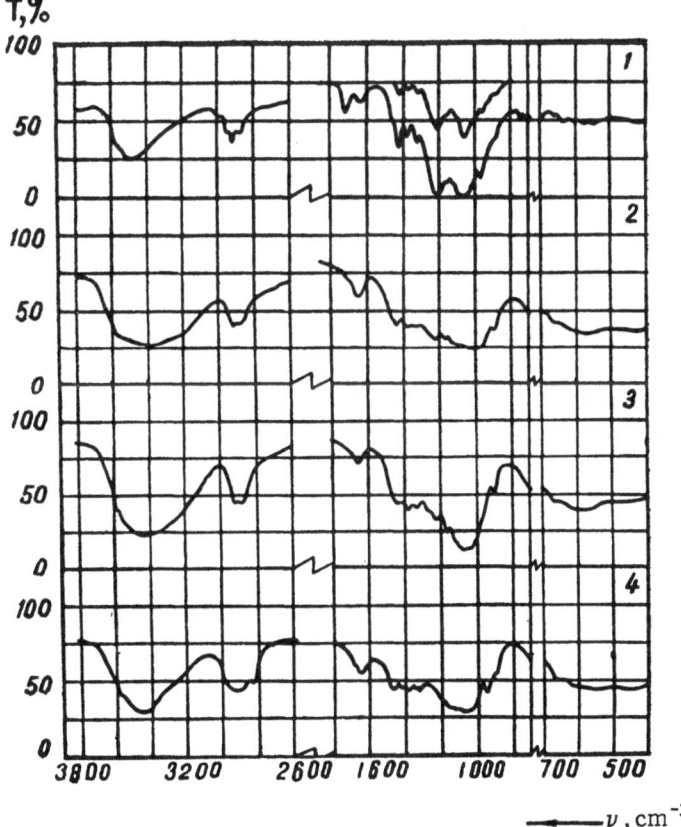

Fig. 69. Spectra of methyl derivatives of cellulose xanthate with γ equal to: 1) 160; 2) 30; 3) 8; 4) spectrum of methylcellulose.

In the spectrum of cellulose methylxanthate with γ = 160 (Fig. 69), the maximum of the hydroxyl band is located at 3480 cm^{-1}, i.e., in the region of hydroxyl groups involved in hydrogen bonding. However, the band differs from that in the cellulose spectrum in that it is clearly asymmetrical and the absorption maximum is displaced to longer wavelength. This may indicate the predominance of definite types of hydrogen bonds.

In the region of the CH group stretching vibrations, the spectrum of cellulose methylxanthate with γ = 160 shows some reduction in absorption in the vicinity of 2900 cm^{-1}, the appearance of a sharp band at 2920 cm^{-1}, with a step at 2950 cm^{-1}, and a weak band at about 2990 cm^{-1}.

Considering the high intensity of the bands corresponding to the stretching vibrations of CH$_2$ groups, and the weakness of those corresponding to the stretching vibrations of CH groups, we can assume that the integral absorption in the 2800-3000 cm^{-1} region of the cellulose spectrum is largely due to the CH$_2$ groups. A reduced absorption in the 2900 cm^{-1} region has been observed in the spectra of a number of cellulose esters and provides indirect evidence for the existence of an ester grouping at the sixth carbon atom of the elementary unit of the cellulose macromolecule [25, 26].

The bands at about 2920 and 2990 cm^{-1} in the spectrum of the methyl xanthate can be attributed to the stretching vibrations of CH in the grouping $-S-CH_3$. Indeed, the spectra of such compounds as dimethyl sulfide (CH_3-S-CH_3), dimethyl disulfide ($CH_3-S-S-CH_3$), and dimethyl trisulfide ($CH_3-S-S-S-CH_3$) show strong bands in the region of the CH stretching vibrations at about 2910-2920 cm^{-1} and weak bands in the region 2980-2990 cm^{-1} [180, 181]. On the other hand, it is known that, in the case of aliphatic hydrocarbons, the methyl group stretching vibration frequencies are highly characteristic and are located at about 2960-2870 cm^{-1} [24, 53].

It should be noted that, in spite of the high degree of esterification of some cellulose methyl xanthate preparations, the spectral bands corresponding to the stretching vibrations of the CH$_3$ groups are not very intense. A similar phenomenon has been observed with acetylated cellulose products. The low intensity of the bands corresponding to the stretching vibrations of the methyl groups in the spectra of these esters is not fortuitous; it can probably be explained by the effects of the double bonds in the ester groupings.

The spectrum of cellulose methyl xanthate with γ = 160 shows a low-intensity band at about 1720 cm^{-1}. The existence of this band indicates the presence of a small number of C=O groups. It is difficult to interpret this. The appearance in the xanthate spectrum of a weak band at 1640 cm^{-1} can be explained by the presence of a small amount of adsorbed water.

The existence of sharp bands at 1420, 1380, and 1320 cm^{-1} is characteristic of the spectrum of a cellulose methylxanthate with a high degree of esterification. The band at 1420 cm^{-1} can be attributed to the asymmetric deformation vibrations of methyl groups. For instance, dimethyl sulfide and dimethyl disulfide also show strong bands in the region 1400-1440 cm^{-1} [181].

The absorption at about 1380 cm^{-1} cannot be explained by the existence of symmetrical deformation vibrations of methyl groups, by analogy with their frequencies in the spectra of aliphatic hydrocarbons. It is well known [24] that the frequency of these vibrations is mainly determined by the electronegativity of the element to which the groups are bound. For example, in the spectra of methyl halides, the frequencies of the symmetrical deformation vibrations of the methyl groups vary within the range 1255-1470 cm^{-1}; in the case of CH_3-O, the frequency is 1456 cm^{-1} for methanol, and 1466 cm^{-1} for dimethyl ether [24], and so on. When the methyl groups are attached to sulfur, the frequencies of their symmetrical deformation vibrations are 1323 cm^{-1} (dimethyl sulfide CH_3-S-CH_3) and 1300 cm^{-1} (dimethyl disulfide $CH_3-S-S-CH_3$) [24, 181].

It thus appears that the band at 1320 cm^{-1} is the most convenient for analyzing the spectra of methylxanthates with respect to the symmetrical deformation vibrations of methyl groups.

According to a number of authors [24, 53, 183], the frequencies of the stretching vibrations of $C=S$ bonds lie in the region 1300-1400 cm^{-1}. If this is correct, then the band at 1380 cm^{-1} in the spectra of the products which we have investigated may be tentatively attributed to the $C=S$ bond.

It should be noted that the spectrum of potassium methylxanthate

$$CH_3-O-C\overset{\displaystyle S}{\underset{\displaystyle SK}{\diagup\diagdown}}$$ also shows a weak absorption band in the region of 1380

cm^{-1} and this can hardly be attributed to symmetrical deformation vibrations of CH_3 in the grouping CH_3-O-.

It has been shown [24, 53] that the spectra of esters are characterized by the strong absorption bands of $C-O-C$ ester bonds in the range 1050-1300 cm^{-1}. According to the literature [24, 53, 181], the frequency of $C-O-C$ stretching vibrations for ethers does not exceed 1150 cm^{-1}, while the spectra of most compounds containing ester groupings with unsaturated carbon atoms usually show intense bands in the region 1150-1300 cm^{-1}. A similar phenomenon has been observed in the spectra of cellulose esters. Thompson and Torkington [162] studied a number of the simple esters and found that their spectra showed two intense bands in the region 1000-1250 cm^{-1}. It has already been noted that the spectrum of cellulose methylxanthate, with $\gamma = 160$, shows a second band at about 1060 cm^{-1}. From a comparison of the spectra of cellulose methylxanthate and of acetylcellulose, both with a high degree of substitution, it is clear that the spectra are very similar in the range 1050-1250 cm^{-1}.

All this provides a basis for attributing the bands at 1220 and 1060 cm^{-1} in the cellulose methylxanthate spectrum to the stretching vibrations of

$$\cdots \overset{/}{\underset{\diagdown}{C}}-O-C\overset{\diagup S}{\diagdown} \quad \text{ester bonds.}$$ The correctness of this interpretation is con-

firmed by the fact that the spectra of low-molecular xanthates [184-186] always show not less than two bands in the region 1040-1220 cm^{-1}, which are attributed to the grouping R−O−C [184].

The intensity of the 1220 cm^{-1} band can be used for direct determination of an increase in the amount of ester grouping in the composition of cellulose xanthate. The same analysis can be carried out indirectly by measuring the decrease in integral intensity of the hydroxyl group bands at 3200-3500 cm^{-1}.

The spectrum of cellulose methylxanthate shows a band at 970 cm^{-1}, which can very probably be attributed to the CH$_3$−S− grouping. Confirmation of this is provided by the intense bands at about 950 cm^{-1}, which are characteristic of the spectra of dimethyl sulfide and dimethyl trisulfide [181].

Weak bands at 750, 730, and 700 cm^{-1} are observable in the spectra of cellulose methylxanthates with a high degree of substitution. There are known reasons for attributing these bands to the stretching vibrations of C−S [24, 181].

As in the case of other cellulose derivatives, there is reduced absorption in the 400-700 cm^{-1} region as substitution of the hydroxyl groups increases.

The spectrum of cellulose nitrophenylxanthate shows steps at 3100, 3060, and 3030 cm^{-1} against a background of the OH band, as well as sharp bands at 1600, 1580, and 740 cm^{-1}, both of which indicate the presence of an aromatic ring in the structure.

It has been shown [24] that a distinct band at 1580 cm^{-1} shows the presence of a benzene ring attached to an unsaturated grouping, and that a band in the region 735-770 cm^{-1} shows the existence of a benzene ring with four unsubstituted hydrogen atoms. The strong bands at 1520 and 1350 cm^{-1} in the spectrum of cellulose nitrophenylxanthate can be attributed, respectively, to the asymmetric and symmetric stretching vibrations of the nitro group [24, 53].

Absorption at 1600, 1500, and 750 cm^{-1} in the spectrum of cellulose phenylthiourethane can also be attributed to the presence of aromatic rings.

The band at 1550 cm^{-1} in the spectra of these products can very probably be attributed to deformation vibrations of NH.

The strong band at about 1650 cm^{-1} in the spectrum of cellulose di-ethylacetamidoxanthate can be attributed to stretching vibrations of C=O, while the bands at 2970, 2920, and 1450 cm^{-1} can be attributed to the methyl and methylene groupings of the diethylacetamido radical.

The spectra of these cellulose xanthate derivatives also show bands at 1200 and 1300 cm^{-1}, which must be attributed to the existence of ester bonds.

It is interesting to note the very high stability of some of these com-pounds. For example, the spectrum of cellulose methylxanthate, with γ = 30, obtained after storing the sample for a year, showed very little change.

Investigation of the Products of Thermal Decomposition of Cellulose Methylxanthate

The synthesis of unsaturated cellulose compounds has extended the pos-sibilities of obtaining new cellulose derivatives and, in the case of their com-plete dehydration, has led to the production of a new class of aromatic poly-mers with systems of conjugated double bonds [84]:

Until recently, the only such compounds known were the so-called cellulosene compounds having a double bond between the fifth and sixth car-bon atoms; these were first produced by Kaverzneva, Ivanov, and Salova [85]. A new means for dehydrating cellulose was discovered in the process of thermal decomposition of some cellulose xanthate derivatives, in accordance with the Chugaev reaction [86]. Polyakov, Dervitska, and Rogovin [84] showed that the thermal decomposition of cellulose methylxanthate gave rise to volatile products containing methyl mercaptan and carbonyl sulfide, and that the solid residue consisted of partly dehydrated cellulose, containing double bonds. In view of the difficulty of precise elucidation of the com-position and structure of this product by chemical means, it was of interest to use spectroscopic methods for this purpose [204].

Cellulose methylxanthate was decomposed in a series of Wurtz flasks, at various temperatures and for various times, in a fume cupboard, under an atmosphere of nitrogen. The nitrogen was first freed from traces of water by passage through sulfuric acid and solid NaOH. Figure 70 shows the spectra of the initial cellulose methylxanthate and of the products after thermal decom-position to various extents. When the temperature was increased from 70 to

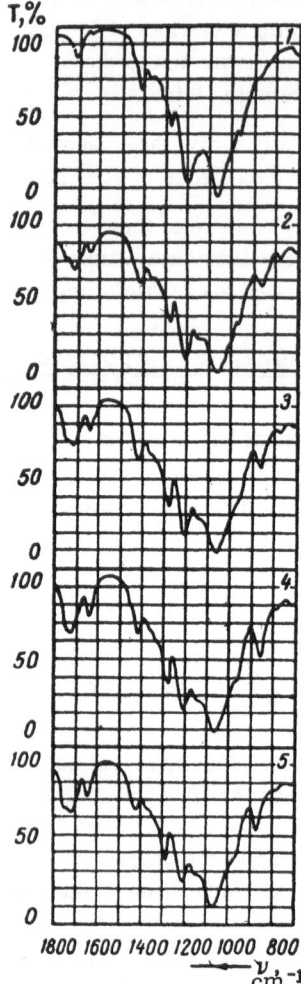

T, %

1800 1600 1400 1200 1000 800
ν, cm^{-1}

Fig. 70. Spectra of cellulose methylxanthate (γ = 160): 1) before (1) and after thermal treatment at 200°C for: 2) 0.4 h; 3) 3 h; 4) 10 h; 5) 15 h.

200°C, and the decomposition time from 15 min to 15 h, there was an increase in absorption in the regions 1720 and 1650 cm^{-1}, new bands appeared at 870 and 790 cm^{-1}, and the ester bond band at 1220 cm^{-1} became less intense. The band at 1720 cm^{-1} should undoubtedly be attributed to the stretching vibrations of the carbonyl group $C=O$ [24, 53].

The reaction, leading to the formation of carbonyl groups in the products from thermal decomposition of cellulose methylxanthate, can evidently be explained as follows:

$$S=C-SCH_3$$

$$\begin{array}{cc} H & O \\ -C-C- \\ OH & H \end{array} \longrightarrow \left[\begin{array}{cc} C=C \\ OH & H \end{array} \right]$$

$$\longrightarrow \begin{array}{cc} & H \\ C-C- \\ O & H \end{array}$$

At the same time, the increased absorption in the region of $C=O$ stretching vibrations can be attributed to a regrouping, occurring at the elevated temperature, as follows [189]:

$$\begin{array}{c} S \\ \| \\ -C-O-C-SCH_3 \end{array}$$

$$\begin{array}{c} O \\ \| \\ \longrightarrow -C-S-C-SCH_3 \cdot \end{array}$$

The increased absorption in the 1650 cm^{-1} region should be attributed to the accumulation of $C=C$ double bonds. For example, the frequencies of

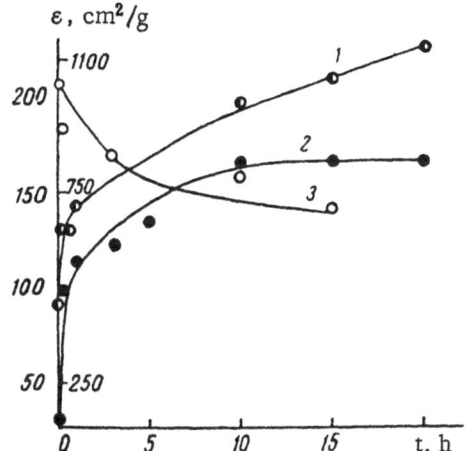

Fig. 71. Changes in intensity of the C=C, C=O,

and C-O-C bands as functions of the time of

thermal treatment at 200°C: 1) C=O at 1720 cm^{-1};

2) C=C at 1650 cm^{-1}; 3) -C-O-C at 1220 cm^{-1}.

the C=C bonds in olefins are in the region 1620-1680 cm^{-1} [24, 53]. The accumulation of C=C double bonds in the products from thermal decomposition of cellulose methylxanthate is confirmed by the appearance of new bands at 870 and 790 cm^{-1}, which can be attributed to deformation vibrations of the CH bonds in conjugated C=C double bonds [24, 53].

Thus, the presence of double bonds in the products of thermal decomposition of cellulose methylxanthate confirms the mechanism proposed [84] for the thermal decomposition of this compound.

As shown above, the strong band in the 1150-1250 cm^{-1} region of the spectra of cellulose esters is due to the stretching vibrations of the -C-O-C ester bonds. The frequency of this band in the cellulose methylxanthate spectrum lies at about 1220 cm^{-1}. The reduction in intensity of this band shows that the ester bonds are broken down in the process of thermal treatment.

It follows from Fig. 71 that, although the thermal decomposition of the methylxanthate groupings is completed after 10-15 h, the intensity of the 1720 cm^{-1} band continues to increase with further heating. It must be presumed that further formation of C=O groups occurs as the result of oxidation of partly dehydrated cellulose by traces of oxygen, contained in the nitrogen passed through the reaction flask.

Cellulose Compounds Containing New Functional Groups

It has already been mentioned that one way of modifying the properties of cellulose is to synthesize new derivatives containing different functional groups. Most of the possibilities in this direction have been discovered by introducing an aromatic amino group into the cellulose macromolecule, followed by diazotization and replacement of the diazonium group by a halogen atom or other grouping. In this way a number of new cellulose derivatives have been synthesized [89], containing iodine, thiocyanate, oxime, hydrosulfide, and arylhydrazine-N',N"-disulfonic acid groupings.

We give below some data on the investigation of these products by infrared spectroscopy [87].

The starting material was a mixed ether of cellulose and 4-β-hydroxyethylsulfonyl-2-aminoanisole (γ = 25), obtained by alkylating cellulose with the sulfuric ester of 4-β-hydroxyethylsulfonyl-2-aminoanisole. The amino group in this cellulose ether was then replaced by various functional groups as the result of chemical conversions (Table 11).

$$C_6H_7O_2(OH)_3 \longrightarrow C_6H_7O_2(OH)_{3-x}\left(OCH_2CH_2SO_2\underset{NH_2}{\underset{}{\bigcirc}}OCH_3\right)_x \longrightarrow$$

$$\longrightarrow C_6H_7O_2(OH)_{3-x}\left(OCH_2CH_2SO_2\underset{R}{\underset{}{\bigcirc}}OCH_3\right)_x .$$

It is clear from Fig. 72 that the spectra of these compounds are specific, showing that they differ in chemical structure. The spectrum of the initial ester (Fig. 72, 2) shows a number of bands, characteristic of the new functional groups introduced into the cellulose structure, at 1635, 1600, 1520, 1470-1375, 1315, 1290, 1250, 785, and 750 cm^{-1}. The bands at 1600, 1525, 785, and 750 cm^{-1} can be attributed to the presence of an aromatic structure [24, 53]. The existence of bands in the region 750-800 cm^{-1} is specific for benzene rings with three hydrogen atoms substituted. Increased absorption in the range 1375-1470 cm^{-1} can be attributed to deformation vibrations of methylene and methyl groups. The intense absorption between 1250 and 1320 cm^{-1} in the spectra of these compounds can be attributed to asymmetric stretching vibrations of the SO_2 group, and the absorption band at about 1635 cm^{-1} can be assigned to deformation vibrations of NH_2 [24, 53].

A comparison of the spectrum of the initial ether, which contains a large number of amino groups, with the spectra of the other products, which, according to chemical analysis have fewer of these groups, shows a reduction in intensity of the 1635 cm^{-1} band. This provides a direct indication of the disappearance of NH_2 groups in the course of the various reactions, and agrees well with the results of chemical analyses.

Table 11. Characteristics of the Cellulose Products under Consideration

Product	R	Extent of substitution of amino groups
3	$-\overset{\overset{\displaystyle H}{\textstyle \mid}}{C}=N-OH$	30
4	$-SH$	70
5	$-I$	95
6	$-S-C\equiv N$	80
7	$-N=N-SO_3Na$	75
8	$-\underset{\underset{\displaystyle SO_3Na}{\textstyle \mid}}{N}-NH-SO_3Na$	60

The spectra of all the products showed a band in the region 1470-1530 cm^{-1}: at about 1490 cm^{-1} for products 5 and 6 (Fig. 72, curves 5 and 6); 1500 cm^{-1} for products 3 and 4 (Fig. 72, curves 3 and 4); 1515 cm^{-1} for products 2, 7, and 8 (Fig. 72, curve 2). A band at about 1500 cm^{-1} has been observed in the case of compounds containing an asymmetric triply substituted benzene ring [24]. This indicates that all the above cellulose derivatives contained an asymmetric triply substituted benzene ring. The individual features of the spectra of these compounds in the aromatic group frequency range reflect the differences between the substituents in the benzene rings.

The experimental data show that the band in the 1470-1530 cm^{-1} region is more sensitive, as regards frequency and intensity, to a change in the nature of the third substituent, than is the band in the region 1580-1600 cm^{-1}.

All the compounds considered here contain a benzene ring conjugated to an unsaturated sulfonyl group. All their spectra show an intense band in the region 1580-1600 cm^{-1}, a fact which confirms a previous suggestion [24] as to the effect of conjugation on the intensity of bands in this spectral range.

The frequency range of the sulfonyl group in the spectra of these compounds is somewhat displaced toward longer wavelength, as compared with the frequency range given in the literature. This fact provides a basis for supposing that there is an association between the sulfonyl groups and the unsubstituted OH groups of the cellulose. A number of the absorption bands in the spectra of these ethers are characteristic of the nature of the substituent replacing hydrogen in the benzene ring. For example, the band at 2160 cm^{-1} in the spectrum of cellulose $-O-(CH_2)_2\,SO_2-\!\!\!\!\langle\!\!\!\!\!\!\bigcirc\!\!\!\!\!\!\rangle\!\!\!\!-OCH_3$ can
$$SCN$$
undoubtedly be assigned to the stretching vibrations of the bond $C\equiv N$, and so on.

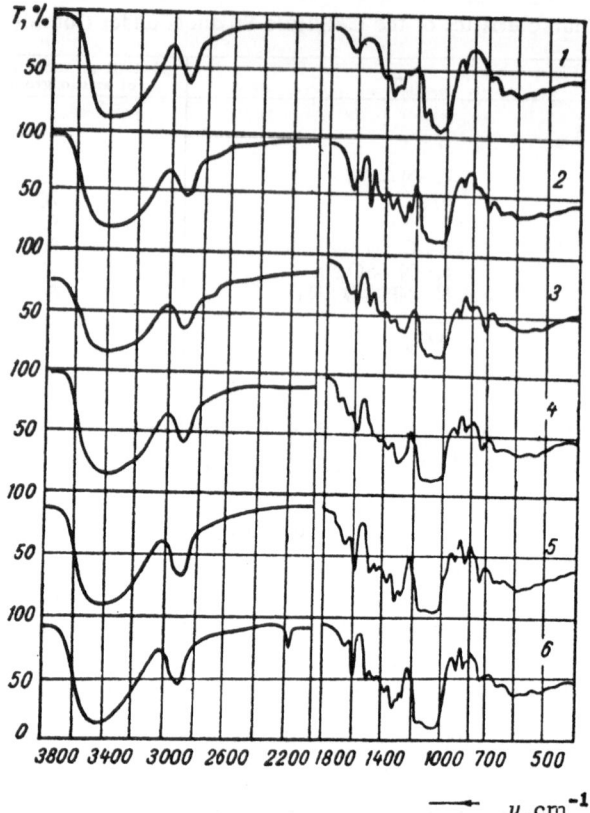

Fig. 72. Spectra of: 1) mercerized cellulose; 2) the ether of cellulose and 4-β-hydroxyethylsulfonyl-2-aminoanisole; 3) the corresponding cellulose derivative containing an oxime group; 4) the corresponding hydrosulfide derivative; 5) the corresponding derivative containing iodine; 6) the corresponding thiocyanate derivative.

Graft Copolymers of Cellulose with Poly-2-methyl-5-vinylpyridine

The synthesis of block and graft copolymers provides a new and promising method for the chemical modification of polymers. Unfortunately, this method of chemical modification has not as yet come into sufficiently general use. This is because of the lack of available methods for carrying out the copolymerization reaction, under conditions which give strict control of the composition and structure of the product, without requiring special equipment and expensive reagents.

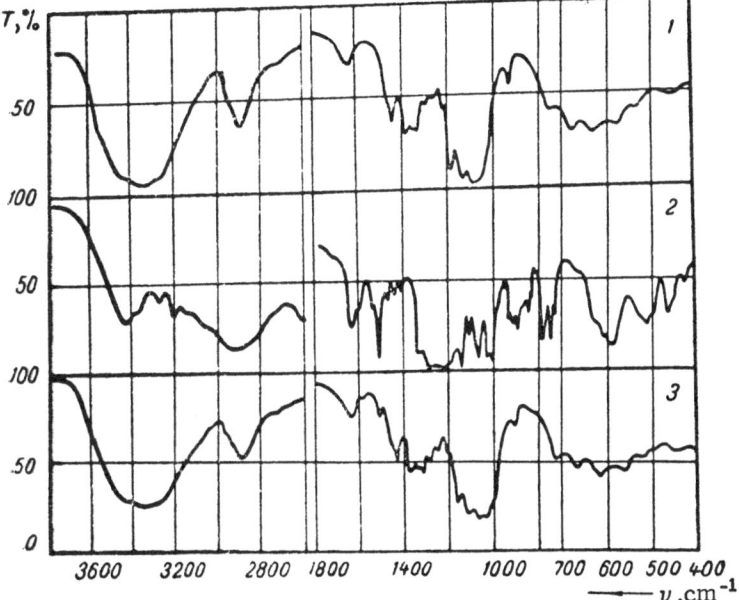

Fig. 73. Spectra of: 1) cotton cellulose; 2) 4-β-hydroxyethylsulf-onyl-2-aminoanisole; 3) cellulose alkylated by means of the sulf-uric ester of 4-β-hydroxyethylsulfonyl-2-aminoanisole.

One of the available monomers, whose polymers have a number of technologically interesting properties, is 2-methyl-5-vinylpyridine. Investigations have shown [90, 91] that it is possible to synthesize graft copolymers of cellulose with poly-2-methyl-5-vinylpyridine (PMVP). The graft polymerization has been achieved in two ways: a) by the chain transfer method; b) by a method involving formation of a macroradical, by decomposition of a diazo group previously introduced into the cellulose macromolecule.

The infrared spectra of graft copolymers of cellulose with PMVP have been investigated [92]. Table 12 shows the characteristics of the compounds investigated. These compounds were prepared for infrared investigation by direct pressing [6, 17, 18], and in other cases by the KBr-pressed disc method.

Infrared spectra were obtained for the following: graft copolymers; PMVP obtained from the graft copolymers by hydrolysis of the original cellu-lose; cellulose, alkylated by means of the sulfuric ester of 4-β-hydroxyethyl-sulfonyl-2-aminoanisole in order to introduce aromatic amino groups, re-quired to form macroradicals for the reaction of graft copolymerization by decomposition of diazo groups (Figs. 73 and 74). From a comparison of the spectra of the original cellulose and the alkylated product, it is obvious that

Table 12.   Characteristics of Products Investigated

Sample No.	Product investigated	Preparation conditions	Analyses, %		Chemical structure
			N	S	
1	Initial cotton cellulose		—	—	
2	4-β-Hydroxyethylsulfonyl-2-aminoanisole		6.10	13.80	$HOCH_2CH_2SO_2$— (ring with —OCH_3, —NH_2)
3	Cellulose alkylated by means of the sulfuric ester of 4-β-hydroxyethylsulfonyl-2-aminoanisole		0.40	0.91	cell.—$OCH_2CH_2SO_2$— (ring with —OCH_3, H_2N—)
4	Graft copolymer of cellulose with PMVP	Decomposition of the diazo group in the alkylated cellulose in presence of MVP phosphate [91]	2.80	0.18	cell.—$OCH_2CH_2SO_2$— (ring with —OCH_3); $(—HC—H_2C)_n$— (pyridine ring with N, CH_3)
5	The same	The same in presence of MVP acetate [91]	3.35	0.19	The same

No.		Reaction			Formula
6	The same	Polymerization of MVP in presence of mercerized cotton cellulose (chain transfer method) [90]	3.42	—	$cell.-O-(-CH_2-CH-)_n$; pyridinium ring, N, CH_3
7	The same	The same in presence of natural cotton cellulose	2.44	—	The same
8	PMVP, formed from product 4	Acid hydrolysis of the cellulose contained in the graft polymer [91]	11.60	0.76	$C_6H_7O'-OCH_2CH_2SO_2$— , OCH_3 ; $(-HC-H_2C-)_n$ pyridinium ring, N, CH_3
9	PMVP, formed from product 5	The same	11.60	0.68	The same
10	PMVP, formed from product 6	The same	11.70	—	$C_6H_7O(OH)_4-O-(-CH_2-$ $-CH-)_n$, pyridinium ring, N, CH_3
11	Homopolymer of MVP, obtained in synthesis of product 6		11.70	—	$(-CH_2-CH-)_n$, pyridinium ring, N, CH_3

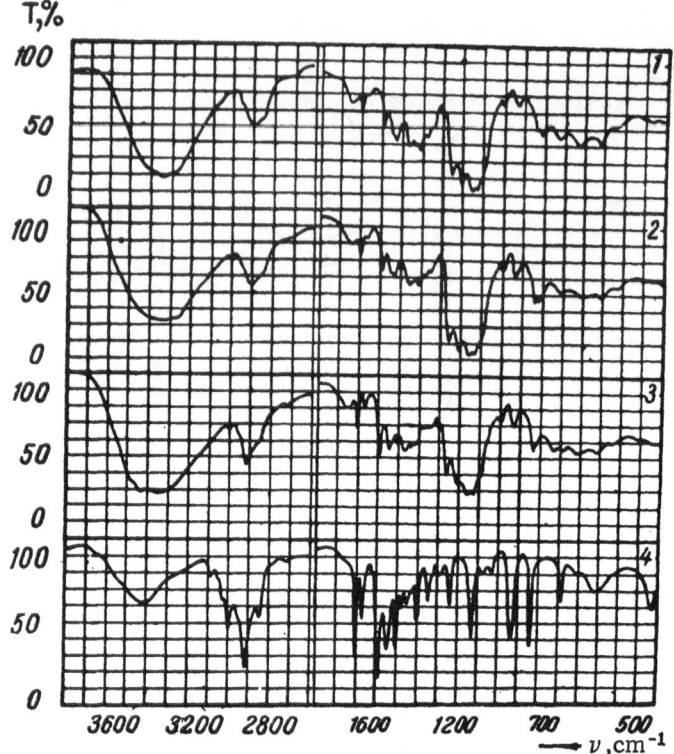

Fig. 74. Spectra of graft copolymers of cellulose with PMVP. 1) Product 4; 2) product 5; 3) product 6; 4) homopolymer of MVP.

the latter shows new absorption bands at 1590 and 1520 cm^{-1} and increased absorption in the region 1200-1300 cm^{-1}; both changes are characteristic of the presence of 4-β-hydroxyethylsulfonyl-2-aminoanisole. The weak intensity of the new bands shows the relatively small amount of new functional groups in the macromolecule of partially alkylated cellulose and is in accordance with the chemical analyses. The new bands at 1590 and 1520 cm^{-1} in the spectrum of the alkylated cellulose can be assigned to the stretching vibrations of the C $=$ C bonds in the aromatic rings [24,53]; the increased absorption between 1200 and 1300 cm^{-1} may be attributed to the presence of sulfonyl groups. Indeed, an absorption band between 1200 and 1300 cm^{-1} has been observed in the spectra of other compounds containing sulfonyl groups [24, 53], but in the present case the interpretation is not quite so definite.

The spectra of relatively thick samples, which have been very carefully dried, show a band at 1630 cm^{-1} which is absent from the spectrum of a control sample of cotton cellulose. The reduction in intensity of this band

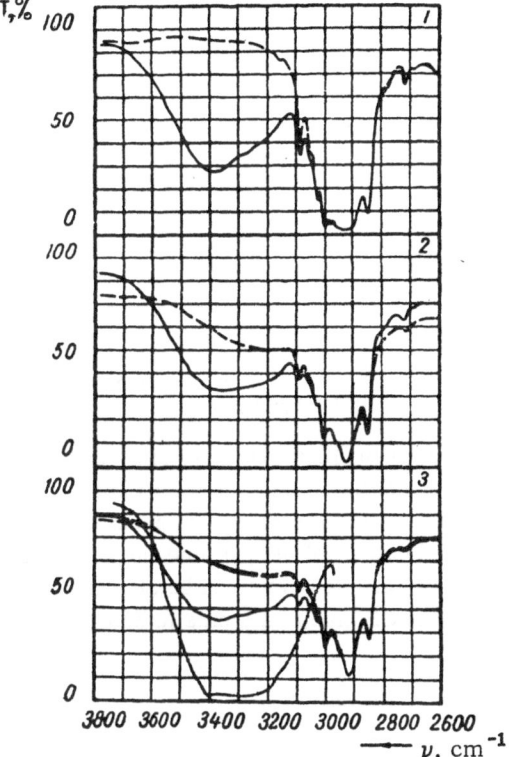

Fig. 75. Spectra of PMVP films, isolated from
graft copolymers with cellulose and homopolymer. 1) Product 11; 2) product 9; 3) product 10.
Continuous lines — samples before drying; broken
lines — samples after drying under vacuum conditions. The spectrum of D-glucose is shown for
reference. PMVP films were obtained from solutions in chloroform. Drying times were 6 h, or
6 and 12 h for product 10. Spectra were recorded with samples in a special cell, under
vacuum conditions.

in the spectrum of the graft copolymer indicates that the latter is produced as
the result of formation of a macroradical, by decomposition of the diazo
group, introduced into the cellulose macromolecule by diazotization of the
aromatic amino group in the alkylated cellulose. The spectra of the graft
copolymers 4, 5, and 6 show a number of additional bands or inflections at
3010, 2850, 1600, 1570, 1495, 1450, 1400, 840, and 740 cm^{-1}, and some

increase in absorption between 1240 and 1340 cm^{-1}, which are characteristic of the presence of PMVP in the structure of alkylated cellulose.

The absorption bands at 3010 cm^{-1} in the spectra of graft copolymers
4, 5, and 6 (Table 13) can be assigned to the stretching vibrations of C—H
bonds in a pyridine ring. For example, several authors [24, 53] have investigated the infrared spectra of pyridine and similar aromatic heterocyclics and
have observed absorption bands in the range 3010-3070 cm^{-1}, which they interpreted in the same way.

Some increase in absorption at about 2850 cm^{-1} (products 4, 5, and 6)
and a displacement of the main band maximum in the range 2800-3000
cm^{-1} (product 6), as compared with the spectrum of cotton cellulose, can be
attributed to superposition of PMVP bands.

Absorption bands at 1600, 1570, and 1495 cm^{-1} can be assigned to
stretching vibrations of the pyridine ring. It is well known [24] that the
pyridine spectrum always shows three absorption bands in the region 1480-
1600 cm^{-1}, assigned to vibrations of the pyridine ring. On the other hand,
the absorption bands at 840 and 740 cm^{-1} can be attributed to external deformation vibrations of the CH groups of aromatic heterocyclics [24, 53].

No differences have been observed between the spectra of products 4
and 5, obtained using different salts of MVP.

One of the most important tasks in investigation of the structure of
graft copolymers is to detect chemical bonds between the graft and initial
polymer. In studies on graft copolymers of cellulose with a carbon chain
polymer, the latter can be separated from the graft copolymer by hydrolysis
of the cellulose. If there are chemical bonds between the components of the
graft copolymer, then the polymer isolated by hydrolysis should contain a
definite amount of cellulose hydrolysis product, chemically bound to the
carbon chain polymer (in this case PMVP). Such a product should be absent
in the case of a homopolymer.

The experimental results (Fig. 75) show that, in the spectra of carefully dried films of PMVP there is increased absorption in the hydroxyl group
region on changing from product 11 to products 9 and 10. The presence of
hydroxyl groups can be interpreted, unambiguously, as the result of chemical
bonding of cellulose hydrolysis product to PMVP macromolecules.

It is interesting to note that the spectra of PMVP, isolated from block
copolymers obtained by various methods, differed considerably both in spectral sharpness and in the intensities of some of the bands. Similar differences
were observed between the spectra of the corresponding graft copolymers, and
can be attributed to differences in the structure of the PMVP macromolecules,
produced under different conditions.

A comparison of the spectra of cotton cellulose, mercerized cotton cellulose, and the graft copolymer obtained from cotton cellulose indicates that in the process of synthesis of the graft copolymer there is no change similar to that observed in the transition from natural cellulose to hydrocellulose.

Graft Copolymers of Cellulose and Carbon Chain Polymers, Obtained by Initiating the Graft Polymerization with Pentavalent Vanadium Compounds

One of the most promising methods in the field of synthesis of graft copolymers is by the reaction of radical polymerization. For this it is essential that polymerization of the monomer should be initiated only by a polymer macroradical, and that termination of the growing chain should occur without formation of a low-molecular radical capable of initiating homopolymerization of the monomer. These conditions can be realized by using the method of graft copolymer synthesis in an oxidation—reduction system, where cellulose acts as the reducing agent, and the oxidizing agent is the salt of some metal of variable valency, such as cerium [93] or pentavalent vanadium [94]. This method makes it possible to obtain cellulose copolymer without any homopolymer, and thus reduces the monomer consumption and facilitates the practical realization of the synthesis.

Shown below are some results of the application of infrared spectroscopy to elucidate the mechanism of synthesis of cellulose graft copolymers, with pentavalent vanadium compounds as initiators [95]. Previous investigations [94] led to the development of conditions for carrying out graft copolymerization in an oxidation—reduction system, with modified cellulose containing an aromatic amino group as reducing agent, and pentavalent vanadium as oxidizing agent. The cellulose derivative used for the graft copolymerization was the ether obtained by treating cellulose with the sulfuric ester of 4-β-hydroxyethylsulfonyl-2-aminoanisole [94]. The structure of this ester corresponds to the formula

$$\text{cell.} - \text{OCH}_2\text{CH}_2\text{SO}_2 \overset{\text{NH}_2}{\underset{}{\diagup\!\!\diagdown}} \text{OCH}_3 . \tag{I}$$

The main material selected for investigation was the graft copolymer of compound I with polyacrylonitrile. In order to elucidate the mechanism of the initiation of graft copolymerization, compound I was oxidized both in the presence and in the absence of monomer.

For the oxidation studies we used a modified cellulose in the form of a fabric, consisting of I with γ for the ether grouping equal to 9.02. The oxidant was a solution of vanadic acid in either phosphoric or sulfuric acids. The technique for carrying out the analyses has been described in detail [93, 94]. On the basis of chemical analyses it was accepted that the macroradical, initiating the graft copolymerization, was formed by the interaction of an amino group with one molecule of HVO_3 according to one of the following schemes:

$$\text{cell.} -OCH_2CH_2SO_2 \overset{NH_2}{\underset{}{\diagup\hspace{-0.3em}\boxed{}\hspace{-0.3em}\diagdown}} OCH_3 \longrightarrow VO_2^+ \longrightarrow$$

$$\longrightarrow \text{cell.} -OCH_2CH_2SO_2 \overset{\underset{\cdot N=O}{\overset{H}{|}}}{\diagup\hspace{-0.3em}\boxed{}\hspace{-0.3em}\diagdown} OCH_3 + VO^{2+} + H^+, \tag{II}$$

$$\text{cell.} -OCH_2^\cdot CH_2SO_2 \overset{NH_2}{\diagup\hspace{-0.3em}\boxed{}\hspace{-0.3em}\diagdown} OCH_3 \longrightarrow VO_2^+ \longrightarrow$$

$$\longrightarrow \text{cell.} -OCH_2CH_2SO_2 \overset{\dot{N}H}{\diagup\hspace{-0.3em}\boxed{}\hspace{-0.3em}\diagdown} OCH_3 + VO^{2+} + OH^-. \tag{III}$$

Further oxidation of compounds II and III should occur in accordance with the schemes

$$\text{cell.} -OCH_2CH_2SO_2 - \overset{\underset{N=O}{\overset{H}{|}}}{\diagup\hspace{-0.3em}\boxed{}\hspace{-0.3em}\diagdown} OCH_3 + VO_2^+ \longrightarrow$$

$$\longrightarrow \text{cell.} -OCH_2CH_2SO_2 - \overset{NO_2}{\diagup\hspace{-0.3em}\boxed{}\hspace{-0.3em}\diagdown} OCH_3 + VO^{2+} + H^+, \tag{IV}$$

$$\text{cell.} -OCH_2CH_2SO_2 - \overset{\dot{N}H}{\diagup\hspace{-0.3em}\boxed{}\hspace{-0.3em}\diagdown} OCH_3 + VO_2^+ \longrightarrow$$

$$\longrightarrow \text{cell.} -OCH_2CH_2SO_2 - \overset{NO}{\diagup\hspace{-0.3em}\boxed{}\hspace{-0.3em}\diagdown} OCH_3 + VO^{2+} + H^+. \tag{V}$$

It follows from the above that the determination of NO_2 or NO groups in the oxidation product from compound I should give an unambiguous answer as to the mechanism for oxidation of compound I, and hence as to the mechanism of initiation of the graft copolymerization. The answer was in fact obtained by infrared spectroscopy. The material used for this

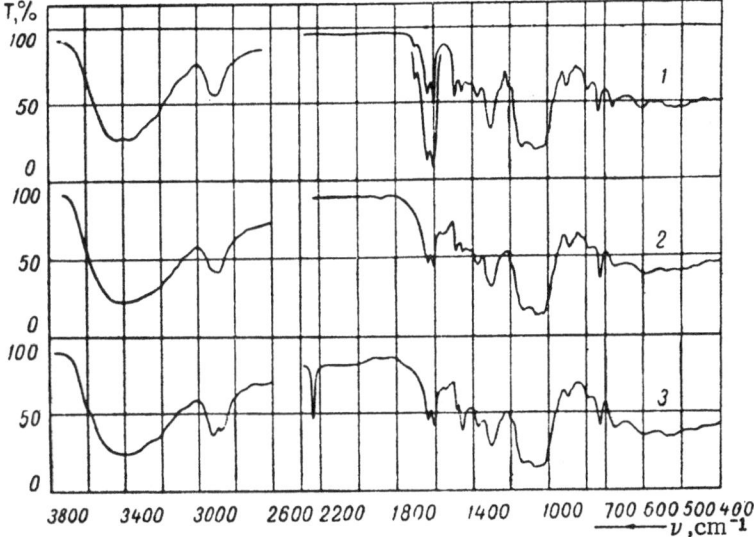

Fig. 76. Spectra of the following: 1) the ether of cellulose and 4-β-hydroxyethylsulfonyl-2-aminoanisole; 2) this ether oxidized by vanadic acid; 3) the graft copolymer of the ether and polyacrylonitrile.

investigation was compound I with γ = 46.5. The extent of conversion in the oxidation was 90.5%, as judged by the consumption of vanadic acid. The content of graft polyacrylonitrile in the copolymer was 85% of the weight of compound I, as determined from the nitrogen content. Figure 76 shows the spectra of the compounds investigated [95].

When compound I was oxidized by vanadic acid, the band at 1700 cm^{-1} disappeared, the 790 cm^{-1} band became weaker, a new band appeared at about 1550 cm^{-1}, and absorption increased in the region 1200-1240 cm^{-1} (Fig. 76). Absorption in the region 1500-1570 cm^{-1} is specific for compounds containing the grouping $C-NO_2$ [24]. The disappearance of the band at 1700 cm^{-1} and the reduction in absorption at about 790 cm^{-1} can be explained by the change in the nature of the substituent in the benzene ring. It is known, for example, that the conjunction of a benzene ring and an NO_2 group leads to a considerable change in the intensity and frequencies of the benzene ring absorption bands, because of the strong interaction of the ring with the NO_2 group [24].

It should be noted that quantitative determination of NH_2 in the oxidation products from compound I is difficult. This is because the bands corresponding to the stretching vibrations of NH_2 are masked by OH group absorption, and the deformation bands of NH_2 are in the region of aromatic ring absorption. Thus spectroscopic determination of the change in NH_2 content is unreliable. The spectrum of the graft copolymer is also characterized by disappearance of the band at 1700 cm^{-1}, a reduction in intensity of the band at 790 cm^{-1}, and the appearance of a weak band in the region 1500-1570 cm^{-1}. The presence of the latter band is associated with the partial occurrence of the reaction of oxidation to give an NO_2 group together with the graft copolymerization; in confirmation of this, the band is less intense with the graft copolymer, as compared with the oxidation product. The spectrum of the graft copolymer shows a number of new bands at 2920, 2880, and 2240 cm^{-1} and increased intensity of the band at about 1450 cm^{-1}. The bands at 2920, 2880, and 1450 cm^{-1} can be assigned to deformation vibrations of methylene groups [24], and the intense band at 2240 cm^{-1} can be attributed to stretching vibrations of the $C \equiv N$ group.

The above spectral results make it possible to choose between several possible schemes for the reaction and indicate that the structure of the graft copolymer corresponds to the formula

$$\text{cell.} -OCH_2CH_2SO_2 - \underset{}{\overset{}{\left\langle \right\rangle}} OCH_3 \overset{\overset{\displaystyle O}{\overset{\displaystyle \|}{}}}{HN-(CH_2-CHCN-)_n}$$

Products of the Ion Exchange of Graft Copolymers of Cellulose and Polyacrylhydroxamic Acid with Ions of Fe^{3+} and Cu^{2+}

Complex-forming derivatives of cellulose are of considerable interest for the separation of ion mixtures, using fabrics made of modified cellulose fibers.

We present below the results of investigations by infrared spectroscopy on graft copolymers of cellulose and polyacrylhydroxamic acid and ions of polyvalent metals, carried out in cooperation with M.D. Balabaeva, L. S. Gal'braikh, T. V Vladimirova, and Z. A. Rogovin (Table 13). The synthesis of these compounds has been described [96]. The ion-exchange reaction proceeds as follows (see scheme on page 161).

The infrared spectra were obtained by the direct pressing method, with the use of an immersion medium [6, 17, 18]. The samples were prepared as thin fiber films 10-11 μ thick and were immersed in a liquid (tetrachloroethylene) to reduce scattering of the radiation. The spectra of the original compounds (Fig. 77) showed bands in the region 3600-3100 cm^{-1}, and at 2900, 1730, 1670, 1540, 1430, 1370, 1170, 1120, 1060, 1040 cm^{-1}, and others.

Table 13. Characteristics of Graft Copolymers of Cellulose
with Polyacrylhydroxamic Acid

Product	Time for synthesis of graft copolymer, h	Content in graft copolymer, %		Content in graft co-polymer after treat-ment with solution, %	
		nitrogen	OCH$_3$ group	Cu	Fe
1	3	3.35	2.03	5.0	4.79
2	8	4.45	1.10	5.0	5.38
3	12	3.98	0.81	5.20	5.70
4	18	3.03	0.33	7.80	—
5	24	2.94	0	7.61	5.62
6	48	2.89	0	4.44	—

$$\text{cell. } -O-(CH_2-CH-)_n + \underset{\underset{NHOH}{\diagdown}}{\overset{\overset{O}{\diagup}}{C}}$$

$$+ Fe^{+3} \longrightarrow \text{cell. } -O-(CH_2-CH-)_n \quad \underset{\underset{NHOFe\diagdown}{\diagdown}}{\overset{\overset{O}{\diagup}}{C}} \cdot \qquad (VI)$$

A number of these bands are specific for the polyacrylhydroxamic acid residues introduced into the cellulose structure. For example, the band at 1670 cm^{-1} must be assigned to the stretching vibrations of C=O in the group-

ing $-C\diagdown^{\diagup O}_{NHOH}$. Indeed, according to a monograph [24], the bands of car-

bonyl C=O in amide groups are located in the region 1630-1710 cm^{-1}. Bands between 1670 and 1680 (dilute solution), and at about 1667 cm^{-1} (solid phase) have been observed in the spectrum of penicillin [192]. Richards and Thompson [193] investigated the spectra of solutions of monosubstituted amides and observed bands between 1675 and 1680 cm^{-1}. It has also been established [194] that the spectra of hydroxamic acids show bands at about 1656 and 1558 cm^{-1}, which have been attributed to stretching vibrations of the bonds C=O and C=N.

It is evident, from a comparison of Figs. 77 and 78, that when a graft copolymer of cellulose and polyacrylhydroxamic acid is treated with a solution containing Cu^{2+} and Fe^{3+} ions, there is a substantial change in the spectrum in the region 1500-1800 cm^{-1}; the intensity of the 1670 cm^{-1} band

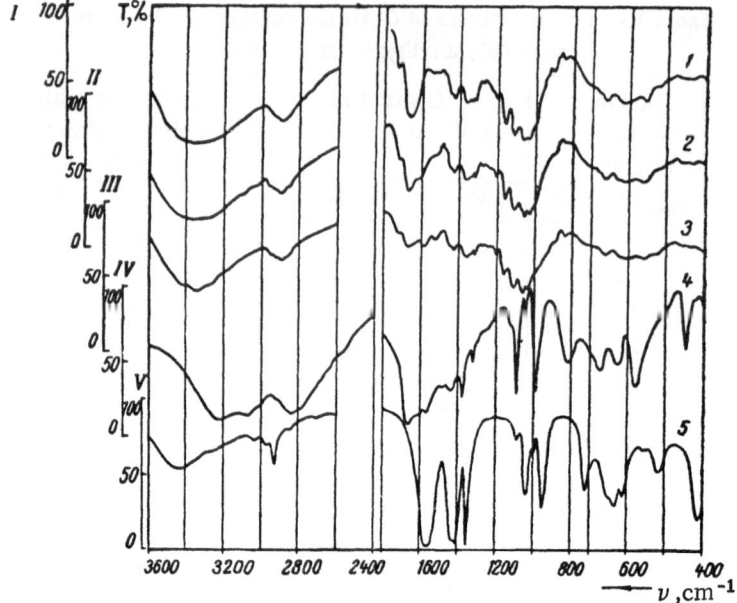

Fig. 77. Absorption spectra. 1) Graft copolymer of cellulose and acrylhydroxamic acid before treatment; 2) after treatment with Cu^{2+}; 3) after treatment with Fe^{3+}; 4) acetohydroxamic acid; 5) copper salt of acetohydroxamic acid.

diminishes, new bands appear between 1580 and 1600 cm^{-1}, and a weak band appears at 1530 cm^{-1}. Similar changes in this spectral region have been observed with model compounds, as in the difference between acetohydroxamic acid and its copper salt; the strong band at about 1670 cm^{-1} disappears, and a new band appears at 1580 cm^{-1} (Fig. 77, curves 4 and 5).

It is known that when salts are formed from carboxylic acids and carboxyl-containing cellulose derivatives there is a reduction in intensity of the carboxyl $C = O$ band, and a new band appears in the region 1580-1650 cm^{-1}.

It is quite clear that the facts enumerated above give grounds for proposing scheme VI as representing the process of ion exchange, when the graft copolymer of cellulose and polyacrylhydroxamic acid is treated with a solution of a metal cation.

The spectral changes observed when the polymer is treated with metallic cations is sensitive to the nature of the cation. It can be seen from Figs. 77 and 78 that, in the case of treatment with Fe^{3+} cations, there is a much greater reduction in intensity of the band at 1670 cm^{-1} and a new band

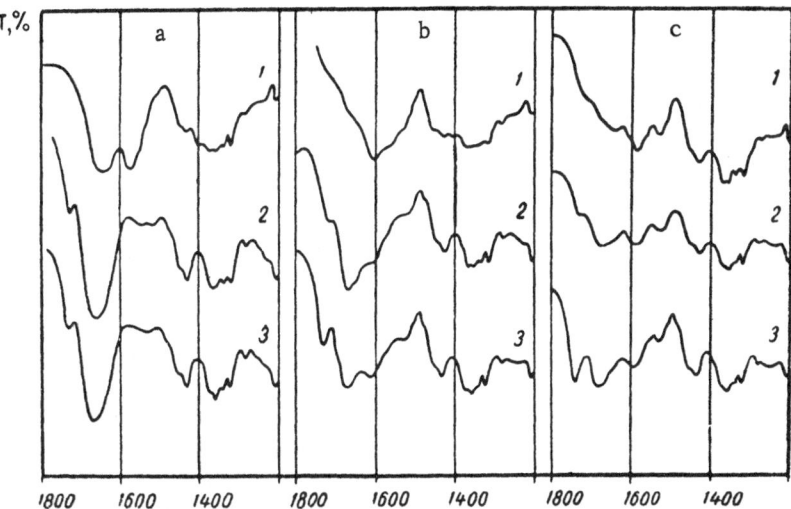

Fig. 78. Absorption spectra of graft copolymer of cellulose and polyacryl-hydroxamic acid: a) before treatment; b) after treatment with Cu^{2+} 3) after treatment with Fe^{3+}. 1, 2, and 3 correspond to samples 5, 2, and 1, respectively, of Table 13.

appears at 1580 cm^{-1}, while, in the case of Cu^{2+} cations the new band is at about 1600 cm^{-1}. Moreover, after treatment with Fe^{3+} cations, the new band at 1530 cm^{-1} is somewhat stronger.

The appearance of two new bands in the double bond frequency region, as the result of ion exchange, is not fortuitous. A similar effect has been observed, for instance, with salts of fatty acids [173] and oxidized celluloses [58], and has been attributed either to differences in the bonding of polyvalent cations, or to the existence of ionized and unionized metal carboxylate groups.

Thus, the experimental material obtained indicates that the graft copolymer of cellulose with polyacrylhydroxamic acid is capable of forming active chemical bonds with polyvalent metal cations, by reaction of its

$$-C\overset{\displaystyle O}{\underset{\displaystyle NHOH}{\big<}}$$

groups. No significant changes have been detected in the spectra of these compounds in the region of alcoholic hydroxyl group absorption, thus suggesting that there are no bonds between such hydroxyl groups and metal cations.

It is interesting to consider the effects of the conditions of formation of graft copolymers of cellulose and polyacrylhydroxamic acid on the structure of the product, and on its capacity to interact with metal cations. It can be seen from Fig. 78 that, with increase in the time taken for synthesis of the copolymer, there is increased absorption in the regions 1580 and 1300-1450 cm^{-1}. In this case, obviously, the increased intensity of the band at 1580 cm^{-1} cannot be attributed to ion exchange processes.

It has already been mentioned that the spectra of hydroxamic acids show bands in the regions 1656 and 1558 cm^{-1}, which have been assigned, respectively, to the stretching vibrations of $C=O$ and $C=N$ bonds [194]. Now it is known that hydroxamic acid can exist in two tautomeric forms: the keto $-C\overset{O}{\underset{NHOH}{}}$ and the enol $-C\overset{OH}{\underset{NOH}{}}$. The appearance of a band in the region 1580 cm^{-1} may be attributed to the presence of the enol form. In confirmation of this view, it has been stated [24, 53] that the frequencies of the $C=N$ bond stretching vibrations can lie in this region of the spectrum.

There are published results to indicate that $-C\overset{OH}{\underset{NOH}{}}$ groups can give rise to three-electron bonds of the type $-C\overset{OH}{\underset{NOH}{}}$ [24]. In this case, by analogy with ionized carboxyl groups $C\overset{O}{\underset{O}{}}$, we should expect bands corresponding to symmetric and asymmetric stretching vibrations of $-C\overset{OH}{\underset{NOH}{}}$. This may account for the increased absorption at 1400 cm^{-1} and the greater intensity of the band at 1580 cm^{-1}, shown in the spectra of compounds 5 and 6 (Fig. 78); these bands may be attributed respectively to symmetric and asymmetric stretching vibrations of the type

$$-C\overset{OH}{\underset{NOH}{}} \qquad -C\overset{OH}{\underset{NOH}{}} \; .$$

According to the results of chemical investigation, the equilibrium should be displaced toward the keto form in the process of ion exchange. In fact, the treatment of compounds 5 and 6 with metal cations (particularly iron) tends to level out the spectral differences between these compounds and products 1 and 2, in the region 1300-1400 cm^{-1}.

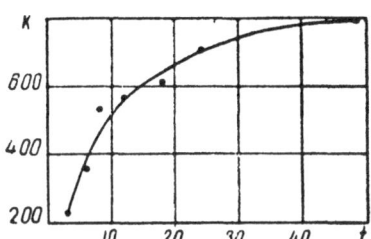

Fig. 79. Variation of the extinction coefficient at the 1580 cm^{-1} band maximum of the initial polymer with the treatment time.

Thus it may be accepted that, depending on the time taken for synthesis of the graft copolymer, and the experimental conditions used, there may be changes in the proportion of keto to enol forms in the polyacrylhydroxamic acid residues. From the spectra shown in Fig. 78 it appears that this factor can have a considerable influence on the process of ion exchange.

Figure 79 shows the variation of the extinction coefficient at the 1580 cm^{-1} band maximum with the treatment time for the product investigated. It is clear that in the initial stages (10-15 h) the enol form of hydroxamic acid is formed at a relatively high velocity; with a further increase in the reaction time the rate of enol formation decreases somewhat, probably owing to a shift in the equilibrium toward the keto form. At the same time there is a reduction in the nitrogen content of the graft copolymer, probably as the result of a similar process of hydrolysis of the hydroxamic acid residue.

In spectra 1, 2, and 3 of Fig. 77, and in spectra 2 and 3 of Fig. 78, there are relatively weak bands in the 1720 cm^{-1} region. The presence of this band in the spectra of samples 1 and 2 can be attributed to small amounts of carboxyl or methylated carboxyl groups; the results of chemical analyses show that these are present in small amounts in the structure (Table 13).

It is clear that, under the conditions of treatment of the graft copolymer (a strongly acid medium), the COOH groups cannot react with metals. Nevertheless, the content of COOH increases in samples 2-6, while the intensity of the 1720 cm^{-1} band decreases. Consequently, this band cannot be attributed to carboxyl groups, but must be assigned to methylated carboxyl groups. A comparison of the data in Table 13 and Fig. 78 shows good agreement between the intensity of the 1720 cm^{-1} band and the content of methylated carboxyl, as determined by chemical analysis.

THE POSSIBILITIES OF THE INFRARED SPECTROSCOPIC METHOD FOR INVESTIGATION OF THE PROPERTIES OF CELLULOSE AND ITS DERIVATIVES

The 2000-4000 cm^{-1} Region

This region of the cellulose spectrum contains the main stretching vibration frequencies of the groups OH, CH_2, and CH.

In the hydroxyl group field we can obtain definite information about hydrogen bonding in cellulose and its derivatives, the extent of substitution of hydroxyl groups, etc. (Fig. 80). For example, we can see the special features of hydrogen bonding in various structural modifications of cellulose [98, 147, 195, 196], in cellulose derivatives, in cellulose nitrates as compared with other cellulose esters, etc. [50, 98]. The considerable reduction in intensity of the cellulose OH group band in the process of alkaline treatment, together with a study of model compounds, has provided a basis for conclusions as to the possibility of cellulose alcoholate formation [16].

The spectra of celluloses of various age and origin, and of their derivatives, show no bands which can be attributed with certainty to free hydroxyl groups.

Cellulose derivatives, depending on the extent of substitution and the nature of the new groups introduced, can be distinguished only by the character of the hydrogen bonds. There is insufficient evidence for accepting the view, advanced by several authors [46], that the introduction of bulky groups into the cellulose structure leads to the appearance of unassociated hydroxyl groups, as a result of the rupture of intermolecular hydrogen bonds.

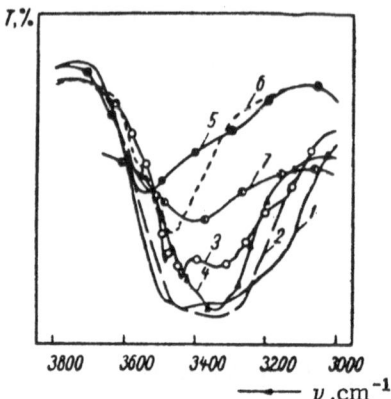

Fig. 80. Spectra in the hydroxyl-group hydrogen-bond region of: 1) hydrocellulose; 2) cellulose; 3) hydrocellulose treated with D_2O; 4) cellulose treated with D_2O; 5) cellulose nitrate ($\gamma = 220$); 6) benzylcellulose ($\gamma = 200$); 7) cellulose treated with 17.5% NaOD in D_2O.

In the author's view, it is correct to state that, not only in cellulose but also in its derivatives, all — or at any rate most — of the hydroxyl groups are involved in hydrogen bonding.

Interpretation of the possible types of hydrogen bonding in cellulose is a very complex task, requiring preliminary elucidation of the most probable conformations of its molecules and individual units. Up to now models have been constructed only for the stereochemical distribution of cellulose molecules whose parts have a three-dimensional state of order [197] and, even here, an accepted point of view has not yet been completely established.

Of the papers dealing with the elucidation of the type of hydrogen bond, we must first consider the work of Marrinan and Mann [147, 195] and of Marchessault and Liang [148, 196]. Marrinan and Mann used the deuteration technique to determine differences in hydrogen bonding between ordered and unordered parts of the cellulose macromolecule. They identified the undeuterated with the crystalline parts and attempted to elucidate the geometry of the hydrogen bonds in the crystalline parts of various structural modifications of cellulose. In the spectrum of cellulose II, they established the presence of bands at 3484 cm^{-1} ($\parallel$) and 3444 cm^{-1} ($\parallel$), and of a double band at about 3350 cm^{-1} ($\perp$), and in the spectrum of cellulose III$_1$ they observed a sharp band at 3484 cm^{-1} ($\parallel$), and bands at 3300 ($\perp$) and 3150 ($\perp$) cm^{-1}, etc. In the case of cellulose I, these authors observed two somewhat different spectra, and therefore refrained from suggesting proposed structures for the hydrogen bonds. Their spectroscopic data enabled Marrinan and Mann to suggest the following schemes: for intramolecular hydrogen bonding $O_3-H \ldots$ $\ldots O_5'$ and $O_2-H \ldots O_6'$ or $O_2 \ldots H-O_6'$ (cellulose II and cellulose III$_1$); for intermolecular hydrogen bonding, $O_6H \ldots O_3^*$, $O_2H \ldots O_2^*$, $O_3 \ldots HO_6^*$ (cellulose II), and $O_3 \ldots HO_6^*$, $O_2H \ldots O_6^*$, $O_6H \ldots O_3^*$, $O_6 \ldots HO_2^*$, $O_2 \ldots HO_3^*$, $O_3H \ldots O_2^*$, $O_5 \ldots HO_6^*$, $O_6H \ldots O_5^*$ (cellulose III$_1$). *

---

* The atoms marked with a star are in adjacent rings.

Liang and Marchessault [196] investigated doubly oriented films of ramie and bacterial cellulose. They used a somewhat different molecular model from that of Marrinan and Mann [195], consisting of bent cellulose chains [198]. With doubly oriented cellulose films they observed the following bands (Fig. 81b and c): 3405 ($\perp$), 3375 ($\parallel$), 3350 ($\parallel$), 3305 ($\perp$), 3275 ($\parallel$), 3245 ($\parallel$) cm^{-1} (bacterial cellulose), 3405 ($\perp$), 3375 ($\parallel$), 3350 ($\parallel$), 3305 ($\perp$), 3275 ($\parallel$) cm^{-1} (ramie cellulose); 3488 ($\parallel$), 3447 ($\parallel$), 3350 ($\perp$), 3305 ($\perp$), 3175 ($\perp$) cm^{-1} (mercerized ramie cellulose). On the basis of these results, the molecular model used, and the dichroism of the bands attributed by them to the CH_2 groups, Liang and Marchessault suggested the existence of the following: intramolecular hydrogen bonds $O_3H \ldots O_5^!$ (the 3350 cm^{-1} band of natural cellulose, and the 3484 and 3447 cm^{-1} bands of mercerized cellulose); intermolecular hydrogen bonds $O_6H \ldots O_4^*$ (the 3305 and 3405 cm^{-1} bands of natural cellulose, corresponding to bonds in the planes 101 and 10$\bar{1}$) and $O_6 \ldots HO_6^*$ (the 3350 cm^{-1} band of mercerized cellulose).

It follows from the above that, so far, there is no agreement as to the hydrogen bond structure, even in the most ordered parts of cellulose. This is not surprising, seeing that at present there is no accepted view as to the relative conformations of the cellulose macromolecules in these regions.

It should also be noted that the so-called crystalline parts do not alone determine the characteristics of cellulosic materials. A considerable part of cellulose consists of the so-called "amorphous" unordered fraction, where there is a much wider choice of conformation and, consequently, of hydrogen bond type. Thus, elucidation of the conformations of the cellulose macromolecule and its rings constitutes a necessary preliminary to a correct and complete interpretation of the hydrogen bond geometry in cellulose.

It must also be appreciated that hydrogen bonds may affect the conformational characteristics of the cellulose macromolecule and its individual rings by stabilizing unstable conformations. Thus, a change in the hydrogen bond structure may be an important criterion in the conformational conversions of cellulose and its derivatives. This must be borne in mind when estimating the effects of new functional groups introduced into the molecule on the properties of cellulose, and when interpreting changes in the cellulose spectrum.

At present we can draw the following conclusions about hydrogen bonding in cellulose and related compounds.

1. In cellulose, and its derivatives and related compounds, practically all the hydroxyl groups are involved in inter- or intramolecular hydrogen bonding.

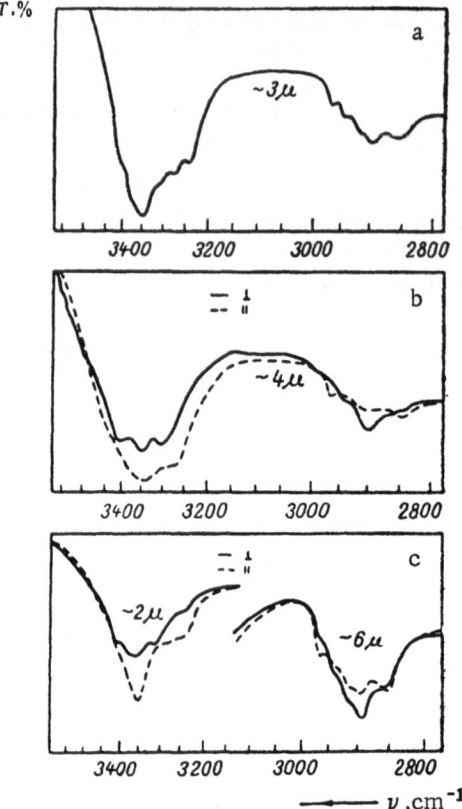

Fig. 81. Spectra of: a) <u>Valonia</u> <u>ventricosa</u> cellulose; b, c) doubly oriented samples of ramie and
bacterial celluloses [196]: ⊥ — electric vector perpendicular to direction of molecular chain; ‖ —
electric vector parallel to direction of molecular
chain.

2. In all these compounds there can exist, in principle, energetically
unequal hydrogen bonds, ranging from very weak to very strong, and covering
a frequency range of 2500-3580 cm^{-1}.

3. Each of the main structural modifications of cellulose is characterized by its own hydrogen bond structure.

4. Introduction into the cellulose structure of a large number of bulky
groups (as in esterification) does not lead to the significant appearance of unassociated hydroxyl groups. In this case, the remaining unsbustituted hydroxyl
groups are still mainly involved in hydrogen bonding, which can, however,

depend on the nature of the substituent, its position in the molecule, and the nature of the reaction occurring.

5. Hydrogen bonds in carbohydrates and similar compounds are inter-dependent, so that a change in one type of hydrogen bond may lead to a change in the nature of the other hydrogen bonds.

It has been pointed out above that the stretching vibrations of the CH_2 and CH groups give a band in the 2900 cm^{-1} region of the cellulose spectrum. The spectra of cellulose fibers show, superimposed on the band as background, steps at 2960, 2940, and 2860 cm^{-1}; these are particularly noticeable in the case of cotton cellulose. An attribution of frequencies in this region, based on one sample of cellulose [143, 148, 196], is not, in the author's view, convincing. For example, Tsuboi [143] investigated the po-larized spectra of cellulose fibers and attributed the weak bands at 2967 (‖) and 2851 (‖) cm^{-1} to symmetric and asymmetric stretching vibrations of CH_2, and the main band at 2907 cm^{-1} (⊥) to the stretching vibrations of CH. Tsuboi considered that absorption in the region 2800-3000 cm^{-1} of the cellu-lose spectrum was due mainly to CH, since the ratio of CH to CH_2 groups in the cellulose structure is 5:1. Liang and Marchessault investigated the spec-tra of doubly oriented cellulose films and established the existence of the following absorption bands in the spectra of natural celluloses: bacterial cellulose 2970 (‖), 2945 (⊥), 2914 (⊥), 2891 (⊥), 2870 (⊥), and 2853 (‖) cm^{-1}; ramie cellulose 2970 (‖), 2945 (⊥), 2910 (⊥), 2870 (⊥), and 2953 (‖) cm^{-1}; mercerized ramie cellulose 2981 (⊥ ?), 2968 (‖), 2955 (‖), 2933 (⊥ ?), 2904 (⊥), 2891 (⊥), 2874 (⊥), and 2850 (‖) cm^{-1} [196].

These authors attributed the bands at 2914, 2910, 2891, and 2870 cm^{-1} (natural cellulose), and at 2968, 2955, 2904, 2891, and 2874 cm^{-1} (mer-cerized cellulose), to the stretching vibrations of CH, and the bands at 2945 and 2853 cm^{-1}, respectively (natural cellulose), and 2933 and 2850 cm^{-1} (mercerized cellulose) to the asymmetric and symmetric stretching vibrations of CH_2.

Both Tsuboi [143] and Liang and Marchessault [148, 196] based their attributions of the stretching vibrations of CH_2 and CH on the probable spa-tial distribution of these groups and on literature data for these frequencies (carbohydrates).

It is difficult to accept Tsuboi's conclusion [143] that absorption in the region 2800-3000 cm^{-1} is due mainly to CH groups.

It has been seen above that, in the case of simple compounds related to cellulose, methylene and methyl groups show strong specific absorption bands. Absorption in the 2800-3000 cm^{-1} region in the spectra of sugars is largely the result of absorption by CH_2 groups.

The frequencies of the stretching vibrations of CH_2 in a CH_2OH group can differ considerably from the corresponding frequencies in the case of hydrocarbons. For example, the spectrum of pentaerythritol shows a strong doublet at 2940-2950 cm^{-1}, with a step at 2925 cm^{-1}, and an intense band at about 2890 cm^{-1}; these can undoubtedly be attributed to asymmetric and symmetric stretching vibrations of CH_2.

It must be accepted that it is mainly absorption by CH_2 which determines the absorption intensity in the 2800-3000 cm^{-1} region of the cellulose spectrum. This view is confirmed by the marked reduction in intensity of this band when the CH_2OH groups are oxidized by oxides of nitrogen. A reduction in the intensity of this band is also observed in the spectra of various cellulose esters, a fact which is readily explainable by the known effect of adjacent double bonds on the intensity of methylene group bands.

Because of the strong interaction of CH_2 groups with adjacent structural elements, it is usually difficult to separate the frequencies of the stretching vibrations of CH_2 and CH groups, which can vary over a wide range, depending on the material. It is also necessary to allow for the possible existence of rotational isomers, resulting from rotation or turning of the CH_2OH group around the C_5-C_6 bond, which can lead to an increase in the number of absorption bands. Analysis of a spectrum in the region 2800-3000 cm^{-1} should be carried out with allowance for the possible isomerism of this group. For example, the spectrum of hydrocellulose shows a marked reduction in the intensity of the 1430 cm^{-1} band, attributable to internal deformation vibrations of CH_2 in one of the possible types of isomer involving CH_2OH [17]. This is accompanied by a marked change in the character of the 2800-3000 cm^{-1} band, particularly the splitting at the maximum.

New bands appear in the spectra of cellulose derivatives, in the region of CH stretching vibrations, following the introduction of new functional groups into the cellulose structure: the CH in the benzene rings of tritylcellulose gives bands at 3090, 3060, and 3030 cm^{-1}; ethylcellulose 2970, 2930, and 2870 cm^{-1}; esters of cellulose and chlorinated fatty acids 2960 and 2870 cm^{-1}; acetylcellulose 3020 and 2950 cm^{-1}; cellulose methylxanthate 3000 and 2920 cm^{-1}, etc.

Cellulose itself does not have absorption bands in the region 2000-2600 cm^{-1}. However, useful information can be obtained here from the spectra of a number of cellulose derivatives.

For example, the spectrum of highly oxidized monocarboxycellulose shows a weak band at 2500 cm^{-1} [7, 11], which must be attributed to carboxylic hydroxyl groups, strongly perturbed by hydrogen bonding. Comparison of the spectra of selectively oxidized celluloses in this region provides definite information as to the special characteristics of carboxyl groups in various positions.

The spectra of deuterated celluloses show bands in the region 2450-2600 cm^{-1} (OD). From the intensity of these bands it is possible to judge the rate and extent of deuteration, and hence the structural homogeneity of the cellulose.

It may be assumed that under normal conditions deuterium exchange proceeds mainly in the least ordered parts. The OD bands in the spectrum of cellulose, deuterated under normal conditions by vapor or liquid D$_2$O, are diffuse in character, whereas the hydroxyl bands of undeuterated cellulose are discrete. This circumstance enabled Marrinan and Mann to demonstrate the special features of hydroxyl hydrogen bonds in the "crystalline" parts of various structural modifications of cellulose [147], and to develop spectroscopic methods for determining the content of the amorphous fraction of cellulose [199]. However, it must be appreciated that the use of this method can lead to high results, because of possible partial deuteration in the more ordered parts. Our experiments have shown that, under certain deuteration conditions, it is possible to obtain an OD group band with a structure similar to that of the band of OH in the undeuterated regions. In this case, the spectrum of natural cellulose shows OD group bands at 2520, 2480, and 2450 cm^{-1}, while hydrocellulose shows sharp bands at 2580 and 2550 cm^{-1} and a diffuse band with a complex structure at about 2480 cm^{-1}.

Deuterium exchange can also be used to elucidate some of the other properties of cellulosic materials.

For example, a study of the spectra of deuterated celluloses, obtained from wood of various ages, has shown that the intensity of the OD band is least in the case of wood from the interglacial period [7].

Stepanov, Skrigan, Shishko, and Zhbankov [15] used deuterium exchange to investigate the nature of the bonds between cellulose and the accompanying compounds in plant cells.

For this purpose, they carried out tests on cellulose production under conditions of digestion with soda, using a 12.06% solution of NaOD in D$_2$O. The standard material for comparison was cotton cellulose, the purest cellulose available; this was also subjected to digestion under the same conditions.

Infrared spectroscopy was found to be valuable in elucidating special features of the deuteration process for wood samples of various ages and for certain other plant constituents [100].

The spectra of cellulose derivatives, containing $C \equiv N$ and $C \equiv C$ bonds, show sharp bands in the region 2150-2250 cm^{-1} (at about 2250 cm^{-1} in the case of cyanoethylcellulose and the graft copolymer of cellulose with polyacrylonitrile, and so on). However, it must be appreciated that the intensity of the $C \equiv N$ band varies greatly with the nature of the compound. In some cases the band is too weak to be detected.

The 1500-2000 cm^{-1} Region

This region contains the frequencies corresponding to the stretching vibrations of the double bonds: $N = O$, $C = O$, $C = C$, etc. The symmetric deformation vibrations of the water molecule correspond to a band in the region 1650 cm^{-1}, and this band is convenient for checking the extent of drying of cellulose. Bands at about 1500 and 1600 cm^{-1} appear in the spectra of celluloses with a high content of low-molecular impurities, particularly in samples with a high lignin content, and must be attributed to the aromatic rings of lignin. Absorption in the region 1500 and 1600 cm^{-1} can be used for ligning determination. It is first necessary to determine the sensitivity of the spectroscopic method; this can be increased considerably by using a differential method, with pure cotton cellulose, free from lignin, as a reference material. In this way it is possible to determine the ligning content directly, as opposed to the chemical method.

It is convenient to use bands at 1750 ($C = O$) and 1650 ($N = O$) cm^{-1} for following the kinetics of processes of oxidation, nitration, esterification, etc. Oxidized celluloses may contain, simultaneously, carboxylic, ketonic, and aldehydic groups, all containing $C = O$ bonds. They should therefore show strong bands in the above spectral region. However, dialdehydocelluloses do not display any bands in the double-bond region [7, 10, 166]. This means that the aldehydic groups are not in the free state. This factor must be considered when attempting to carry out spectroscopic analyses for aldehyde content.

Monocarboxycellulose contains a high proportion of carbonyl groups as well as carboxyl groups. The spectrum of a monocarboxycellulose, if obtained with an NaCl prism instrument, shows only one band in the $C = O$ region at 1740 cm^{-1} [7, 10]. Naturally, in this case, and using this band, it is only possible to determine the total number of $C = O$ bonds, and it is not possible, as erroneously proposed by Forziati, Rowen, and Plyler [167], to determine the number of carboxyl groups. However, this spectral region can be used for the separate determination of carbonyl and carboxyl content if the oxidized cellulose is first treated with active metal cations. Experiments have shown that the following substitution occurs: COOH $\rightarrow$ COOMe, whereas the carbonyl groups are not affected [7, 10, 58, 65]. The COOMe group

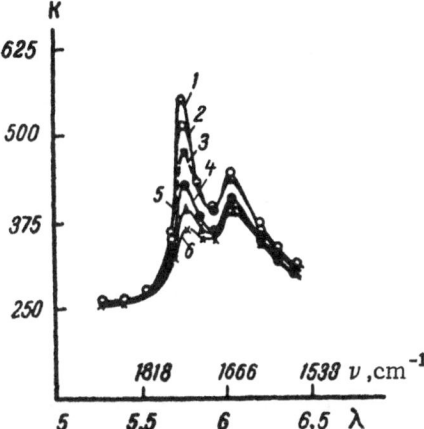

Fig. 82. 1,2,3,4,5, and 6 are spectra of cellulose
which has been treated with oxides of nitrogen for
238, 180, 67, 48, 23, and 6 min, respectively.

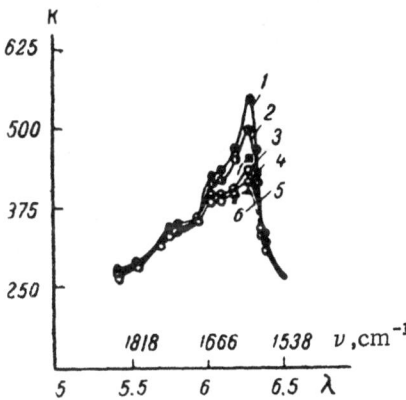

Fig. 83. Spectra of cellulose, oxidized with oxides
of nitrogen for the same times as in Fig. 82, and
then treated with a solution of lead cations. The
numbers have the same significance as in Fig. 82.

gives rise to a band of characteristic frequency, considerably displaced to
longer wavelength at 1550-1650 cm^{-1} [64, 58]. As described above, the frequency and intensity of the COOMe band are sensitive to the nature of the
cation. A spectral method on this basis has been developed for the analysis
of dilute solutions for their metal cation contents [64].

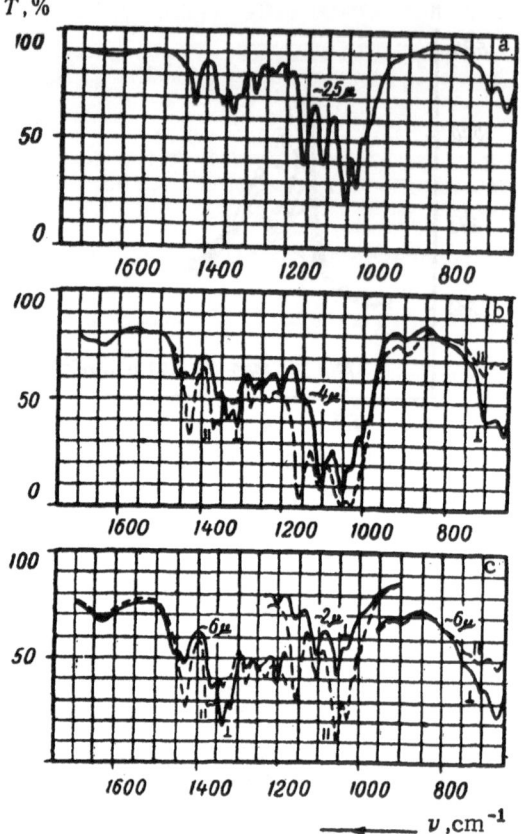

Fig. 84. Spectra of: a) <u>Valonia ventricosa</u> cellulose; b) doubly oriented sample of ramie cellulose;
c) bacterial cellulose [196].

A similar spectroscopic methods has been used to investigate the kinetics of accumulation of carbonyl and carboxyl groups in the process of cellulose oxidation by nitrogen dioxide [55]. The sample of oxidized cellulose is
treated with lead cations, and the analysis of the kinetics of carboxyl group
accumulation is carried out by means of the new band appearing at 1590
cm^{-1} (Figs. 82 and 83). However, it should be noted that the carboxyl group
content determined in this way will be low if the carboxylic hydrogen is in-
completely replaced by the metal cation. This error is small in the case of
Pb, though not with other cations, as has been shown by treating dicarboxycellulose with lead cations. This form of oxidized cellulose contains no
carbonyl groups, and, if it is treated with lead cations, its spectrum shows no
band in the 1740 cm^{-1} region.

Establishment, by direct means, of the presence of definite types of double bonds can be of great value in the analysis of processes involving structural changes in cellulose. Thus, it has been shown above that analysis of the spectrum in this region makes it possible to follow the accumulation of $C = C$ double bonds in the products of thermal decomposition of cellulose methylxanthate, and to show that decomposition proceeds by the Chugaev mechanism [86].

New bands, reflecting special features of the structure, appear at 1500-2000 cm^{-1} in the spectra of cellulose derivatives containing aromatic structures. For example, the spectrum of the trityl ether of cellulose is characterized by a sharp band at 1600 cm^{-1} and a considerably stronger band at 1500 cm^{-1}, whereas, in the case of cellulose benzoate, the intensity ratio of the bands is reversed, and the 1600 cm^{-1} band is a doublet.

The 1200-1500 cm^{-1} Region

In this region of the cellulose spectrum there are a number of relatively sharp absorption bands, which are very sensitive to chemical and structural transformations of the cellulose (Figs. 84 and 85).

Absorption bands, associated with deformation vibrations of bond angles and bonds of CH_2OH, appear in the 1200-1450 cm^{-1} region of the cellulose spectrum. The band at 1430 cm^{-1} should undoubtedly be attributed to internal deformation vibrations of the CH_2 group. When cellulose is selectively oxidized with oxides of nitrogen, the band at 1430 cm^{-1} disappears and there is a marked decrease in intensity of the bands at 1370, 1360, 1340, and 1320 cm^{-1}. On the other hand, oxidation with periodate leads to only a relatively slight change in the spectrum. Reduction of the hydroxl groups in cellulose oxidized by periodate (the appearance of primary hydroxyls at C_2 and C_3) is accompanied by increased absorption in separate parts of the region (1400 and 1240 cm^{-1}).

Several authors, without adequate justification, have attributed some of the bands in the 1300-1400 cm^{-1} region to alcoholic $C-O$ [153], or to CH bonds [137, 143, 149]. The latter attribution can only be partially, if at all, correct, because the total decrease in the number of CH groups on oxidation of cellulose by nitrogen oxides is small, whereas the change in intensity of bands in this region is very considerable.

The attribution of the 1320 cm^{-1} band to external deformation vibrations of CH_2 [149] is highly improbable. In particular, we found that this band vanished completely when cellulose was treated with a concentrated solution of NaOD in D_2O [16], and also when cellulose was isolated from wood by digestion with NaOD in D_2O [15]. However, there is a sound basis for

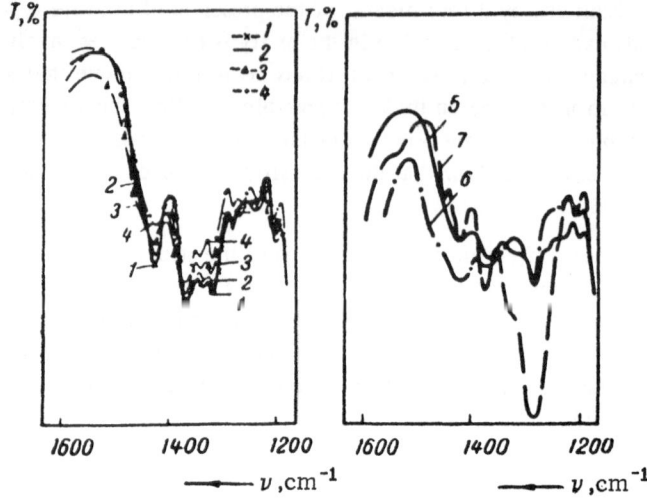

Fig. 85. Spectra of: 1) unhydrolyzed residue of cellulose; 2) cellulose; 3) cellulose regenerated after treatment with 15% NaOH; 4) cellulose regenerated after treatment with 17.5% NaOH; 5) monocarboxycellulose with 9.6% COOH; 6) monocarboxycellulose treated with Al cations; 7) cellulose nitrate.

attributing the 1370 cm^{-1} band to deformation vibrations of CH$_2$, since it is relatively unaffected by deuterium exchange, treatment of the cellulose with alkali, etc.

The bands at 1360 and 1340 cm^{-1} disappear when cellulose is treated, or extracted from wood, with a solution of NaOD in D$_2$O. This fact confirms the attribution of these bands to the deformation vibrations of hydroxyl groups. In view of the abrupt disappearance of the bands at 1360, 1340, and 1320 cm^{-1}, following selective oxidation of cellulose with nitrogen oxides, it may be accepted that the intensity of these bands is largely determined by the primary hydroxyl groups of various rotational isomers.

The weak bands at 1290 and 1240 cm^{-1} can probably be attributed to external deformation vibrations of CH$_2$ groups. These may be the frequencies of CH$_2$ associated with changes in the coordinates φ and χ, as introduced by Stepanov[101] in his calculations on the spectra of normal paraffins, i.e., deformation vibrations of the form

The weak band at 1200 cm^{-1} in the cellulose spectrum has been assigned by some authors to the bond C$-$O [111]. However, this attribution is improbable. Such bands are usually very strong. The disappearance of the 1200 cm^{-1} band when cellulose is oxidized with periodate indicates that it corresponds to the frequency of a ring or of deformation vibrations of secondary hydroxyl groups.

Bands in the 1200-1500 cm^{-1} region are very sensitive to various influences on cellulose. Indeed, cellulose is a good example of a compound where a CH$_2$ group is present only at the sixth carbon atom. In consequence, the intensity of the 1430 cm^{-1} band can give definite information about processes involving the CH$_2$OH group. However, it should always be appreciated that the intensity of this band is also sensitive to structural transformations in cellulose.

From an analysis of a cellulose spectrum in the region 1200-1500 cm^{-1}, it is possible to form definite conclusions as to the nature of the bonds linking cellulose to associated compounds in plant cells. The method used is alkaline digestion of the cellulose with a 12.06% solution of NaOD in D$_2$O[15].

A reduction in intensity of the 1340 cm^{-1} band and disappearance of the 1320 cm^{-1} band are characteristic of the spectra of celluloses obtained by digestion with 12.06% NaOD in D$_2$O, as compared with samples obtained by normal alkaline digestion. Spectra of cotton cellulose subjected to normal alkaline treatment, and of wood cellulose obtained by normal alkaline digestion, are practically identical in the 1300-1400 cm^{-1} region and differ considerably from spectra of wood cellulose produced by digestion with NaOD in D$_2$O. It has been shown [7, 9, 10] that, in the case of selective oxidation of the primary hydroxyl groups, there is a marked leveling of bands in the region 1300-1450 cm^{-1}. Accordingly, the specific character of the spectrum of a cellulose, obtained by digestion with NaOD in D$_2$O, can readily be explained if it is assumed that, in the wood complex, there are chemical bonds between the cellulose and associated compounds, involving the sixth carbon atom of the former. This does not exclude the existence of bonds at other positions in the cellulose. However, the main bulk of the hydroxyl groups in the cellulose macromolecule are not bonded to the associated compounds; this is shown by the existence of intense OH bands in the spectra of plant tissues.

Increased absorption in the 1190-1280 cm^{-1} region is characteristic of celluloses containing carboxyl groups. This is particularly noticeable in the case of dicarboxycellulose and is attributable to deformation vibrations of OH in carboxyl groups. Increased absorption in this region, in the spectra of oxidized celluloses, can provide an additional indication of the presence of carboxyl groups.

The spectra of mono- and dicarboxycelluloses, which have been treated with metal ions, show strong absorption in the range 1350-1450 cm^{-1}; this is attributable to symmetrical stretching vibrations of the COO^{-} group in carboxylate ions [58, 65]. Investigation of other spectral regions can also provide evidence as to the character of the reaction between various cations and the carboxyl groups of cellulose, particularly in the case of organic dyes [7, 68].

When the cellulose structure contains $-ONO_2$ groups, the spectrum shows a very strong band at 1280 cm^{-1} which can be assigned to symmetrical stretching vibrations of the nitrate groups. This band can be conveniently used for following the accumulation of nitrate ester groupings in cellulose derivatives.

Many other cellulose derivatives have their own distinguishing spectral features in the range 1200-1500 cm^{-1}. For example, with the trityl ether of cellulose there is increased absorption at 1450, 1370, 1330, and 1230 cm^{-1}; with cellulose benzoate at 1450, 1420, 1380, 1320, and 1270 cm^{-1}; with cyanoethylcellulose at 1470, 1415, 1370, 1330, 1270, and 1230 cm^{-1}, and so on. These bands reflect the structure and mode of attachment of the new functional groups introduced into the cellulose structure.

The spectra of cellulose esters show intense bands in the region 1150-1250 cm^{-1}. For example, acetylcellulose shows a band at about 1240 cm^{-1} [25], cellulose methylxanthate a band at about 1220 cm^{-1} [27], etc. The appearance of these bands can be attributed to stretching vibrations of ester

bonds $\overset{\diagup}{\underset{\diagdown}{C}}-O-\overset{\diagup\diagup}{C}$. Thus, the spectrum in the 1150-1250 cm^{-1} region can

conveniently be used for the direct determination of the kinetics of accumulation or disappearance of ester bonds, in the course of various reactions of cellulose. For example, the existence of a band in the 1220 cm^{-1} region, in the spectra of stable cellulose xanthate derivatives, can be used to show that the introduction of stabilizing groups does not lead to rupture of the ester

bonding $\overset{\diagdown}{\underset{\diagup}{C}}-O-C\overset{\diagup S}{\diagdown}$ [27].

Infrared spectroscopy provides interesting possibilities for the investigation of various structural modifications of cellulose. It has been seen above that the change to hydrocellulose, regardless of the origin of the cellulose, leads to a marked decrease in intensity of the spectral bands at 1430, 1340, and 1320 cm^{-1} [7, 9, 10, 28, 145]. It has also been suggested [17] that differences between the spectra of natural cellulose and hydrocellulose can be explained by the existence of specific rotational isomers, resulting from rotation or turning of the CH_2OH group around the bond C_5-C_6.

The change in cellulose structure resulting from alkaline treatment, which leads to the above-mentioned spectral change, is achieved very rapidly. Even the spectrum of a product, regenerated after the first minute of interaction of cellulose with alkali, shows all the characteristic signs of the hydrocellulose spectrum.

The absorption spectrum can be used to check the degree of mercerization of cellulose [13], and to determine the concentration of alkali with which the cellulose was treated, on the basis of just a sample of the regenerated cellulose [40].

Comparison of the spectra of sugars and polyhydric alcohols shows that the bands in the 1200-1500 cm^{-1} region depend very much on the location of the individual hydroxyl groups. In particular, it cannot be maintained that the internal deformation vibrations of CH_2 in the CH_2OH groups of such compounds always have the same frequency. The 1430 cm^{-1} band in the cellulose spectrum can be attributed to only the CH_2OH group of a particular isomer.

It is probable that the bands of the cellulose spectrum in the region 1200-1500 cm^{-1} should be associated with a definite complex of vibrations, with participation of CH_2OH groups, determined by a specific force field. The uniform character of the force field, in the vicinity of the groups responsible for these vibrations, is indicated by the uniformity of the physicochemical properties of these groups. In analyzing the spectra of carbohydrates, it is necessary to start from the basis that the same vibration complexes may exist in various compounds. There are good prospects for the use of statistical methods in analysis of the spectra of similar compounds.

Because the bands at 1430, 1340, and 1320 cm^{-1} appear most clearly in the spectrum of unhydrolyzed residual cellulose and disappear when cellulose is hydrolyzed, it must be presumed that their appearance is due to the natural force field in the most ordered parts of the macromolecules of natural modifications of cellulose.

The 950-1200 cm^{-1} Region

It has been seen above that the 950-1200 cm^{-1} region of the cellulose spectrum shows broad diffuse absorption. Interpretation of the absorption bands in this region is practically impossible, owing to the strong interaction of the vibrations of individual groups and bonds, whose frequencies are located in this spectral range. However, several authors have attempted to interpret individual bands in this region [111, 137, 143, 149], which includes the stretching vibrations of $C-O$, $C-C$, ring structures, deformation vibrations of CH_2OH groups, etc.

With mercerized cellulose there is an increase in intensity of the band at 970 cm^{-1}. Well-defined bands in this region appear in the spectra of some cellulose esters, at about 1070 cm^{-1} with cellulose nitrate, 1050 cm^{-1} with acetylcellulose, etc.

Highly oxidized samples of monocarboxycellulose show a weak band at 940 cm^{-1}, which may be due to nonplanar vibrations of OH in carboxyl [7, 10].

Strong bands in this region are characteristic of the spectra of mono- and disacchardies, and of polyhydric alcohols. However, the spectra of straight-chain polyhydric alcohols differ from those of sugars in that there is no sharp band in the frequency range 1120-1170 cm^{-1}. Spectral analysis in this region may be useful for qualitative estimation of the nature of the product.

The 700-950 cm^{-1} Region

The cellulose spectrum shows a weak band at 900 cm^{-1}, which cannot with certainty be attributed, as suggested [137], to vibrations of the ring. Strong bands in this region have been observed in the spectra of all the polyhydric alcohols investigated. It is significant that the spectra of all the sugars investigated show bands at 900 cm^{-1} in cases when the maximum of the stretching vibration band of CH_2 and CH is located at about 2900 cm^{-1}. It has been noted [140] that the transition from tetrahydrofuran to tetrahydrofuranol is accompanied by the appearance of new bands in the region 850-900 cm^{-1}. It has been shown [139] that the doublet band at 900 cm^{-1} in the spectra of ethylene glycol, CH_2OH-CH_2OH, and its derivatives, must be attributed to rocking vibrations of CH_2 in the gauche and trans configurations. Furthermore, there is a band at about 900 cm^{-1} in the spectrum of ethanol, which disappears on deuteration of the alcohol to CH_2CD_2OH [141].

It therefore follows that the band at 900 cm^{-1} in the spectrum of cellulose should be attributed to swinging vibrations of CH_2. This interpretation is confirmed by the fact that the increase in intensity of the 900 cm^{-1} band, on changing from natural cellulose to hydrocellulose, is accompanied by a change in spectral character in the region of the stretching and deformation vibrations of CH_2, while new bands in the regions 850 and 940 cm^{-1} appear in the spectrum of dialcoholic cellulose (with CH_2OH groups in positions 2 and 3).

However, the attribution of the 900 cm^{-1} band to swinging vibrations of CH_2 in CH_2OH groups is insecurely established because of the presence of bands in this region in the spectra of xylan and other simple compounds, such as α-D-xylose, penta- and hexahydroxycyclohexanes, etc. Assignment of the band to one type of vibration of the pyranose ring is also open to doubt, owing to the existence of strong bands in the region 880-900 cm^{-1} in the spectra of polyhydric alcohols, namely erythritol, pentaerythritol, sorbitol, and mannitol. It cannot be excluded that this band may be attributable to a definite stereochemical coupling of CH groups (see pp. 56-57).

When considering the above, it should be borne in mind that the band in the 900 cm^{-1} region, of the spectra of cellulose and other polysaccharides, may be complex in character.

Spectra of wood cellulose show a weak band at about 800 cm^{-1}. This band is absent from the spectra of cotton celluloses, or of celluloses regenerated after removal of alkali in the process of viscose production. Attribution of this band to skeletal vibrations of the whole molecule [137] is unsatisfactory. It is probably due to low-molecular impurities in the cellulose structure. Indeed, the band is considerably stronger in the spectra of celluloses with a high content of low-molecular impurities.

Absorption in the 700-950 cm^{-1} region is specific for many cellulose derivatives. For example, acetylcellulose shows a sharp band at 900 cm^{-1}, with an inflection at 880 cm^{-1}, and bands at 950 and 840 cm^{-1}; ethylcellulose shows bands at 920 and 880 cm^{-1}; trityl ethers of cellulose are characterized by bands at 900, 850, and 700 cm^{-1}; cellulose nitrate shows bands at 840 and 750 cm^{-1}; and so on.

A quite strong diffuse band at 900 cm^{-1} can be observed in the spectrum of dialdehydocellulose, and its intensity increases with the accumulation of aldehydic groups in the combined form. It has been shown that this band can be attributed to aldehyde groupings in the hemiacetal form [7,9,10].

The spectrum of mercerized cellulose differs from that of the original cellulose in showing increased absorption in the 900 cm^{-1} region [7, 9, 10, 28, 145]. However, this band has a much sharper contour than that in the spectrum of dialdehydrocellulose. The increased intensity of the 900 cm^{-1} band in the hydrocellulose spectrum can be attributed to special features of rotational isomers involving the CH$_2$OH group, as compared with natural cellulose.

This region of the spectra of cellulose derivatives and related compounds can contain the vibration frequencies of pyran rings, and the external deformation vibration frequencies of the methylene groups of various rotational isomers, involving C$_{(6)}$H$_2$OH and C$_{(6)}$H$_2$O$-$. According to various authors [125, 127], bands in this region are sensitive to the spatial location of CH groups in the structure of pyranosides. The vibration frequencies of all these groups are very sensitive to changes in the conformation of pyran rings. The

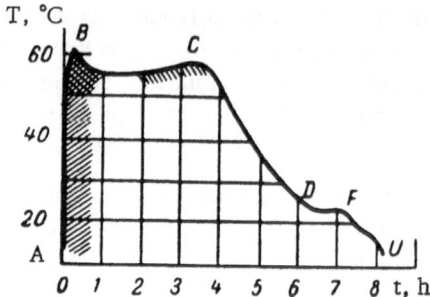

Fig. 86. Determination of the degradation
time by means of an integrating recorder
[102]. AB) Charging; BC) degradation; CD)
cooling; DF) sulfonation; FU) dissolution and
discharge.

relatively small number of quite sharp absorption bands makes this region
very suitable for conformational analysis.

It is highly probable that the spectrum in this region can also reflect
peculiar conformational changes in derivatives of cellulose and related materials.

The use of infrared spectroscopy makes it possible to establish special
features of the interaction of cellulose with alkali, during the various technological processes for the production of alkaline cellulose in viscose manufacture [38, 102].

The spectroscopic results obtained made it possible to draw conclusions as to the predominant role of the first stage of the process (curve ABC
of Fig. 86) on the properties of alkaline cellulose in "VA" plants. The results of this work, carried out in cooperation with colleagues from an artificial fiber plant, have made it possible to recommend automation of the process, not over the whole cycle, which is very difficult, but only over the initial stage ABC (Fig. 86). The same conclusion was reached by workers at the
automation division of the plant, but by a much more difficult method [102].
The application by our colleagues of plant integrators to "VA" equipment
has considerably improved the quality of the viscose [102].

The 400-700 cm^{-1} Region

In the 400-700 cm^{-1} region of the cellulose spectrum, there are bands
at 665, 615, 590, 560, 520, 490, and 435 cm^{-1} against a background of broad
diffuse absorption. The spectrum of hydrocellulose is characterized by the

bands in this region being less sharp, and by disappearance of the bands at 560 and 435 cm^{-1}.

It has been explained above that the spectra of all cellulose derivatives investigated show reduced absorption in this region and the disappearance of the primary structure, following the loss of hydroxyl groups. This spectral interval can include the overtones of the vibration frequencies of the hydrogen bonds themselves and the nonplanar vibrations of hydroxyl groups. The wide range of various types of hydrogen bonds in cellulose may be responsible for the broad diffuse absorption in this region of the cellulose spectrum.

This region can also contain the vibration frequencies of new functional groups introduced into the cellulose structure. For example, the acetyl group gives a sharp band at 600 cm^{-1}, the nitrate group gives bands at 690 and 630 cm^{-1}, while the spectra of cellulose esters of chlorinated fatty acids show a sharp band at 650 cm^{-1} (the C—Cl band) [26].

* * *

In conclusion, we must mention some of the future developments required in connection with the infrared spectroscopy of cellulose. There is a constant need for developments and improvements in the technique of obtaining infrared spectra of cellulose, and for greater precision in the interpretation of frequencies in the spectra of cellulose and its derivatives. It must always be appreciated that we are considering mean spectra of a high polymer, which is inevitably heterogeneous in structure.

There is a need for further spectroscopic analysis of the various structural modifications of cellulose, which will give a better understanding of the specific characteristics of celluloses of various origins, treated in various ways.

For the interpretation of the cellulose spectrum, consideration must be given to various cellulose derivatives, and also to various sugars as model compounds. In deciphering spectra, attention must be directed to peculiarities in polymer structure, particularly conformational features. There is no doubt that, in similar compounds, conformational transformations of the elementary units of the molecule can have a pronounced effect on the infrared spectra. There is a need to distinguish between spectral changes produced by conformational and chemical transformations of the cellulose structure. For this reason, particular interest attaches to the investigation of stereoisomers of cellulose and other polysaccharides, and also of model systems with various conformations of their elementary units.

Depending on conditions, there may be a redistribution of isomeric forms, resulting in a predominance of one or another particular type. It is essential to proceed very cautiously in interpretation of the disappearance of spectral bands, or the appearance of new bands, resulting from various treatments of cellulose, since these changes may be due to structural factors. It is impossible, without allowing for these factors, to give a correct interpretation of changes in the infrared spectra of cellulosic materials, or to carry out quantitative analyses.

The existing experimental material indicates the great possibilities of the infrared spectroscopic method for investigating all the various chemical and structural conversions of cellulose.

There is no doubt that this method will find extensive application in studies on the properties of cellulose, which is one of the most important natural high polymers.

APPENDICES

CONDITIONS FOR RECORDING SPECTRA

Slit program 4, scanning rate 50 cm^{-1}/min, wave number scale 12(32) mm/100 cm^{-1}, amplification 4, time for recording full deflection 50 sec.

SPECTRAL SLIT WIDTHS

KBr Prism

ν, cm^{-1}	400	430	450	500	550	600	700
s[cm^{-1}]	6,0	6.0	5.0	5.0	5.0	5,0	6,0

NaCl Prism

ν, cm^{-1}	700	750	800	900	1000	1300	1500	1800	2000
s [cm^{-1}]	4,0	3.0	3,0	3.0	4.0	5,0	6.0	6,0	6,5

LiF Prism

ν, cm^{-1}	1800	2000	2500	3000	3500	4000
s [cm^{-1}]	2,0	2.0	2.0	3,0	4.0	5.0

SYMBOLS USED TO DESCRIBE STATE OF SAMPLE
WHEN SPECTRUM WAS RECORDED

KBr — Sample pressed with KBr in proportion of 12(6) mg of material to 2 g of KBr.

FP — Cellulose fiber film, obtained by direct pressing; film thickness 9-11 μ.

FM — Cellulose fiber film, obtained by direct pressing, placed in
liquid immersion medium (tetrachloroethylene for 400 to
700 and 1000 to 3800 cm^{-1}, and vaseline oil for 700 to
1000 cm^{-1}); film thickness 9-11 μ.

F   — Film of compound obtained from solution; arbitrary thickness.

FibM— Fiber wound around KBr window and placed in liquid immersion medium.

APPENDIX I

SPECTRA OF MONO-, DI-, and POLYSACCHARIDES
AND THEIR DERIVATIVES, AND SPECTRA OF POLYHYDRIC ALCOHOLS

1. α-D-Glucose (KBr)
2. β-D-Glucose (KBr)
3. α-D-Mannose (KBr)
4. β-D-Mannose (KBr)
5. α-D-Galactose (KBr)
6. α-D-Talose (KBr)
7. β-D-Lyxose (KBr)
8. β-D-Arabinose (KBr)
9. α-D-Xylose (KBr)
10. α-Methyl-D-Glucoside (KBr)
11. α-Methyl-D-Galactoside (KBr)
12. α-Methyl-D-Mannoside (KBr)
13. Rhamnose (KBr)
14. D-Mannoketoheptose (KBr)
15. 2,3-Di-O-Methyl-D-Glucose (KBr)
16. 2,3,4,6-Tetra-O-Methyl-D-Glucose (KBr)
17. 2,3,4,6-Tetra-O-Methyl-D-Mannose (KBr)
18. 2,3,4,6-Tetra-O-Methyl-D-Galactose (KBr)
19. 2,3,4-Tri-O-Methyl-D-Xylose (KBr)
20. Penta-acetate of β-D-Glucose (KBr)
21. Octa-acetate of cellobiose (KBr)
22. Levoglucosan (KBr)
23. Trimethyllevoglucosan (KBr)
24. Triacetyllevoglucosan (KBr)
25. γ-Lactone of gulonic acid (KBr)
26. i-Erythritol (KBr)
27. Pentaerythritol (KBr)
28. Sorbitol (KBr)
29. Mannitol (KBr)
30. Dulcitol (KBr)
31. L-Arabitol (KBr)
32. Maltose (KBr)
33. Cellobiose (KBr)
34. Lactose (KBr)
35. Trehalose (KBr)
36. Raffinose (KBr)
37. Ribose (KBr)
38. Amylose (KBr)
39. Laminarin (KBr)
40. Xylan (KBr)
41. Galactan (KBr)
42. Ethylxylan (γ = 175) (F)
43. Cyanethylxylan (γ = 104; 8% uronic acids) (F)
44. Native dextran, synthesized by the micro-organism Lencostos meseterondes (F)
45. Dextran, partially hydrolyzed (F)
46. Heparin (F)
47. Glycogen (F)
48. Chitin (KBr)
49. Chitosan (KBr)
50. Ethylchitin (KBr)
51. Carboxymethylchitin (KBr)
52. Alginic acid (KBr)
53. Pectinic acid (KBr)

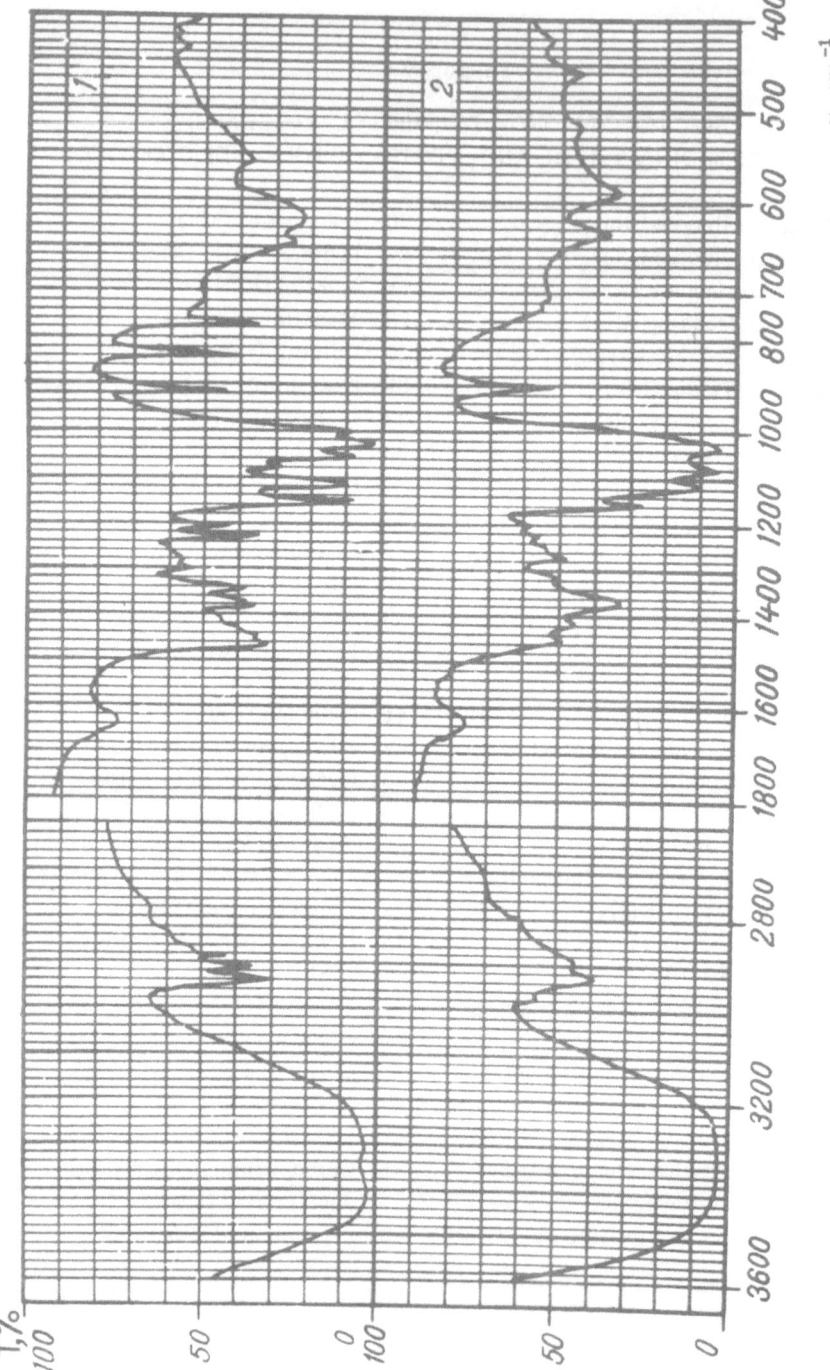

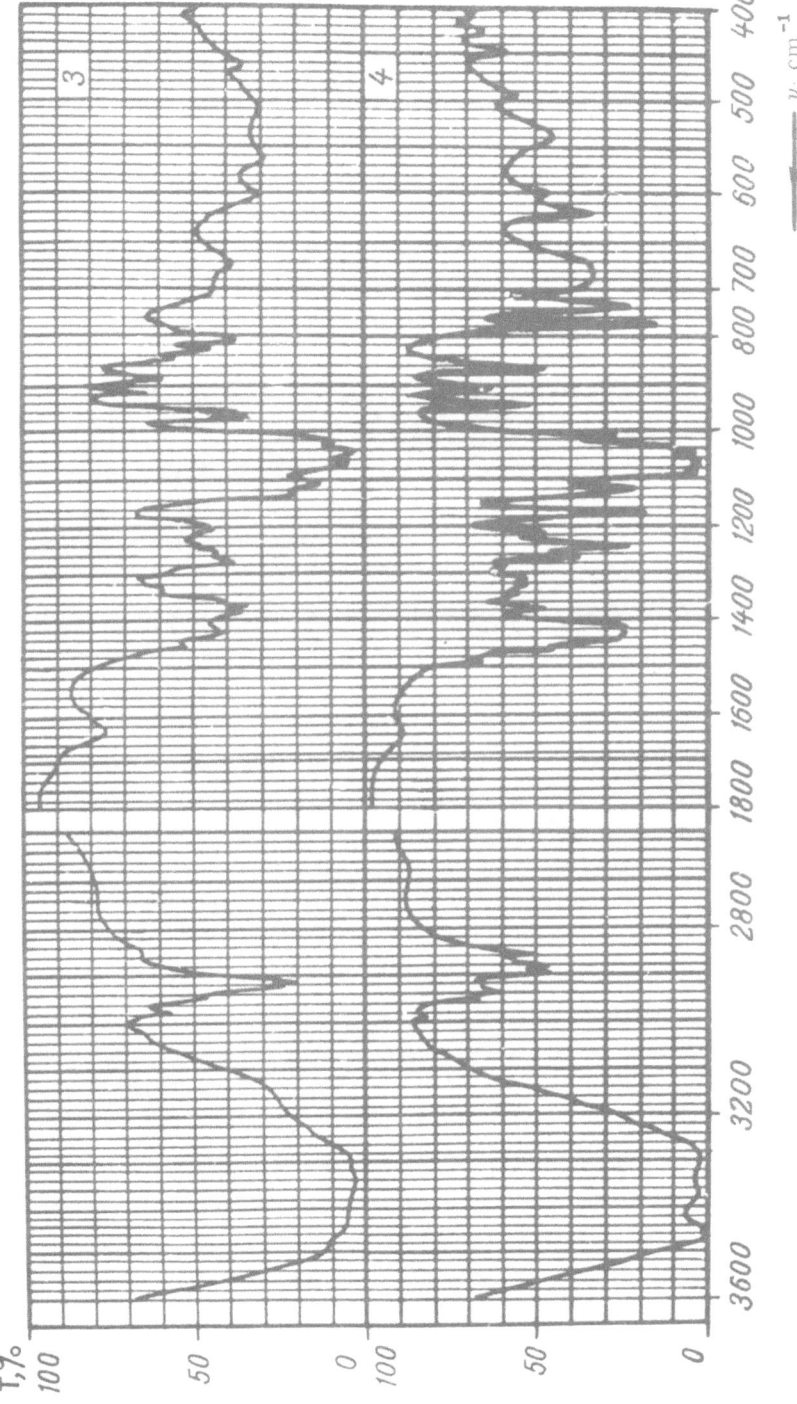

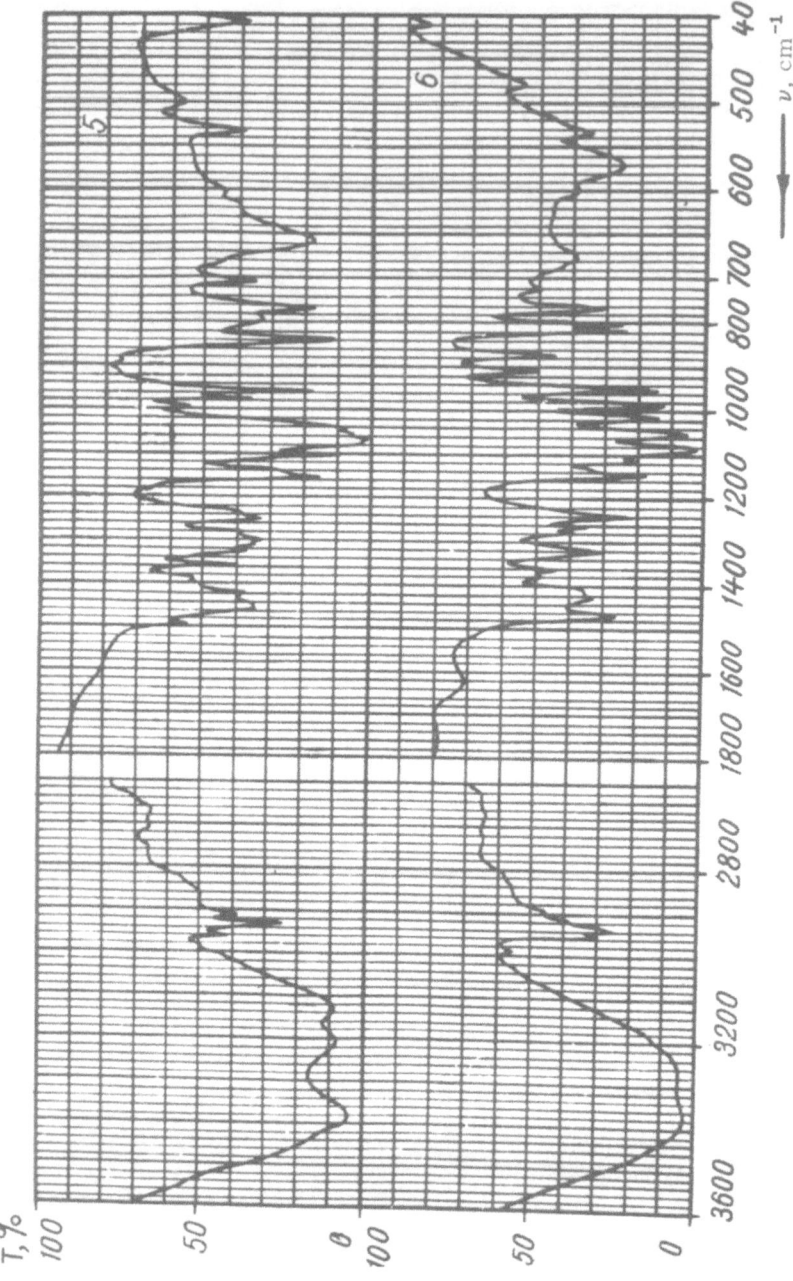

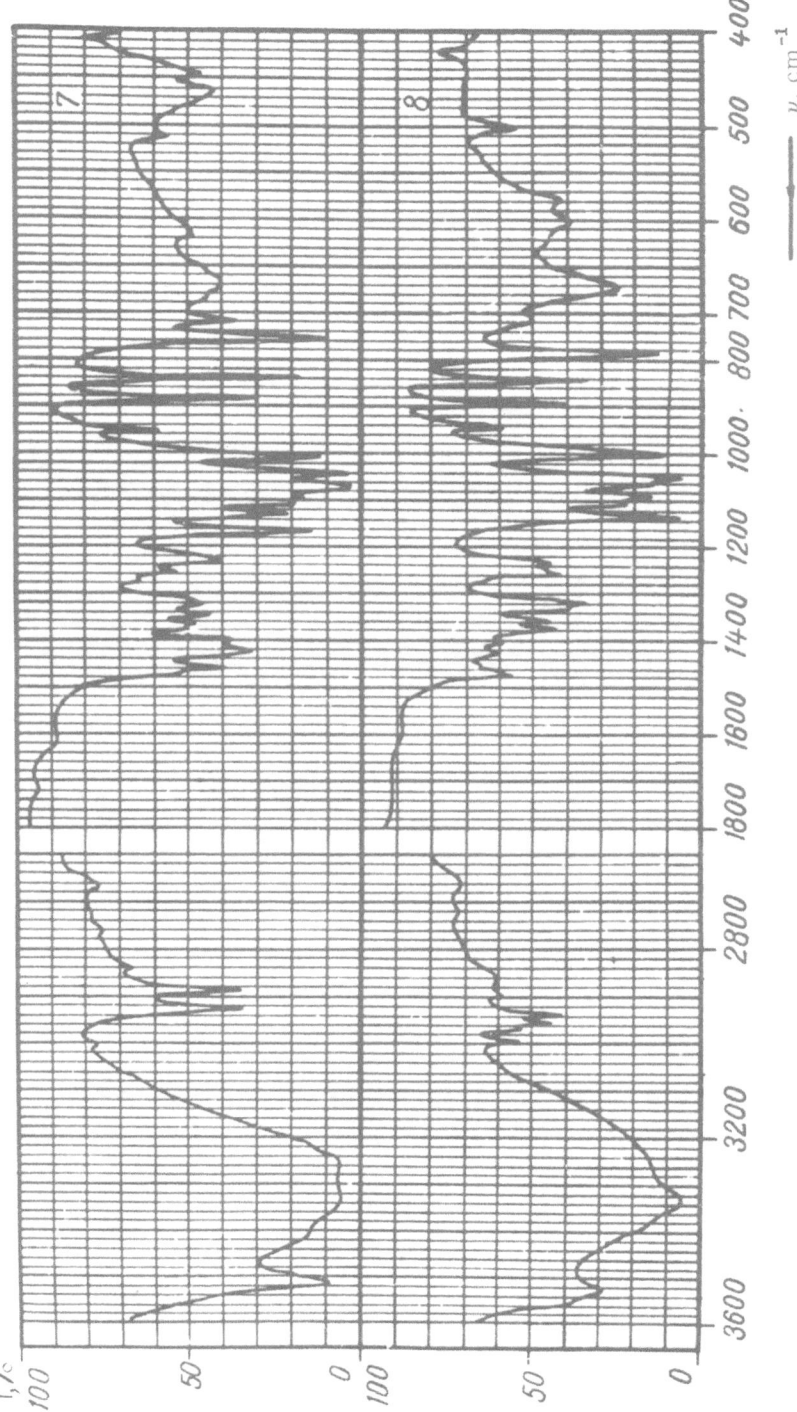

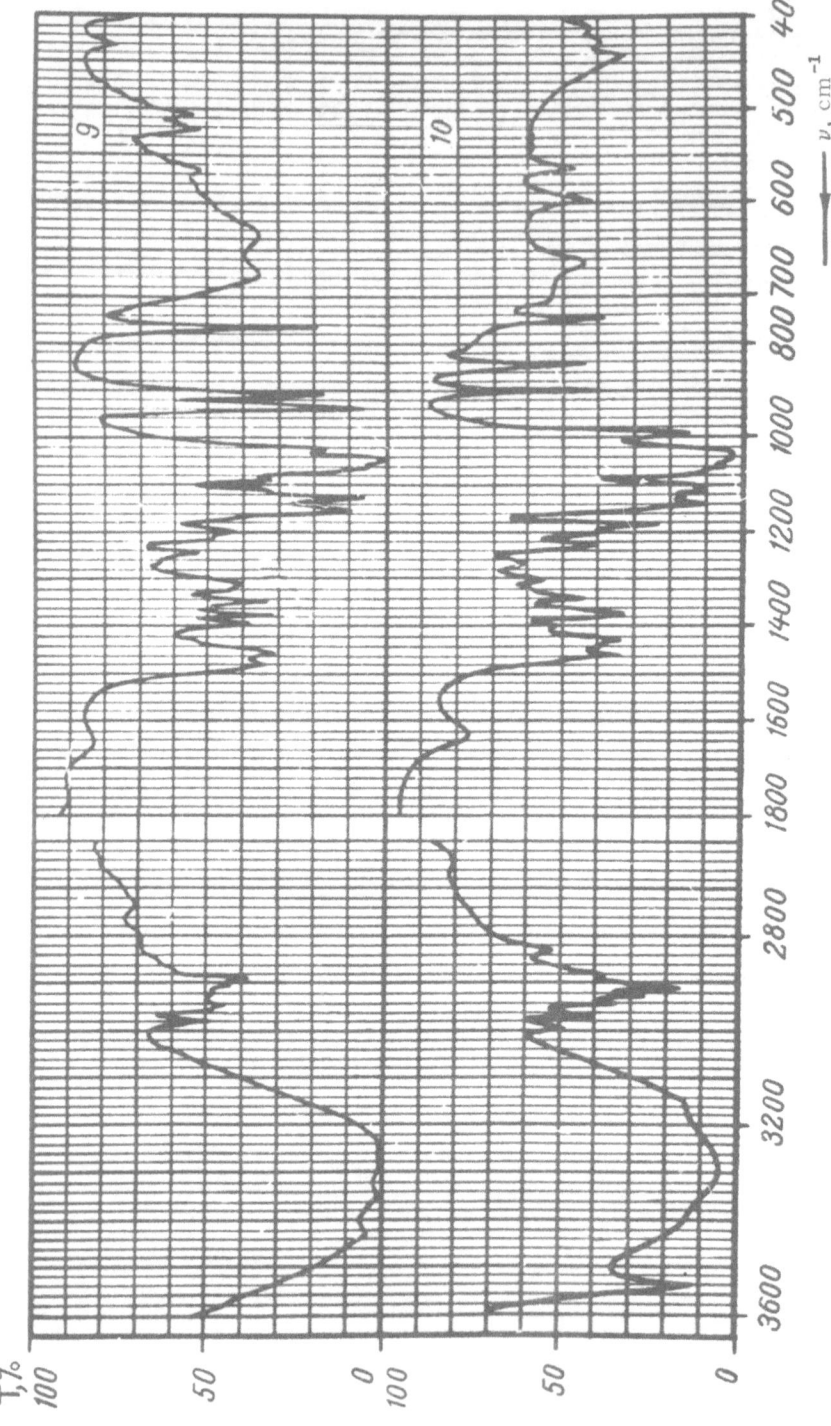

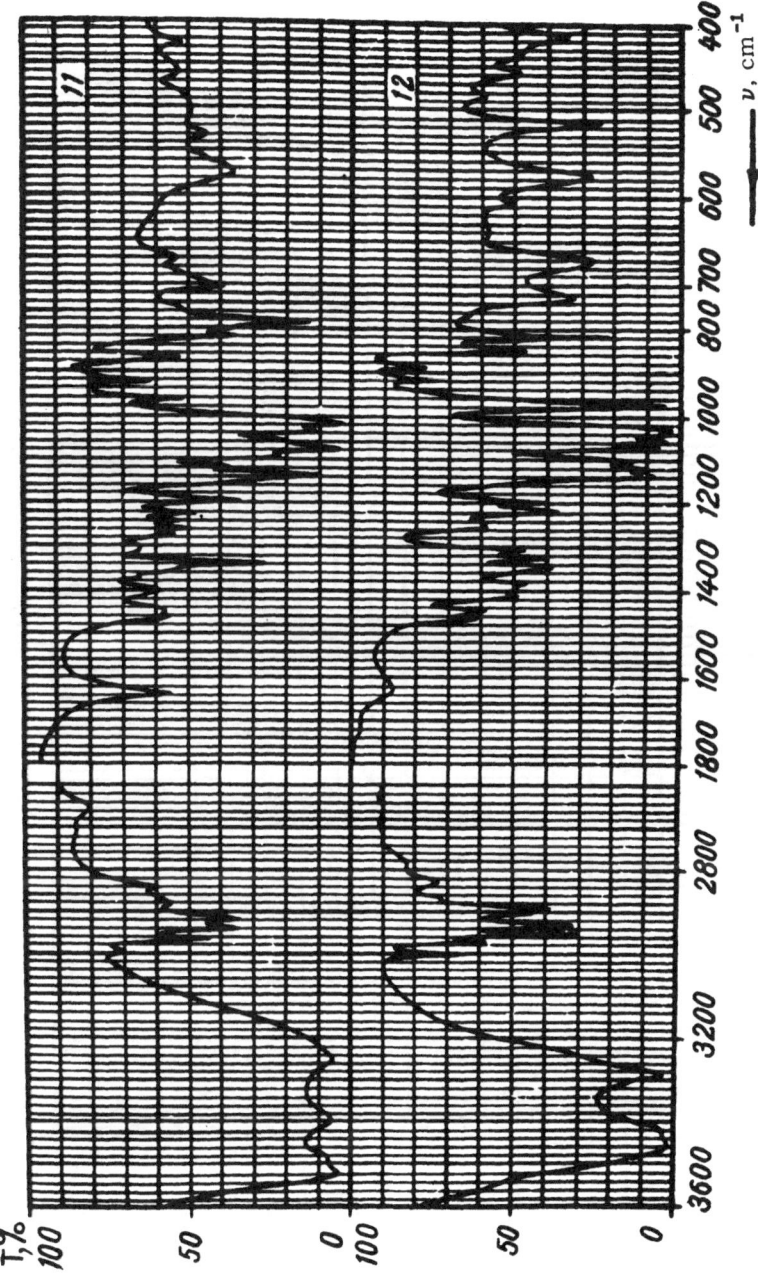

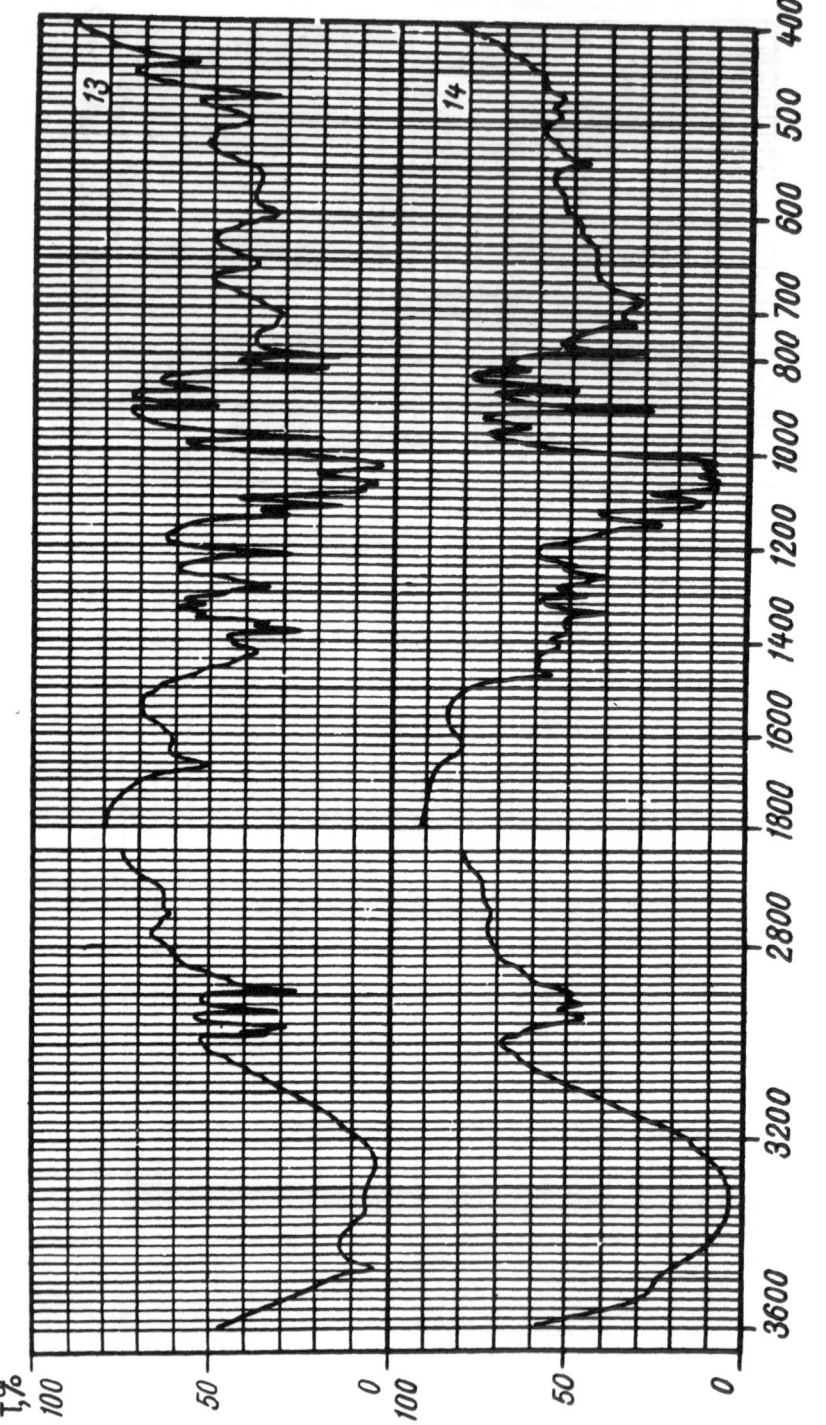

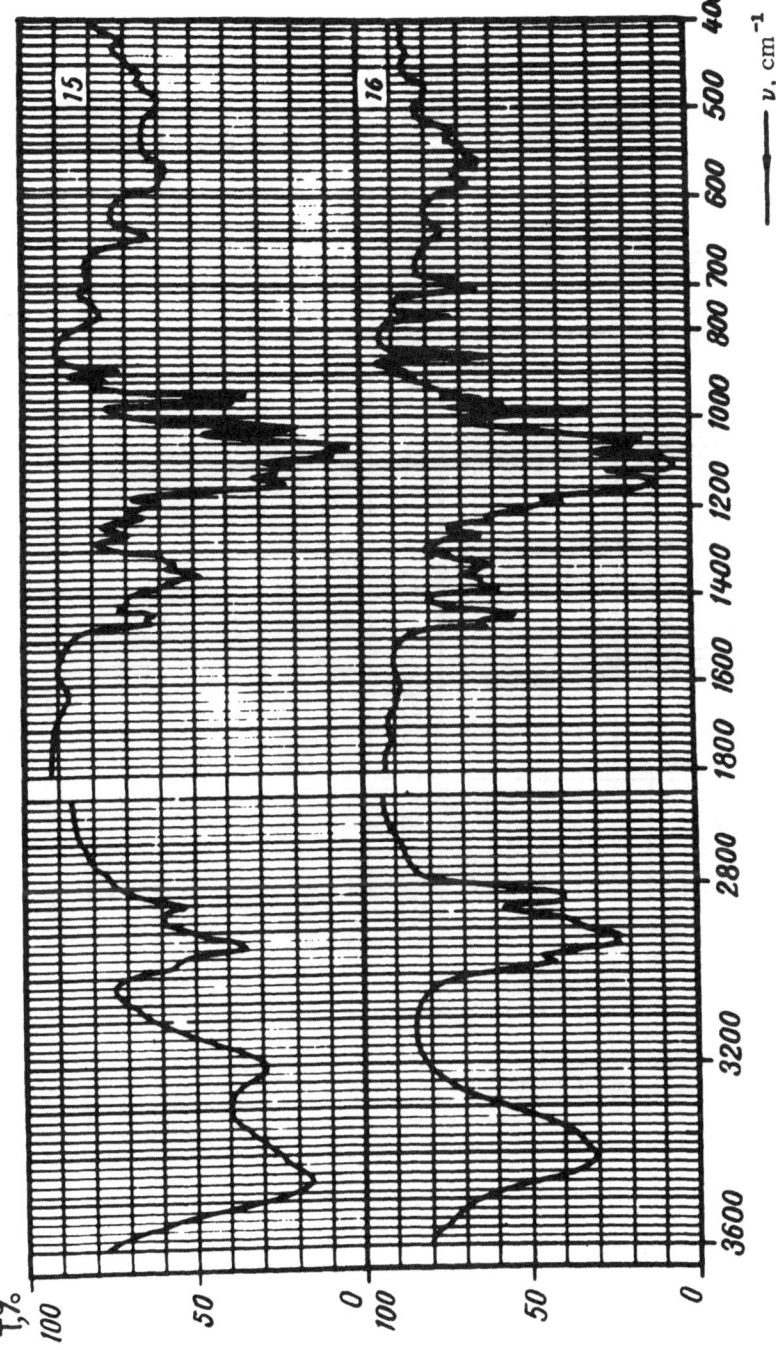

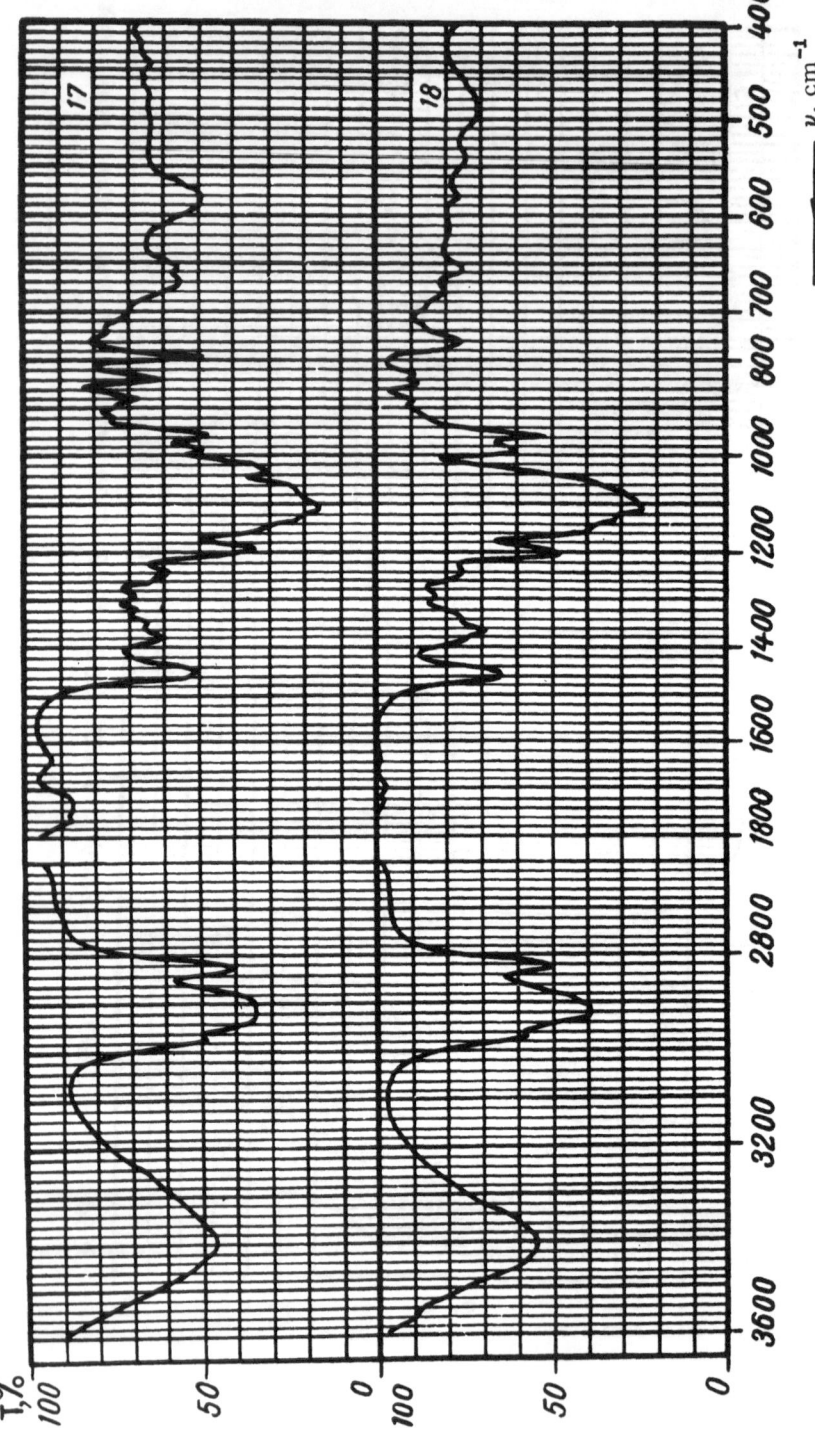

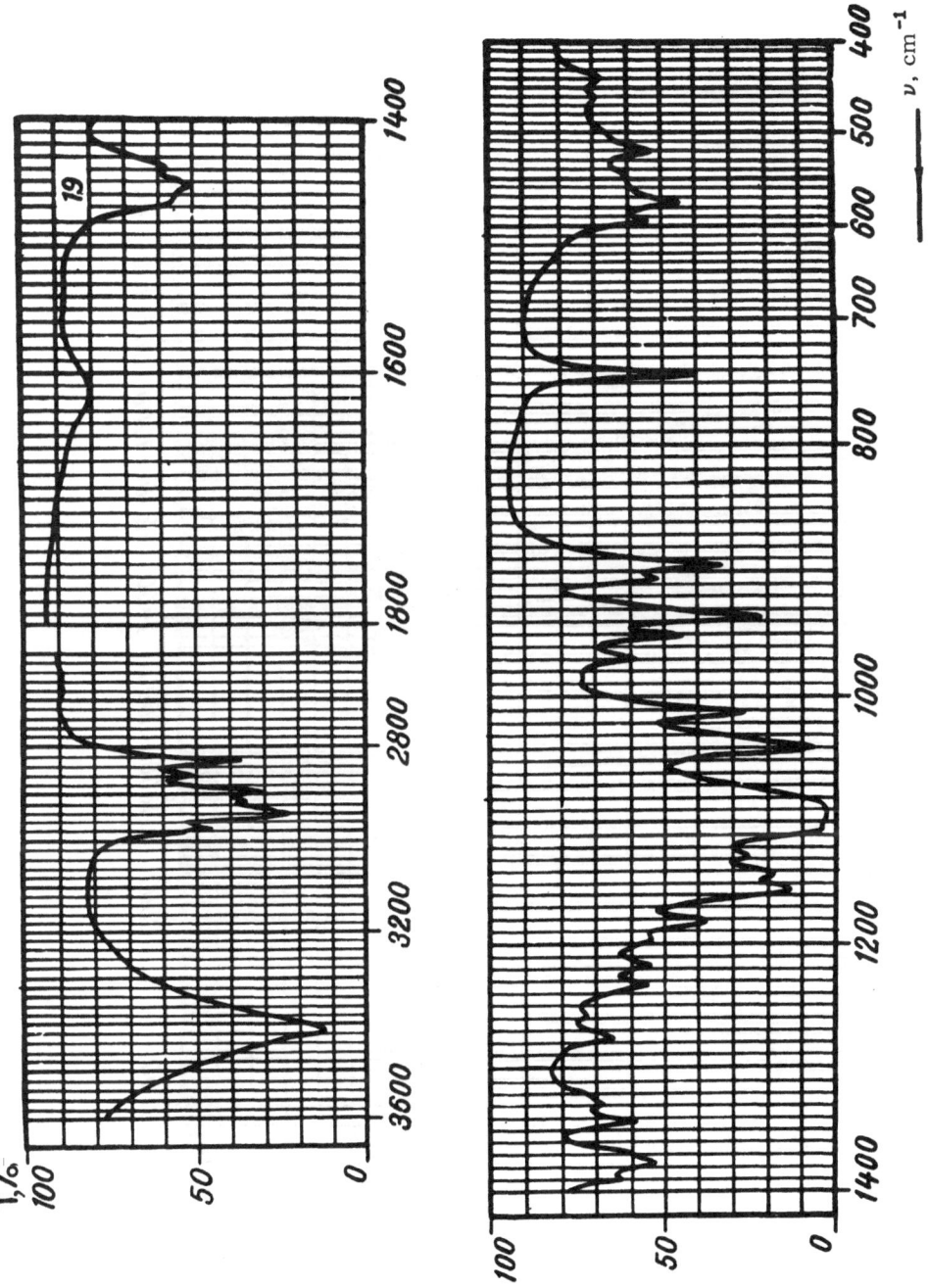

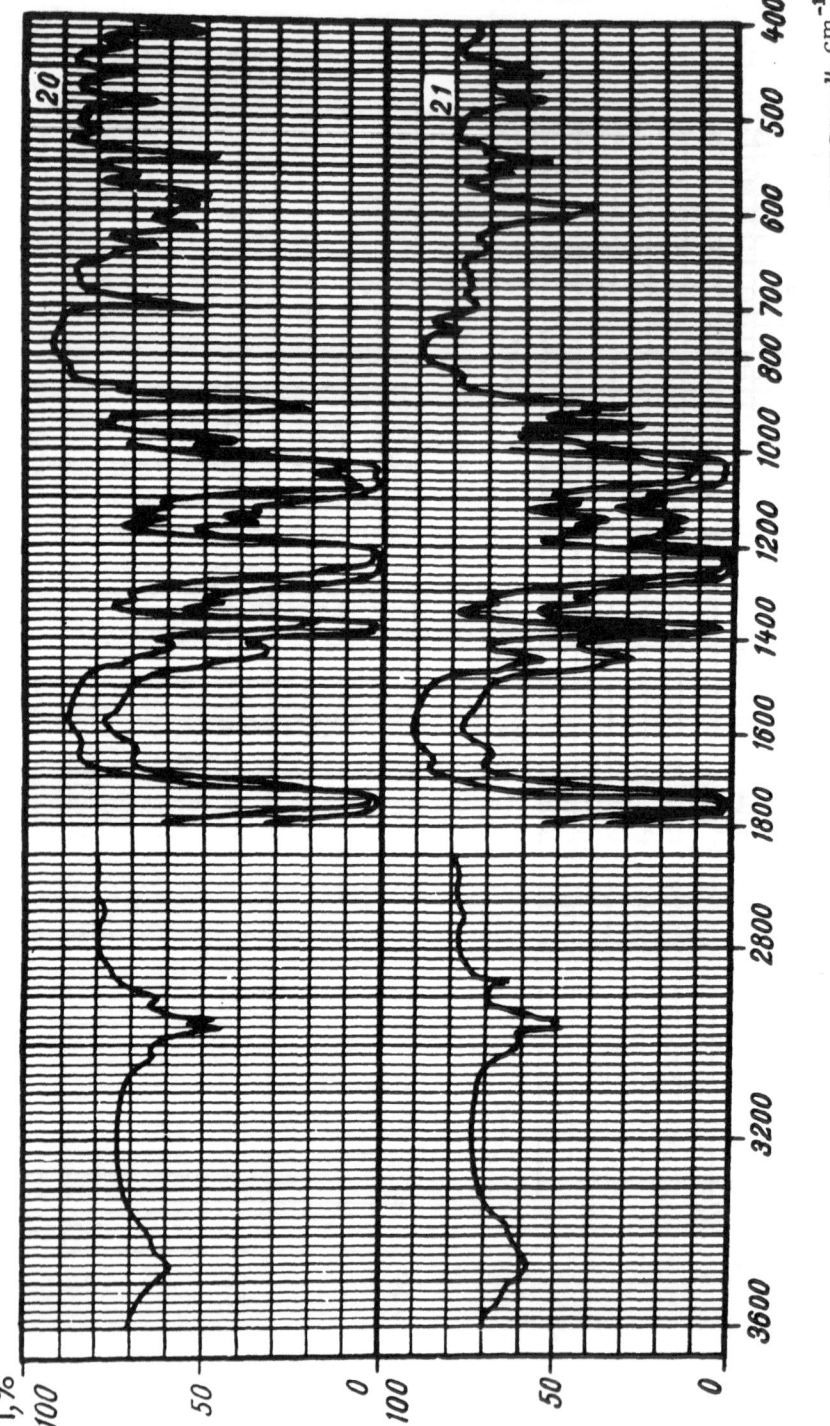

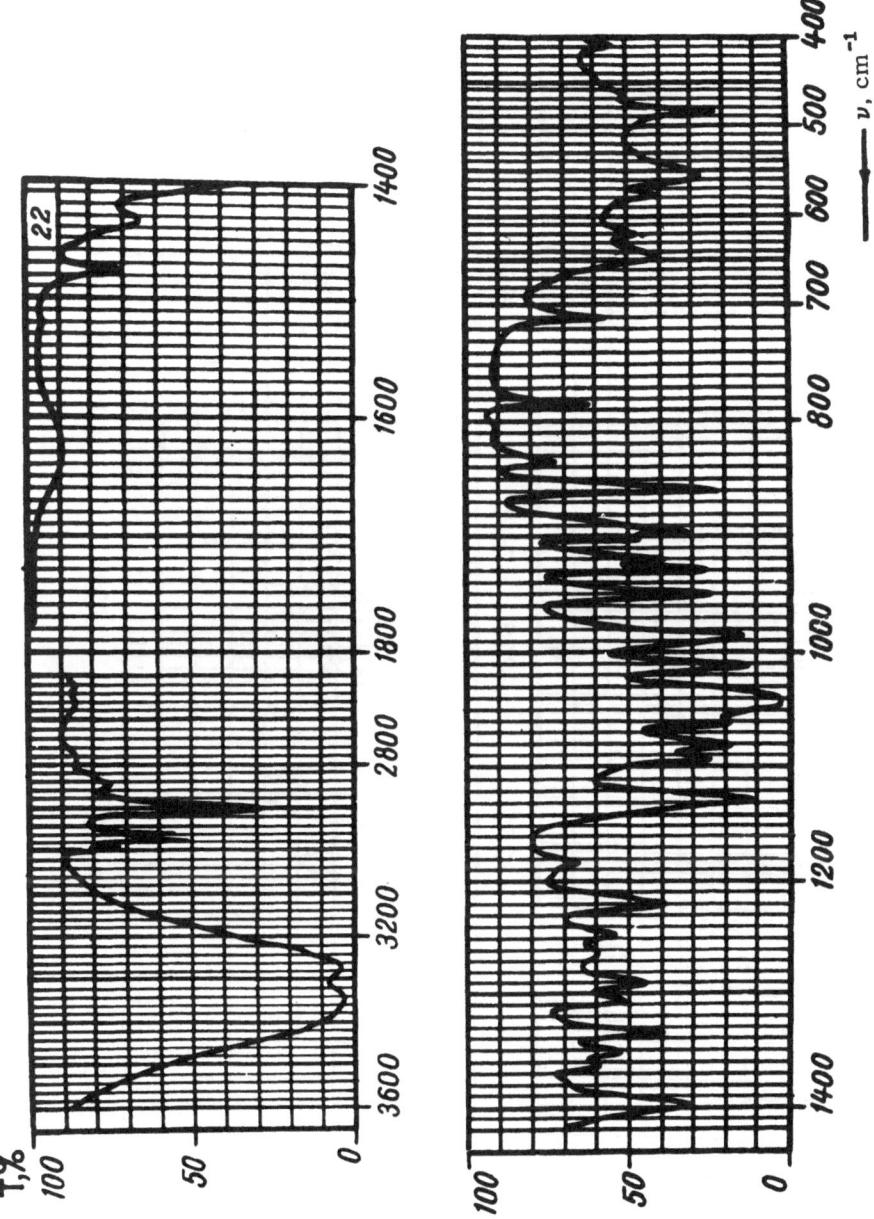

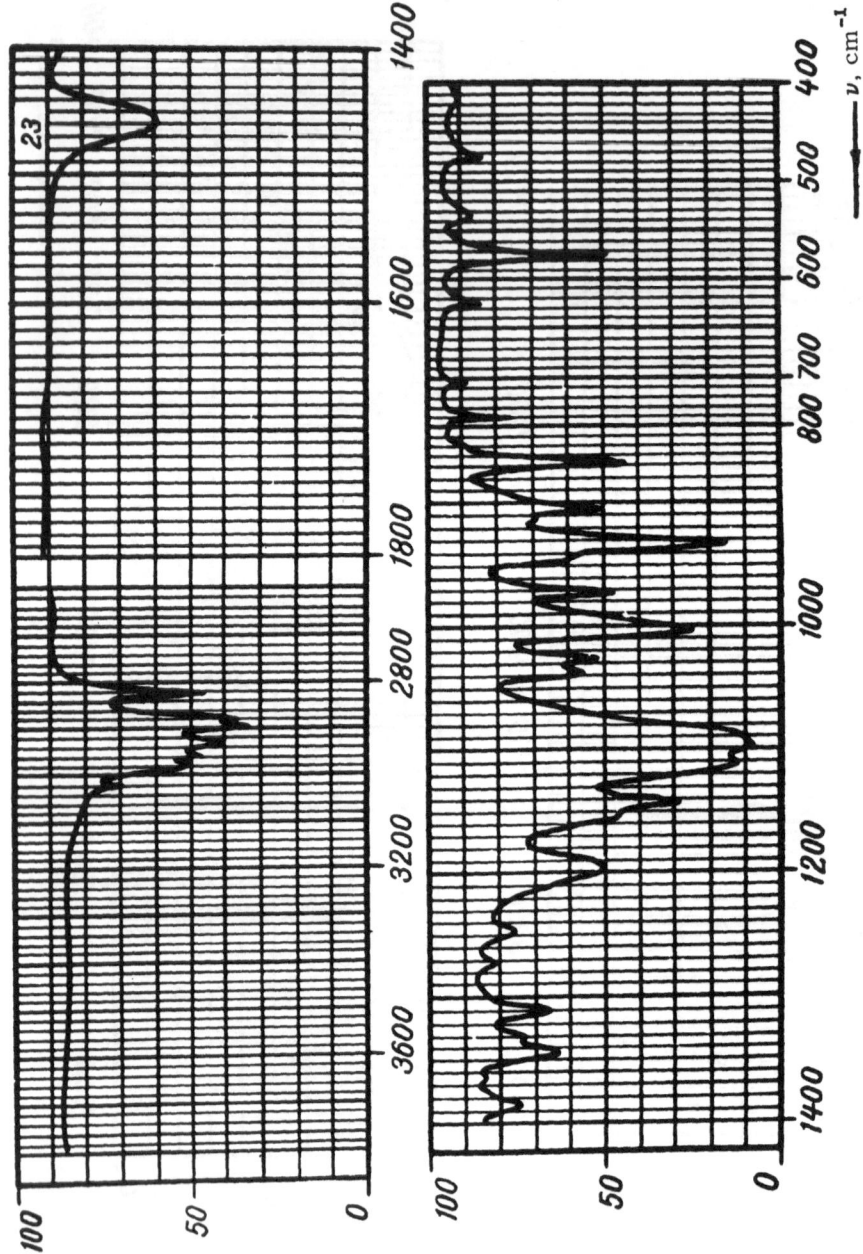

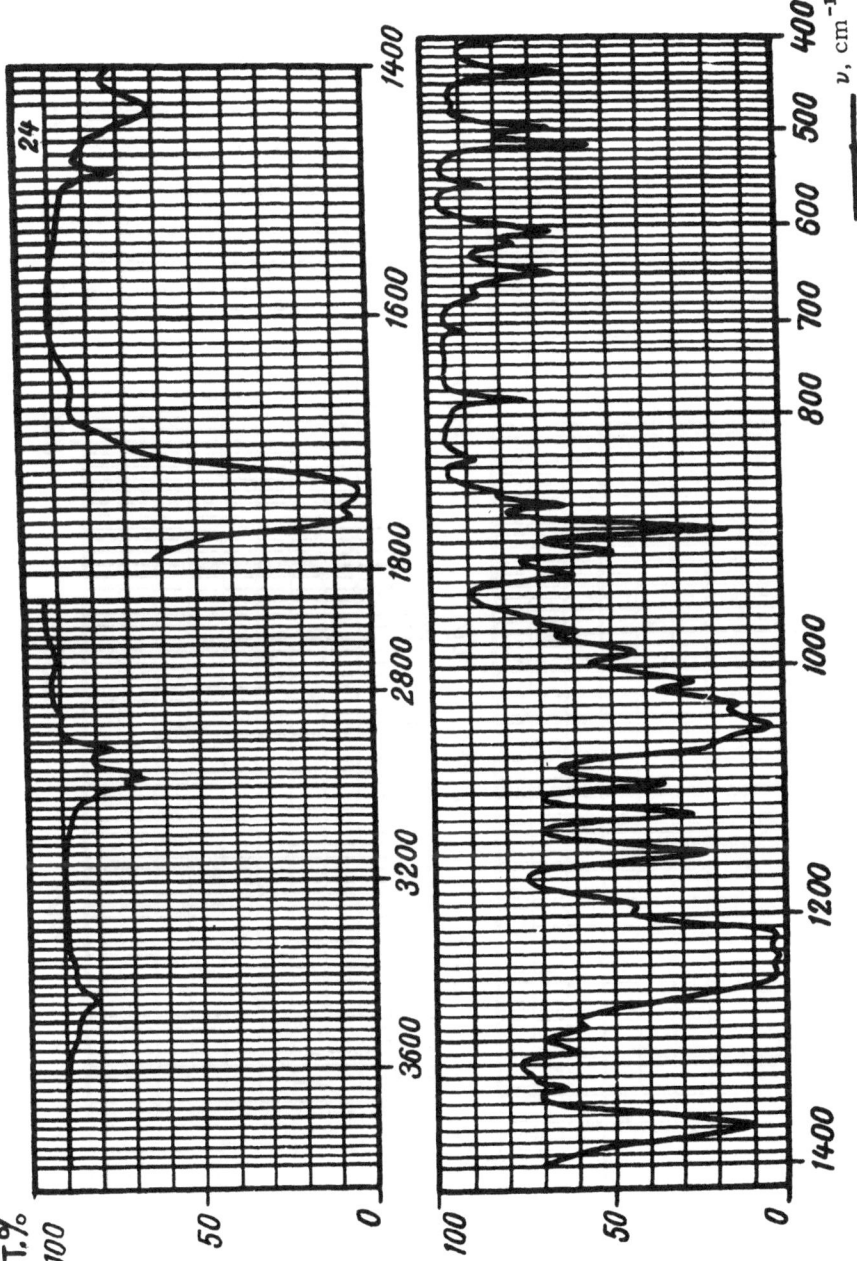

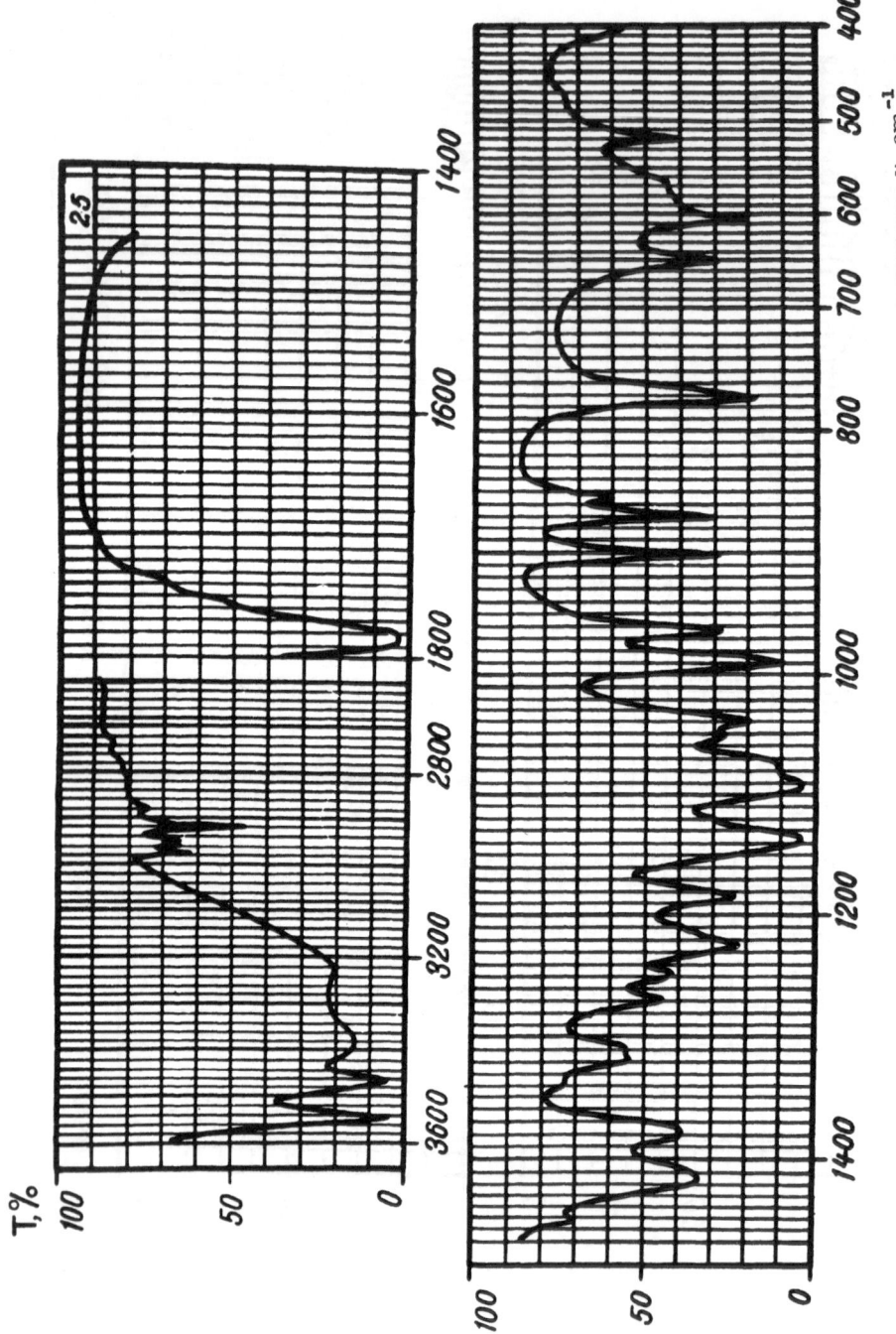

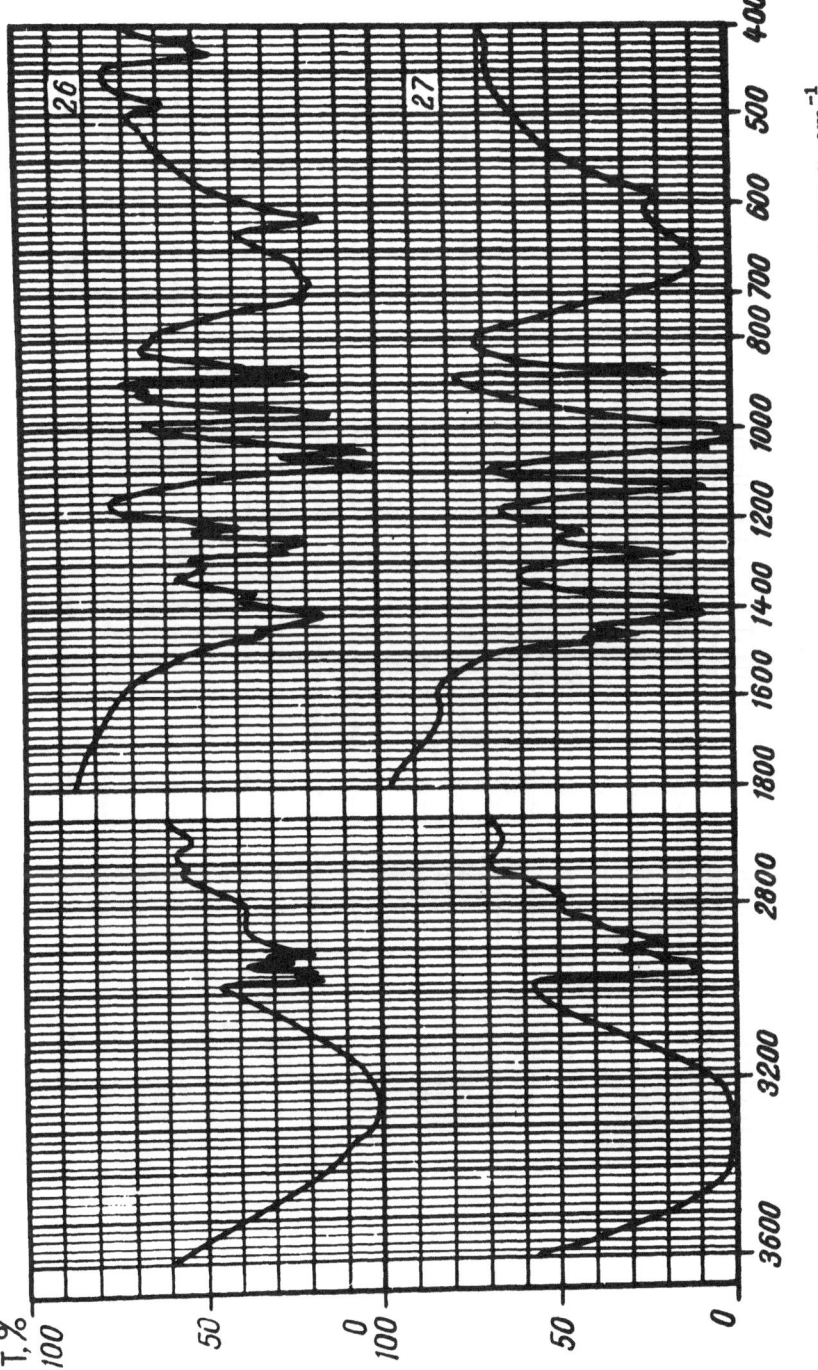

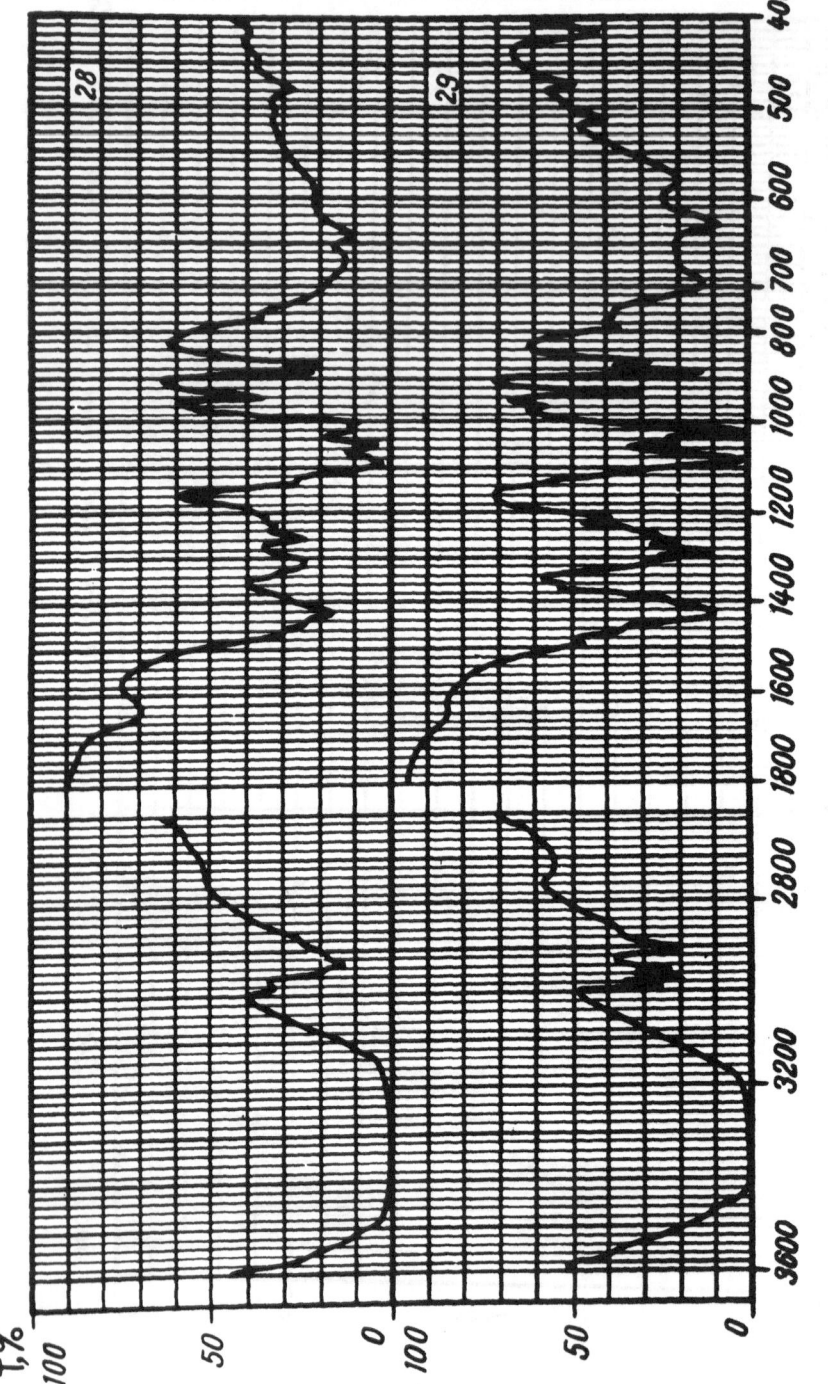

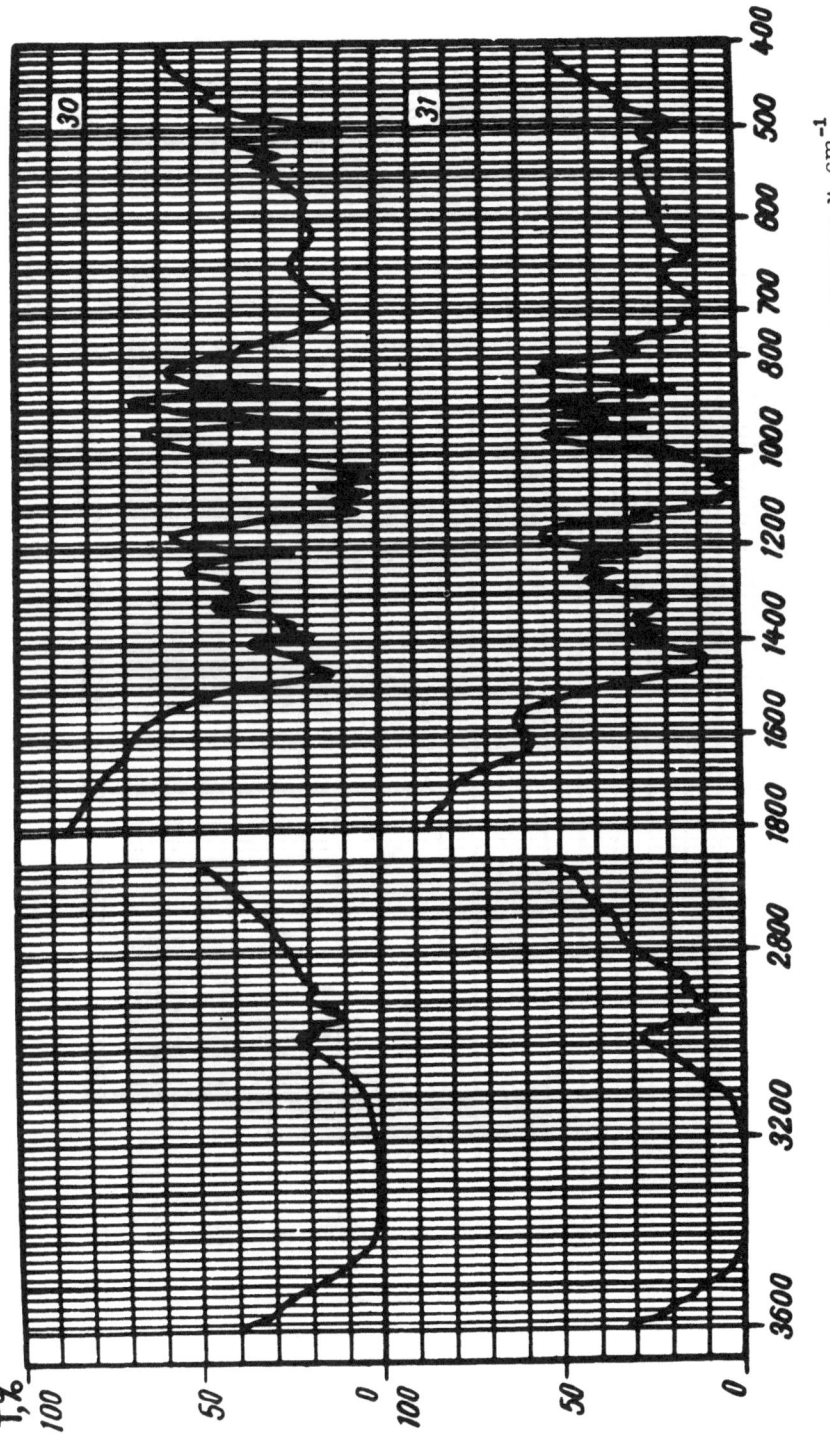

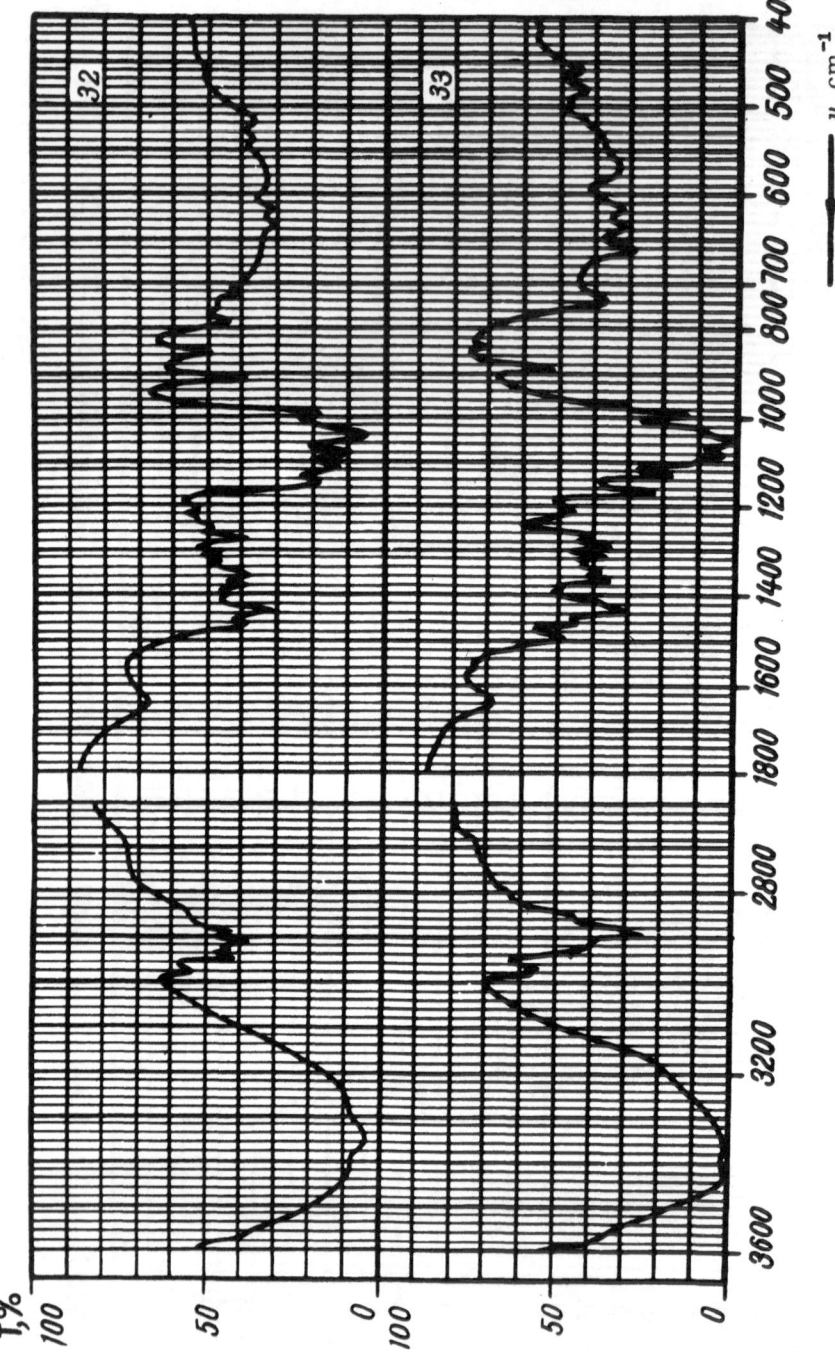

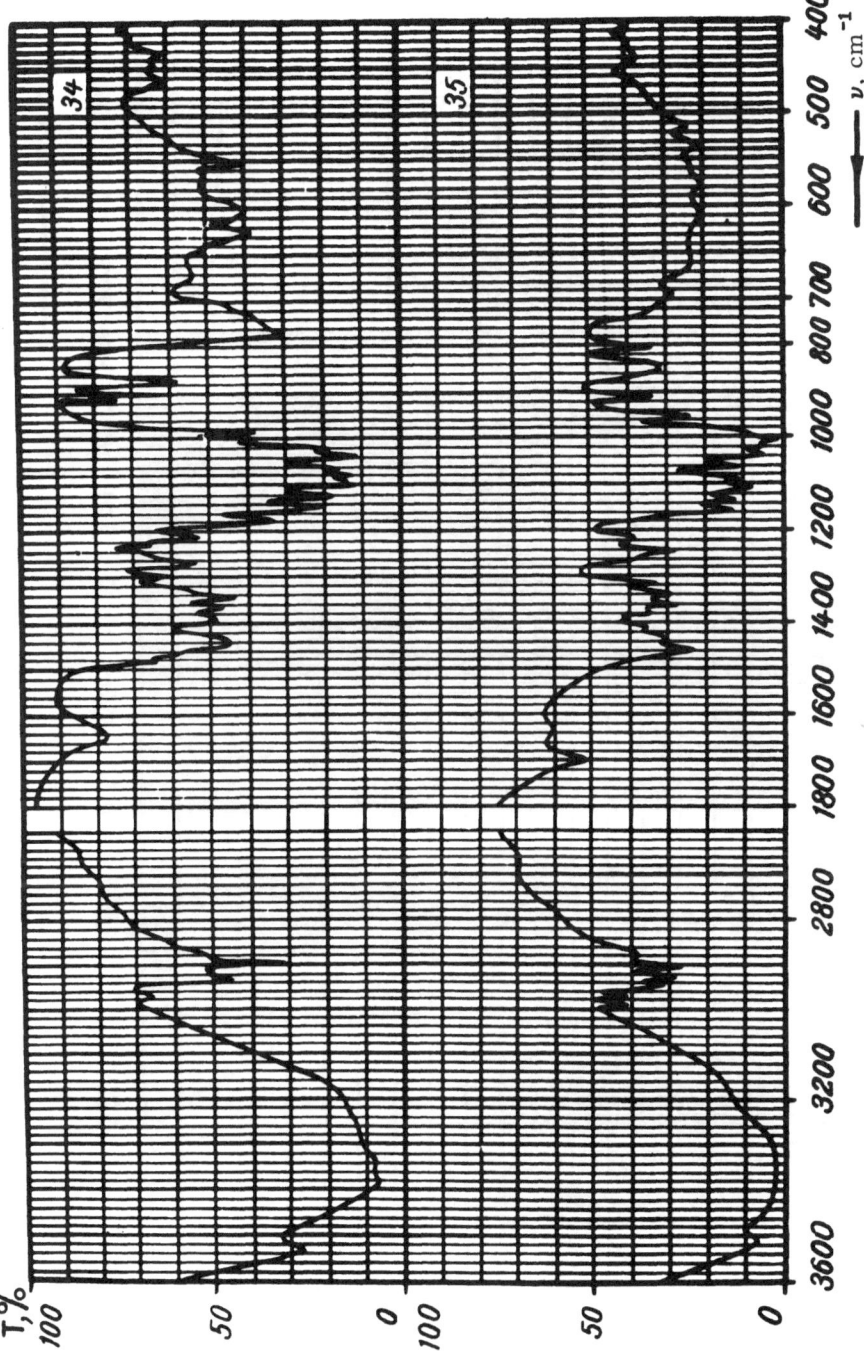

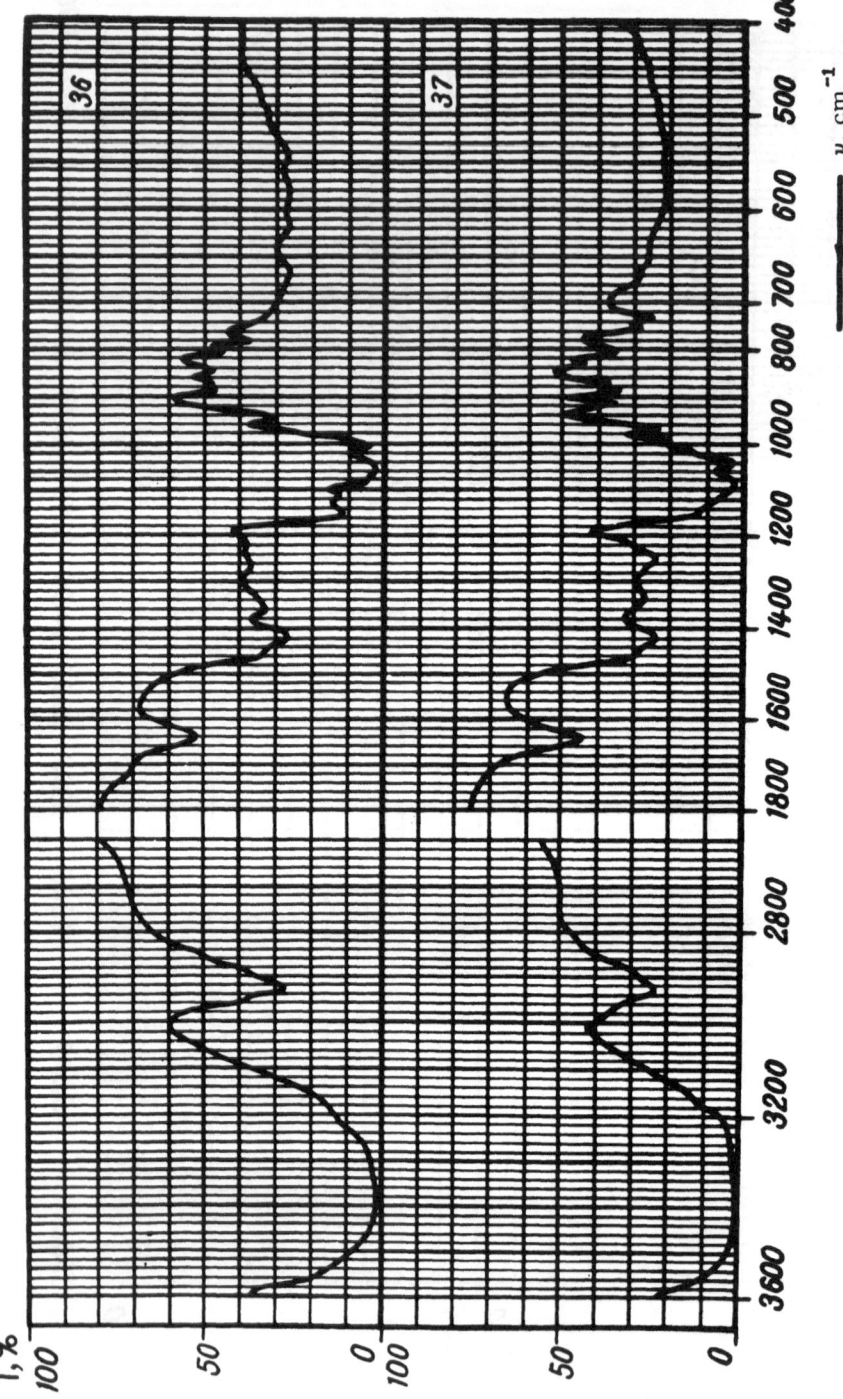

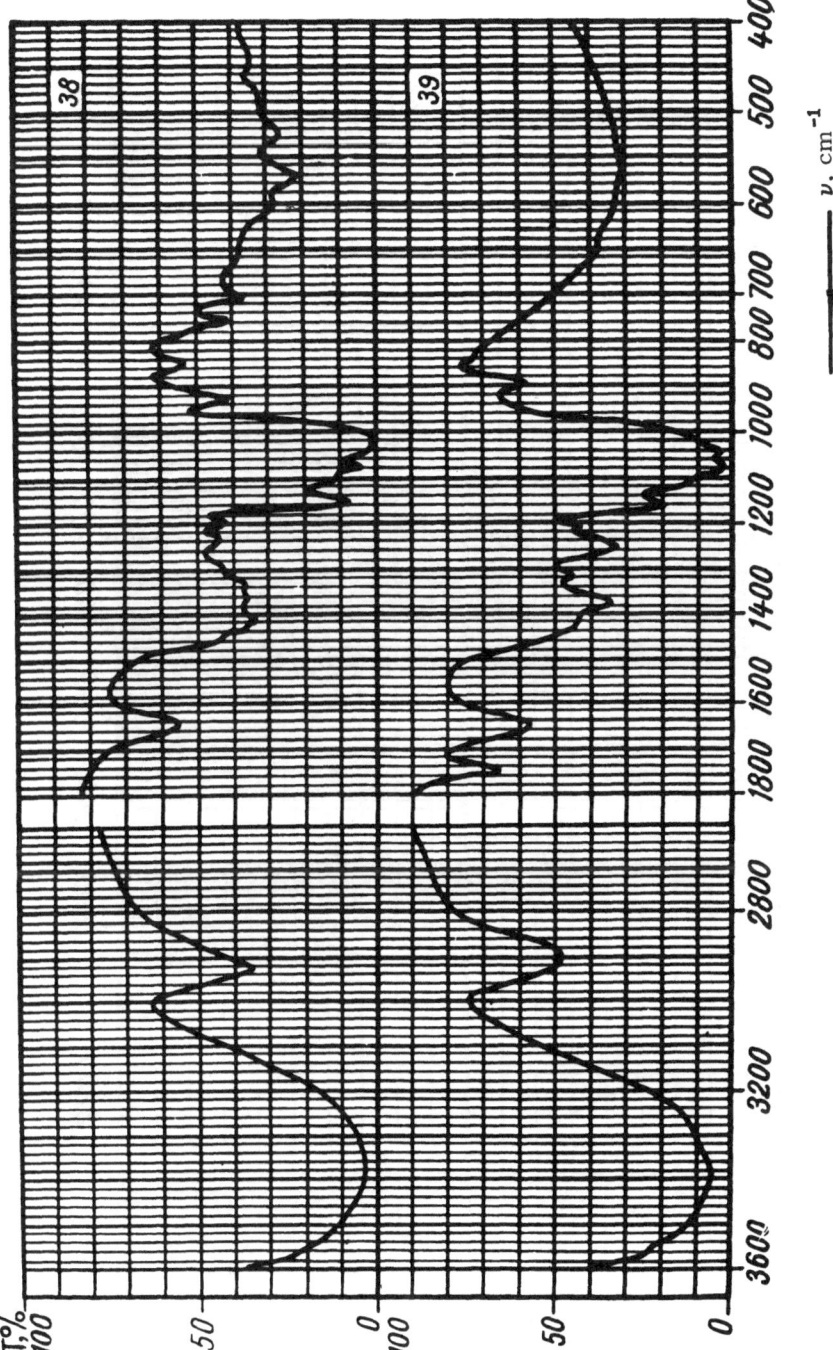

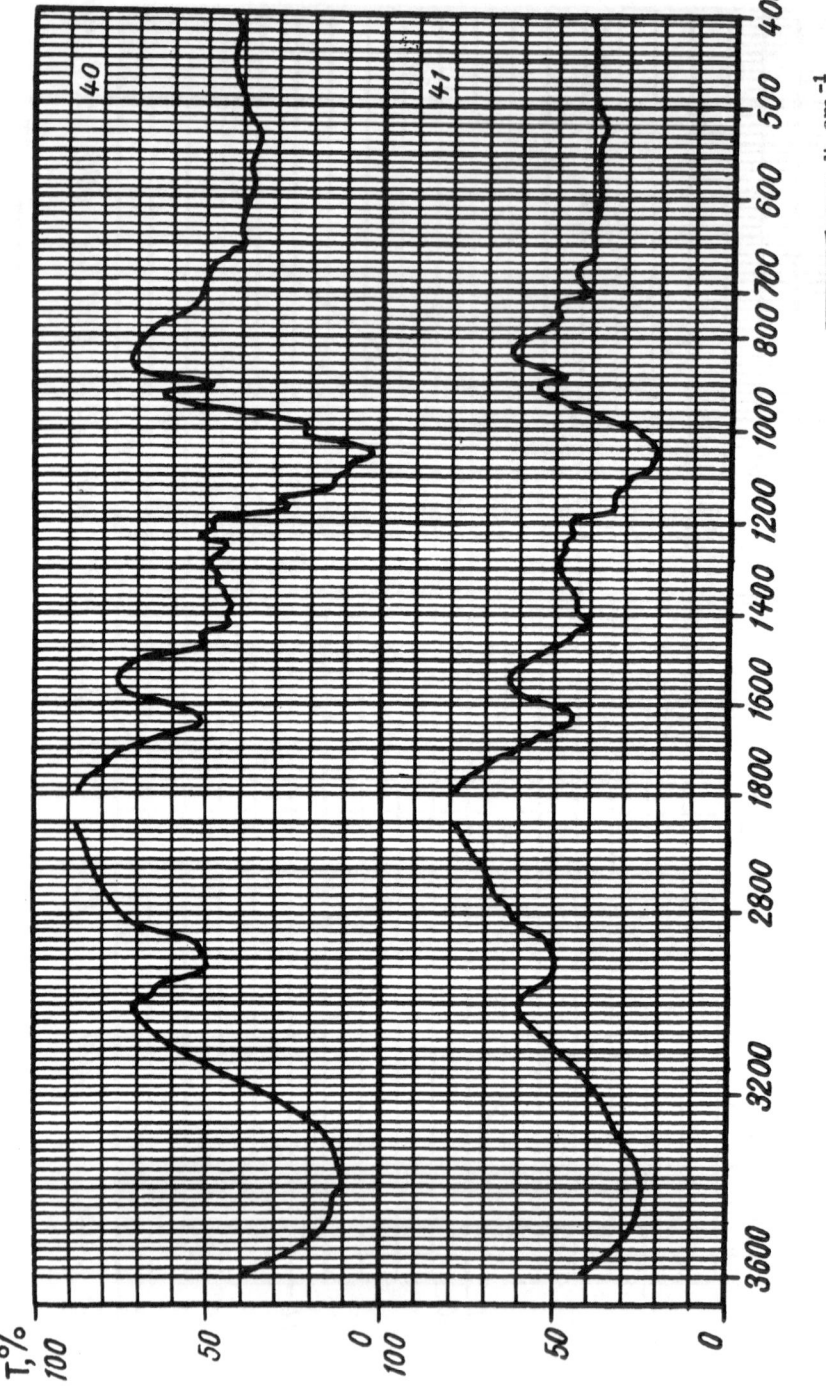

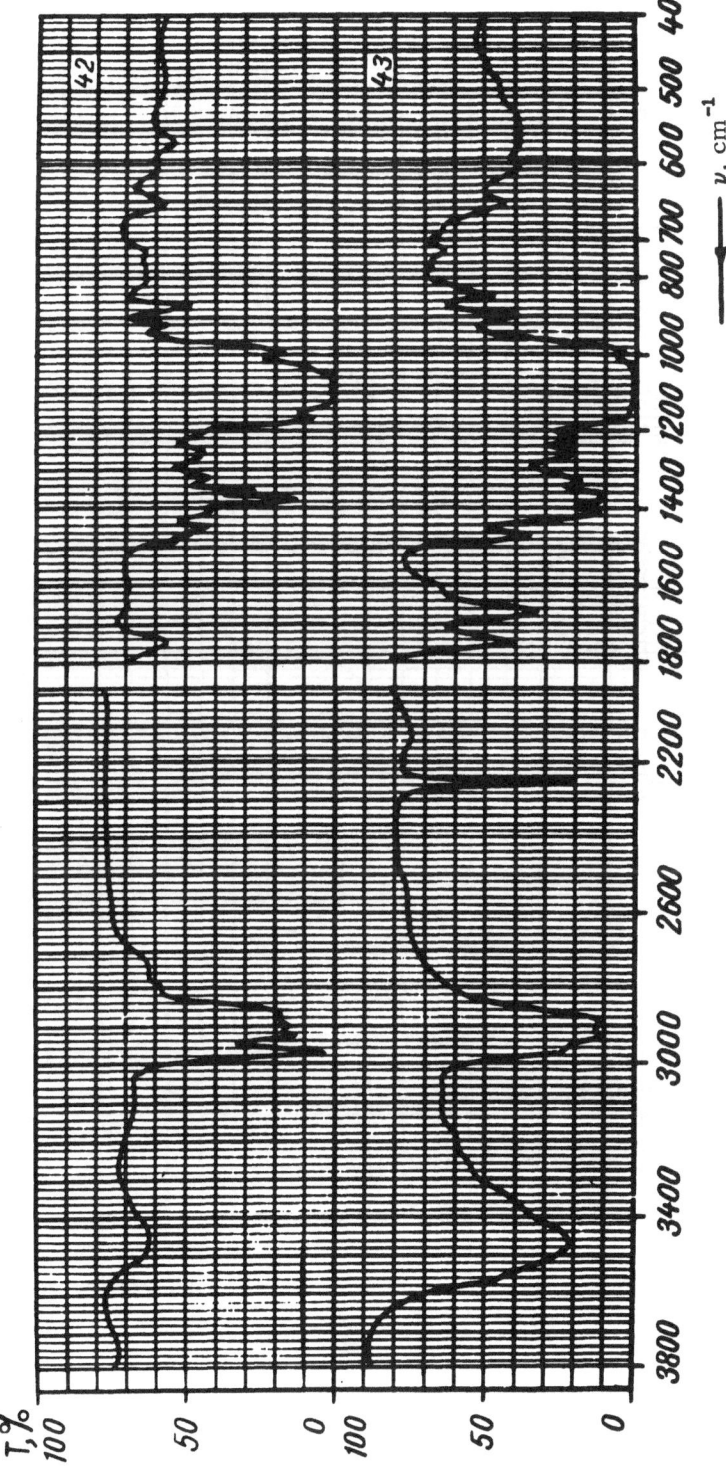

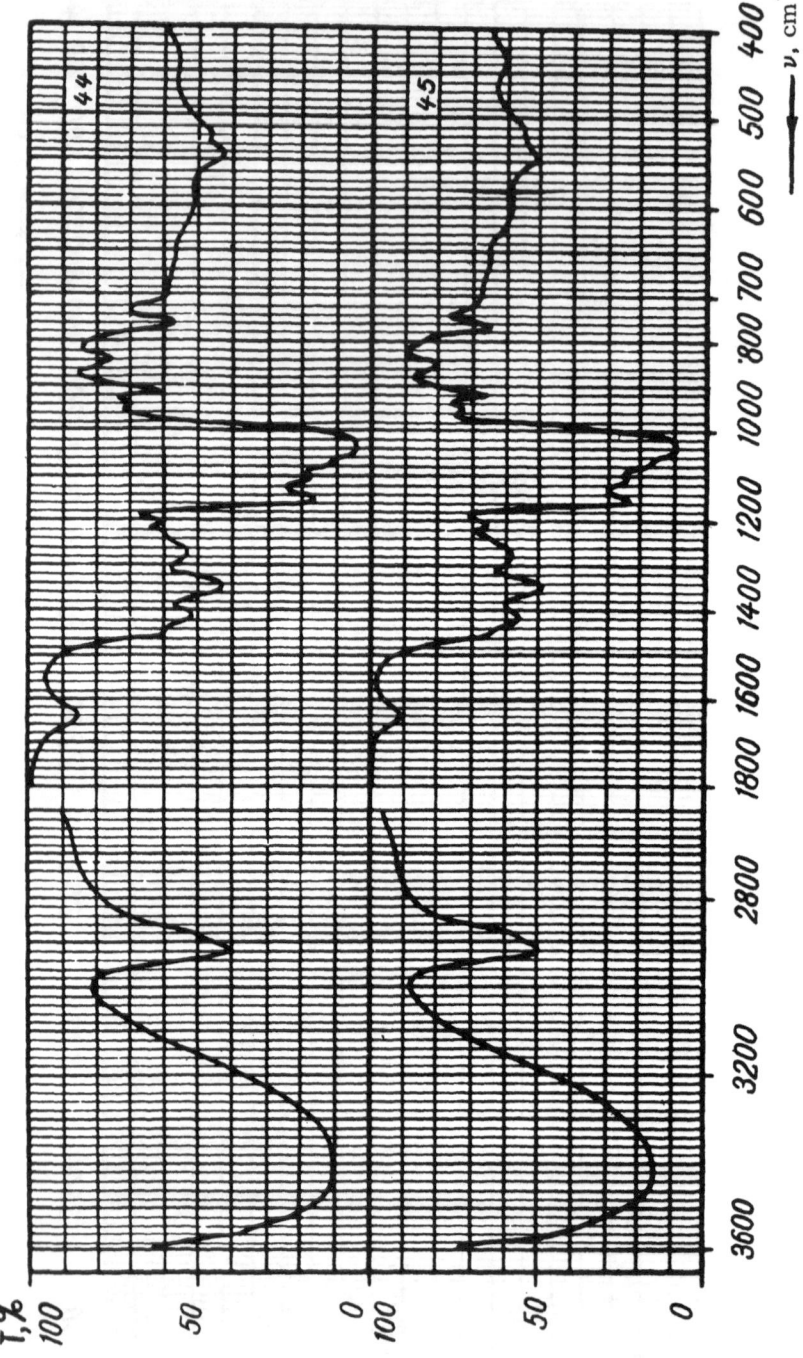

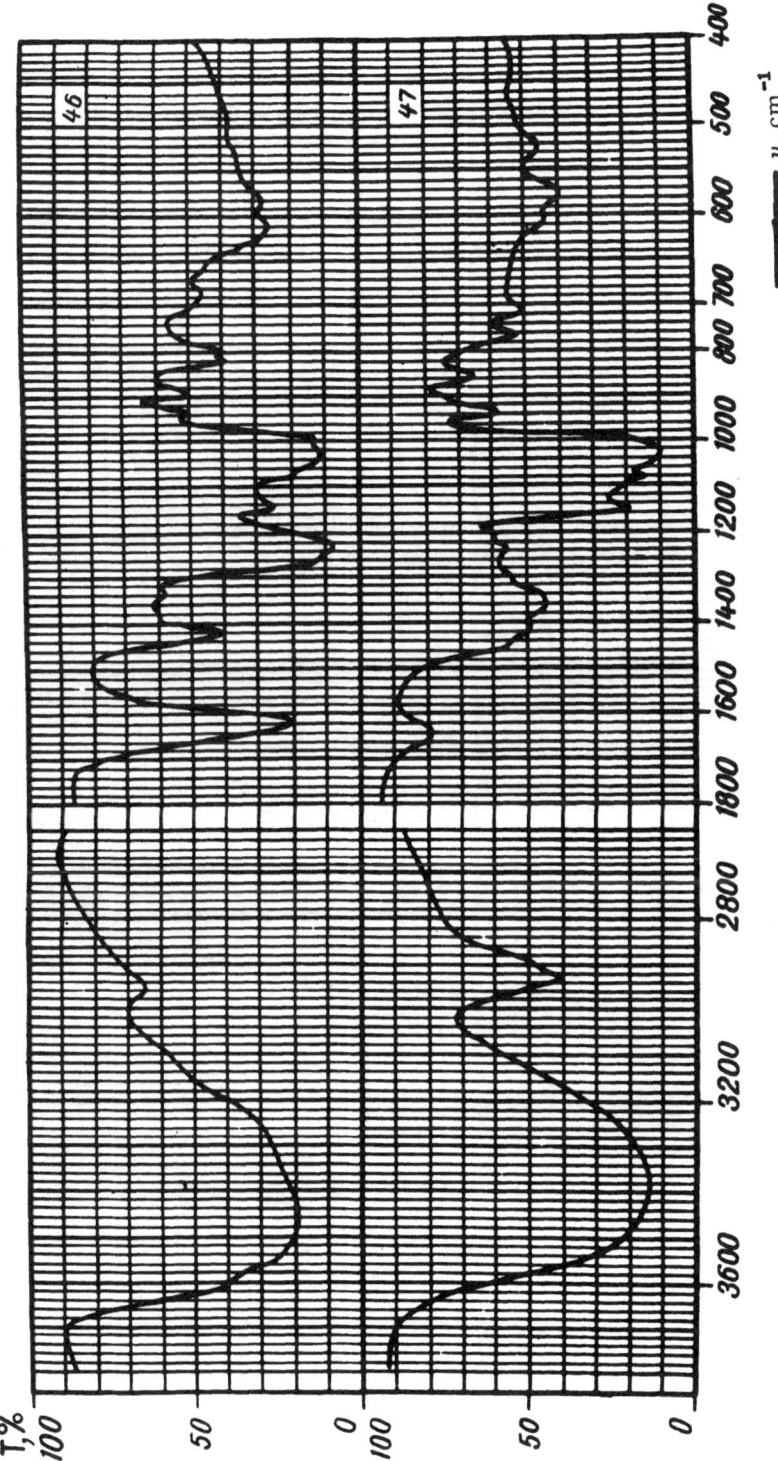

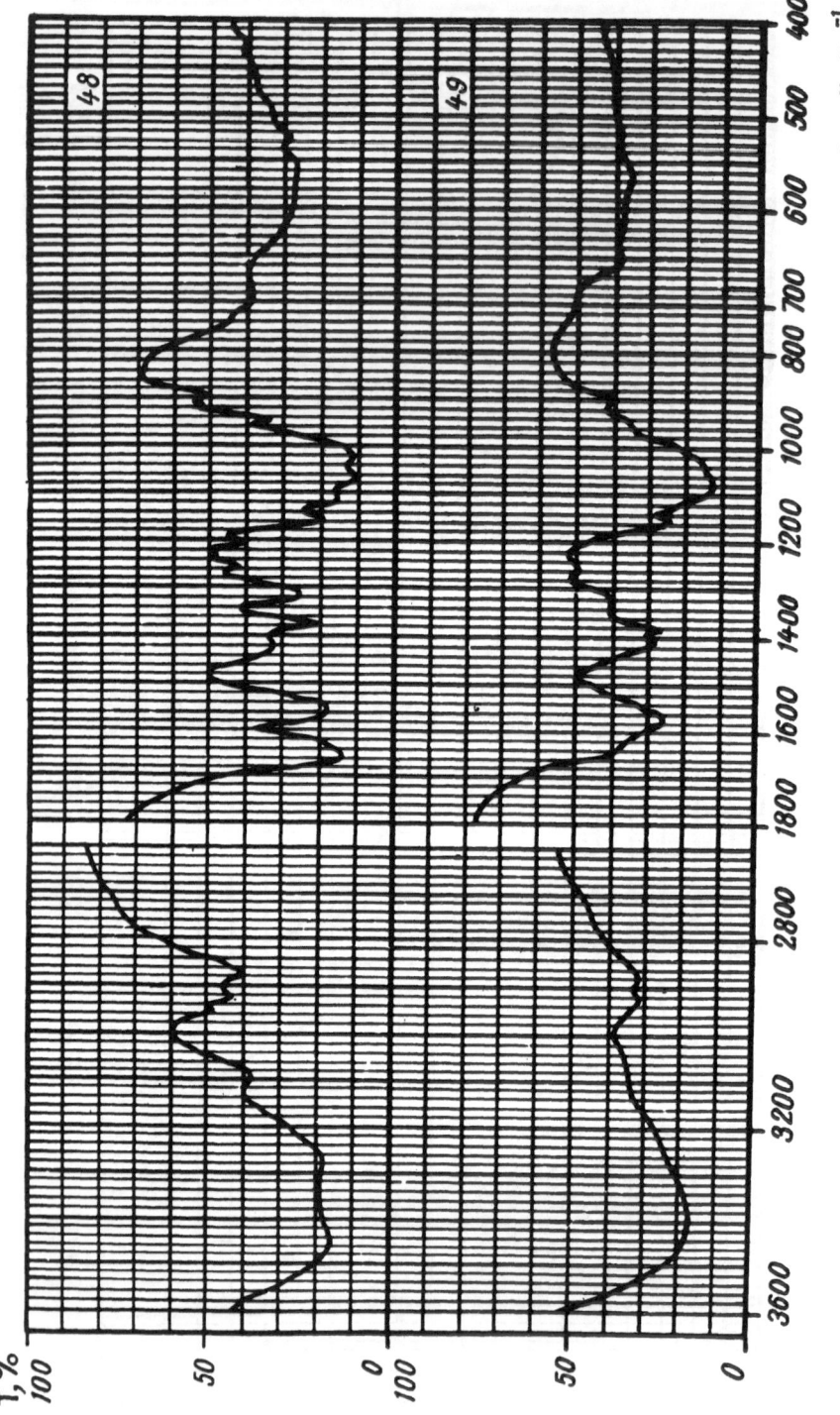

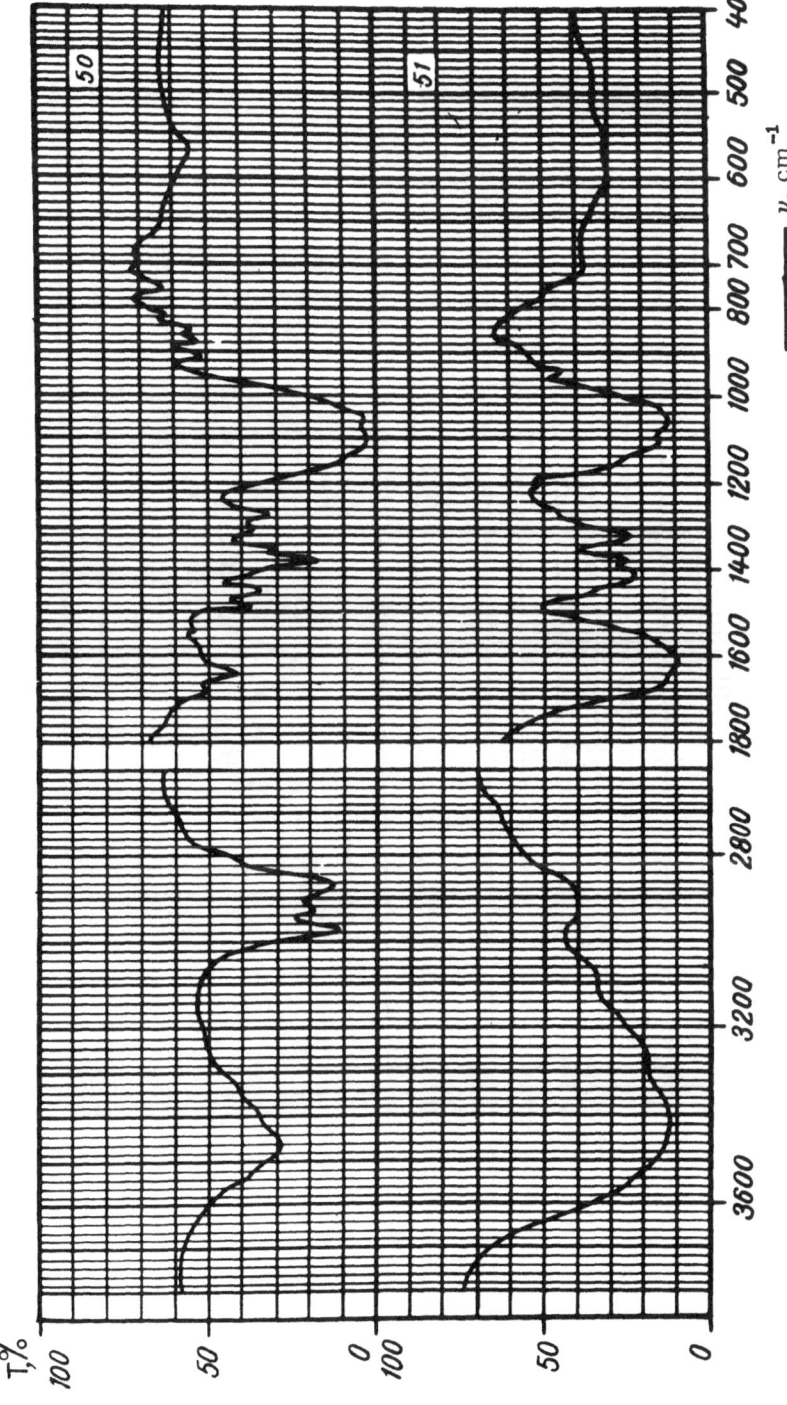

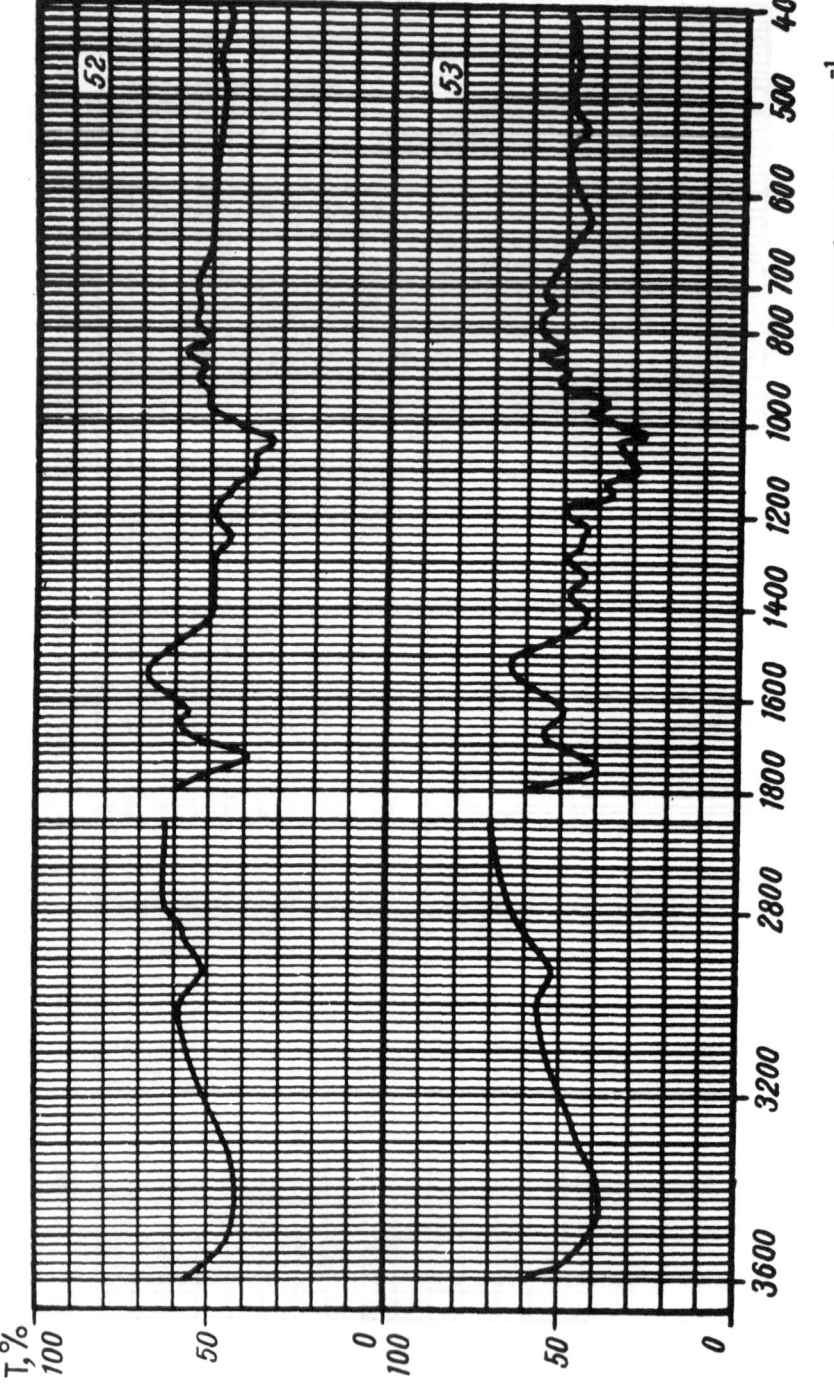

APPENDIX II

SPECTRA OF UNMODIFIED CELLULOSES

54. Cotton cellulose "C" (FM)
55. Wood cellulose "CA" (FM)
56. Flax cellulose (FM)
57. Ramie cellulose (FM)
58. Flax cellulose (unbleached) (FM)
59. Flax tow cellulose (unbleached) (FM)
60. Reed cellulose, cultivated (FM)
61. Cereal straw cellulose (FM)
62. Reed cellulose containing 7.5% lignin (FM)
63. Reed cellulose containing 1.4% lignin (FM)
64. Short-fiber hemp cellulose (unbleached) (FM)
65. Long-fiber hemp cellulose (unbleached) (FM)
66. Maize stalk cellulose (unbleached) (FM)
67. Rye straw cellulose (unbleached) (FM)
68. Viminalis willow shoot cellulose (unbleached) (FM)
69. Salix acutifolia willow shoot cellulose (unbleached) (FM)
70. Birch shoot cellulose (unbleached) (FM)
71. Alder shoot cellulose (unbleached) (FM)
72. Cellulose from June pine shoots (FM)
73. Cellulose from interglacial pine wood, of geological age 140,000 years (FM)
74. Spruce cellulose (FM)
75. Larch cellulose (FM)
76. Cellulose from pine wood 100 years old (FM)
77. Cellulose from pine wood, of geological age 3000 years (FM)
78. Cellulose from wood of sub-Moscovian poplar (unbleached) (FM)
79. Cellulose from bark of sub-Moscovian poplar (unbleached) (FM)
80. Uncultivated wood cellulose (LTI process) (FM).
81. Wood cellulose for high-durability tires (LTI process) (FM)
82. Wood cellulose from the firm Bekai (Canada) (FM)
83. Wood cellulose from the firm Lintra (USA) (FM)

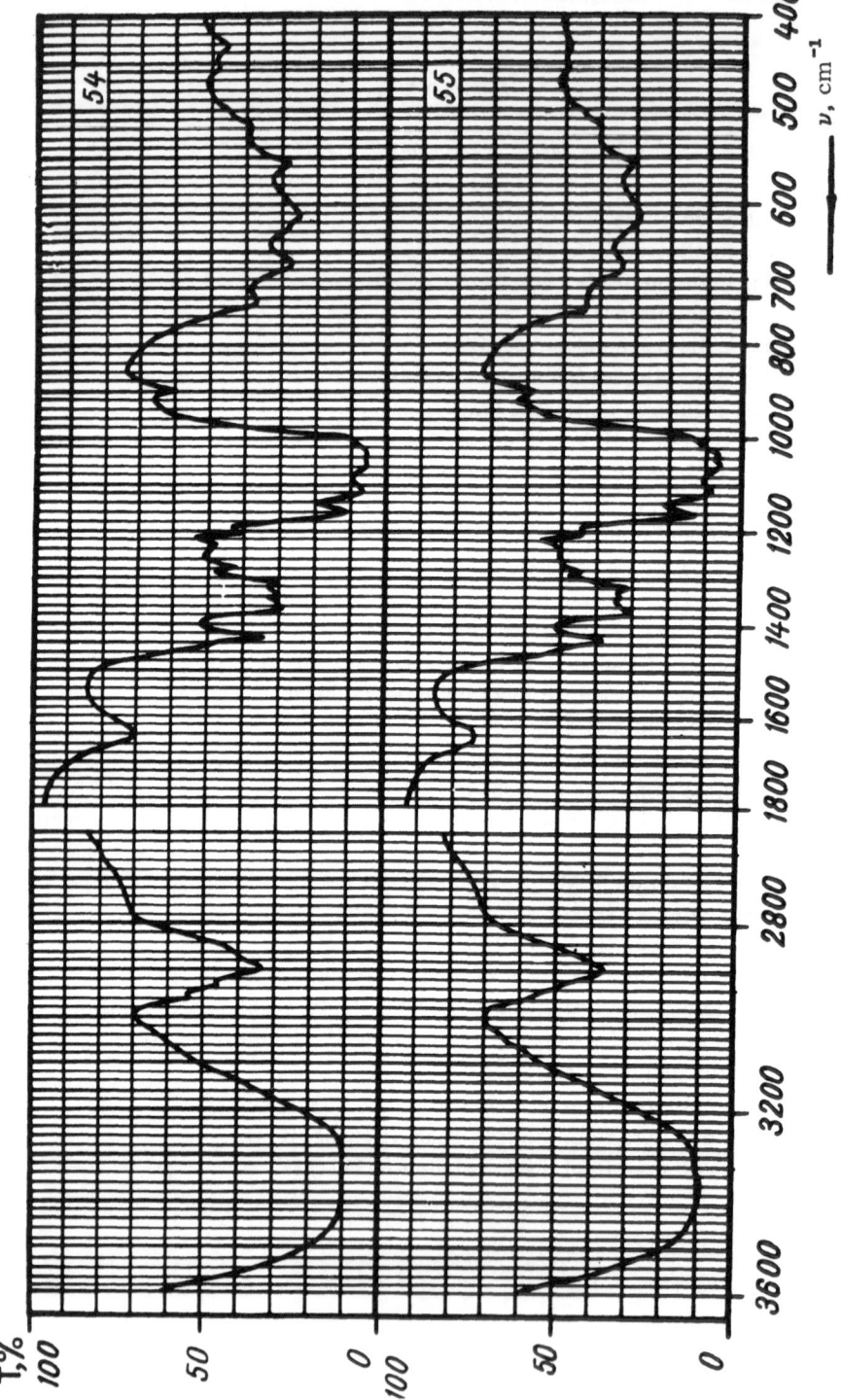

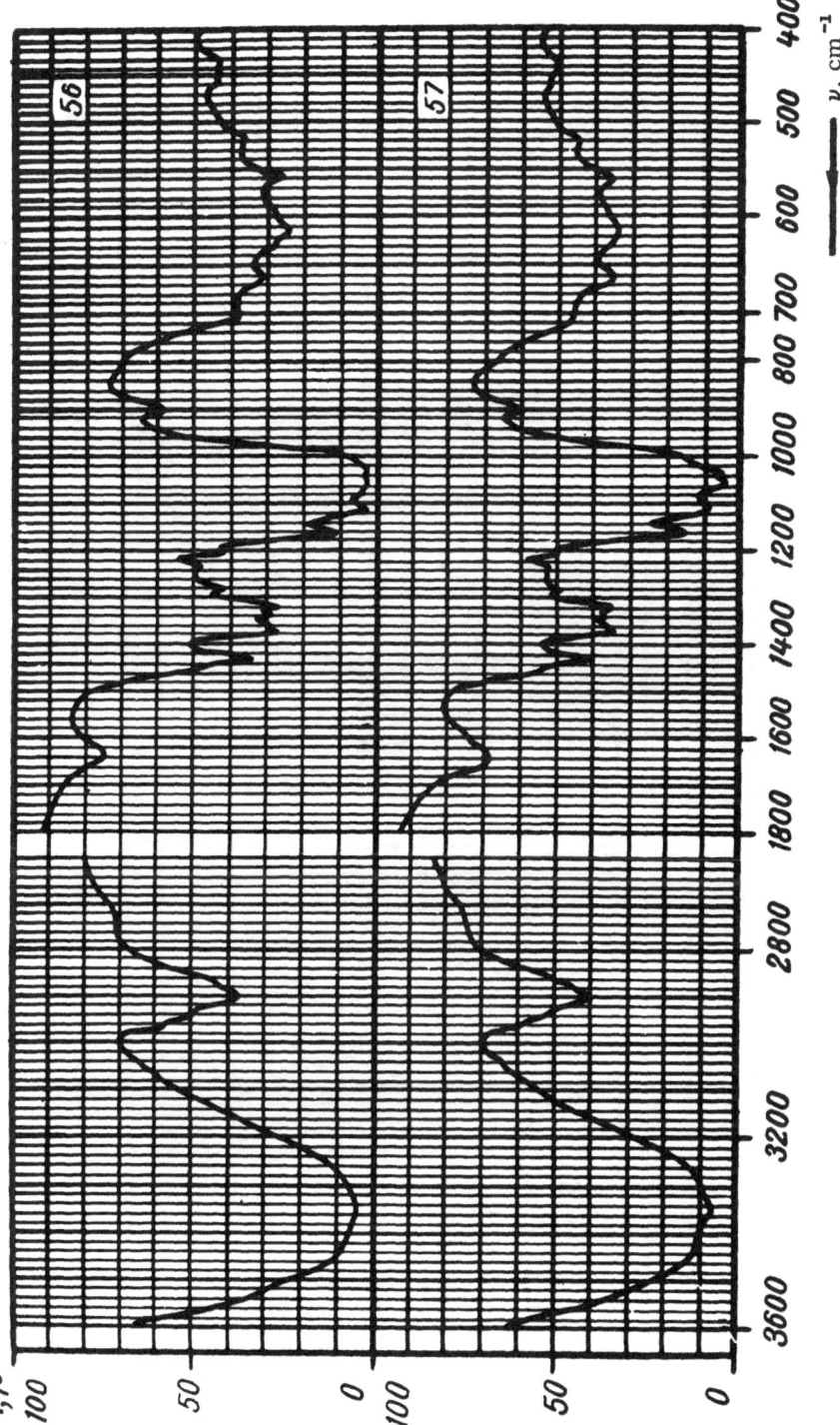

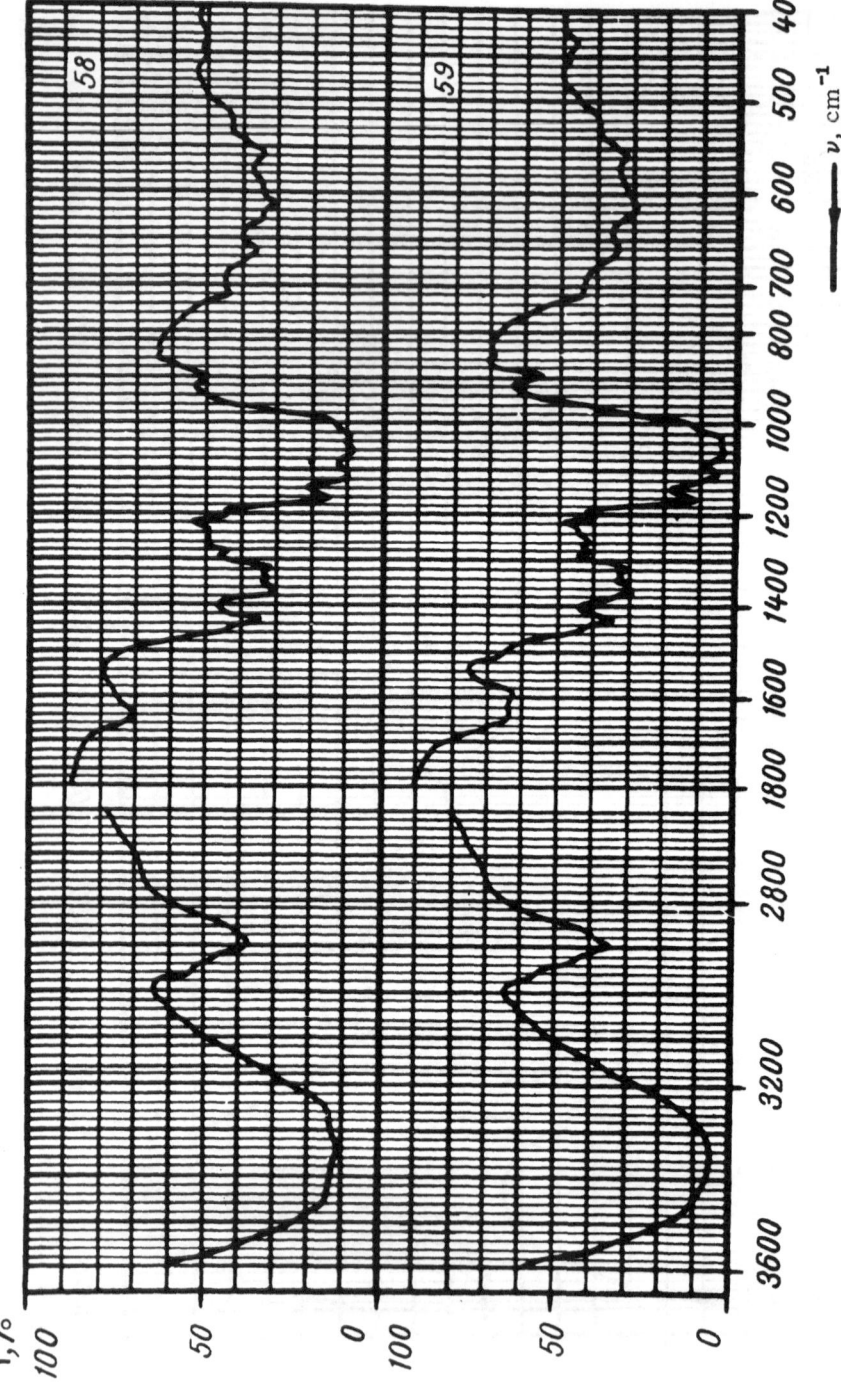

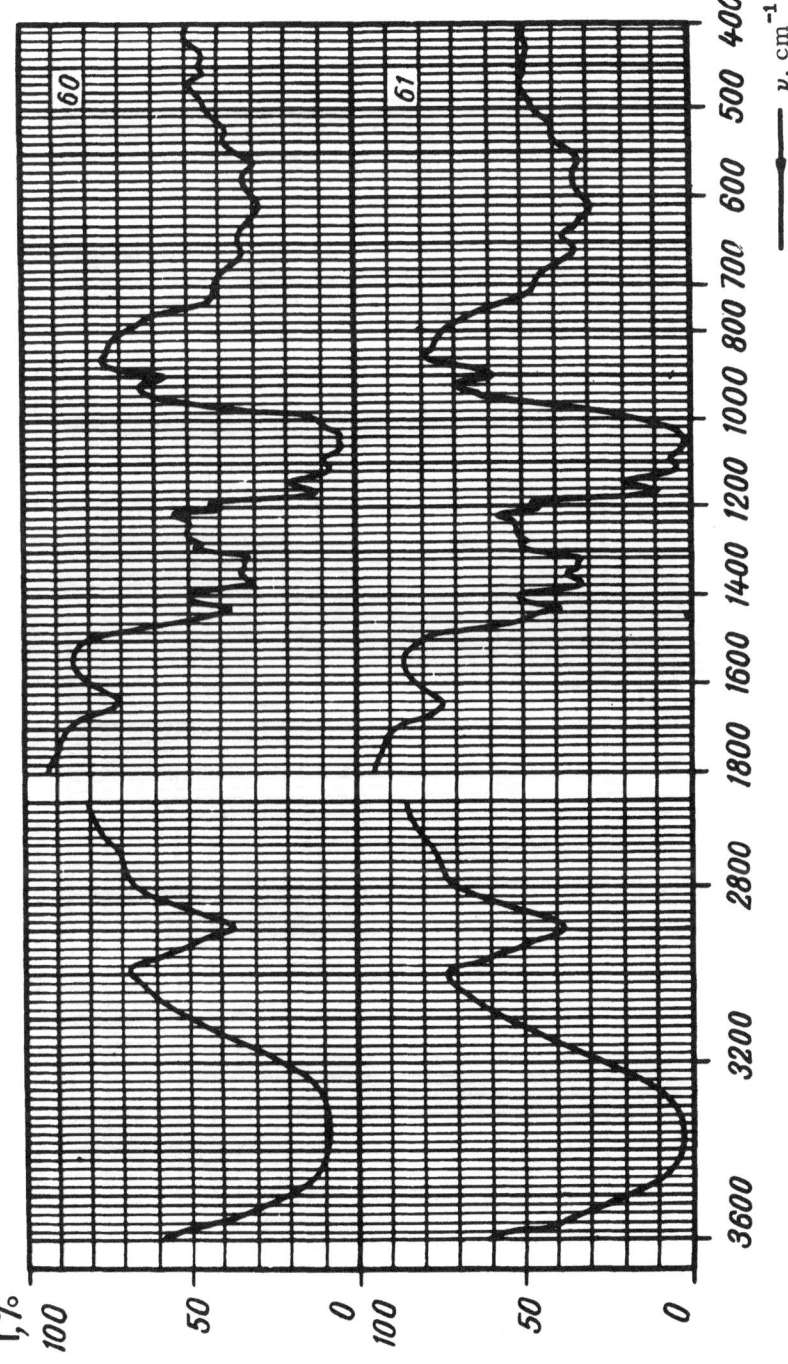

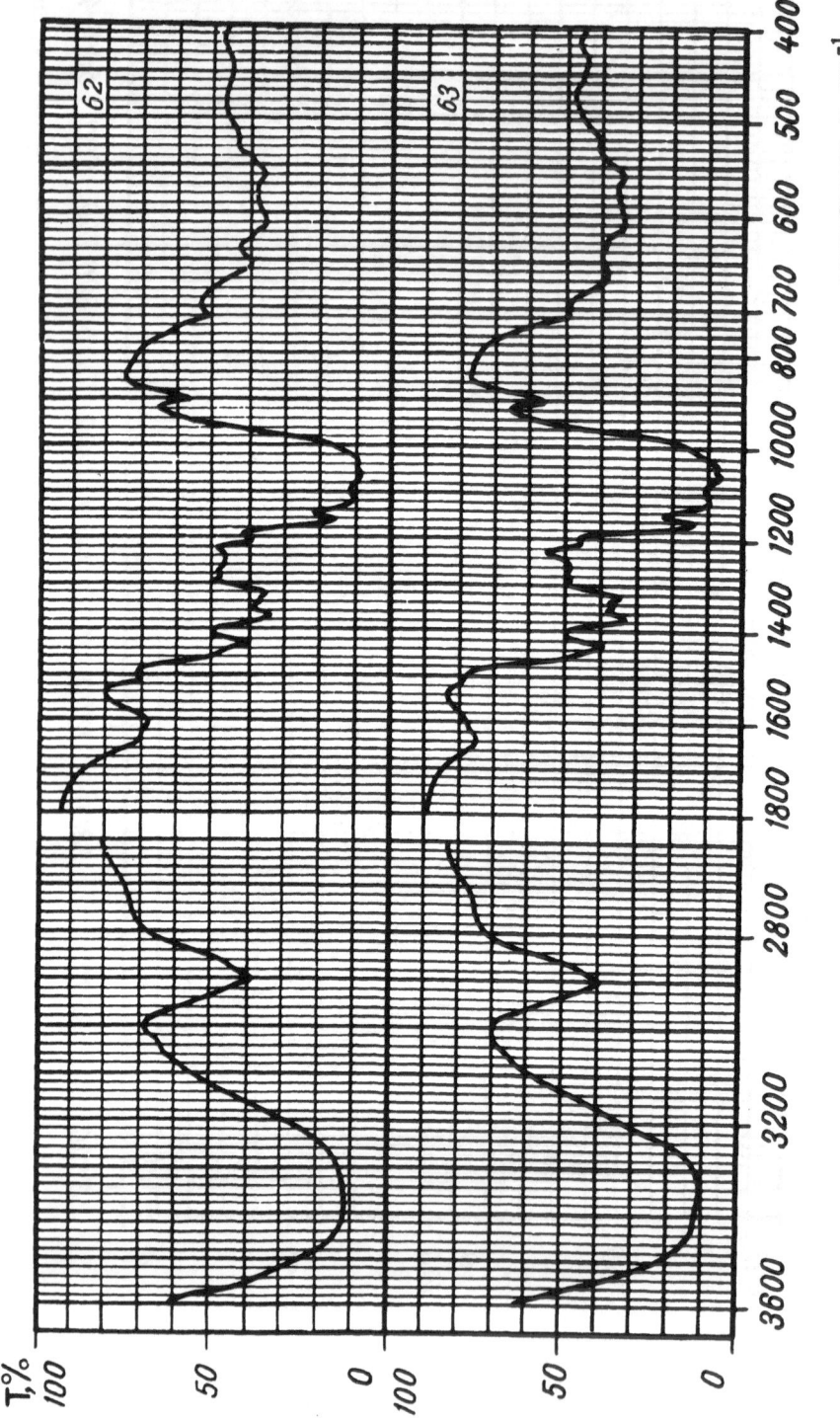

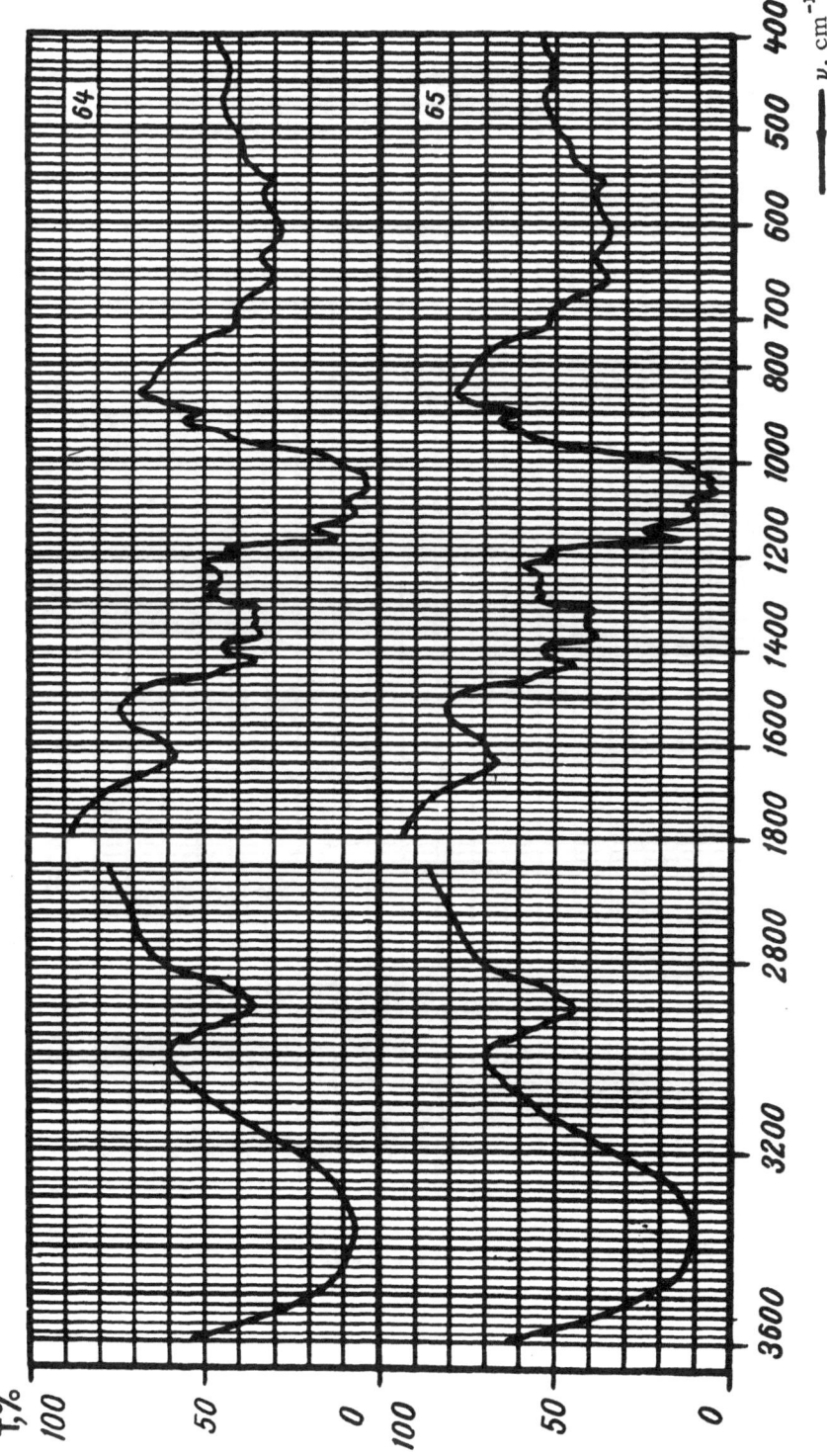

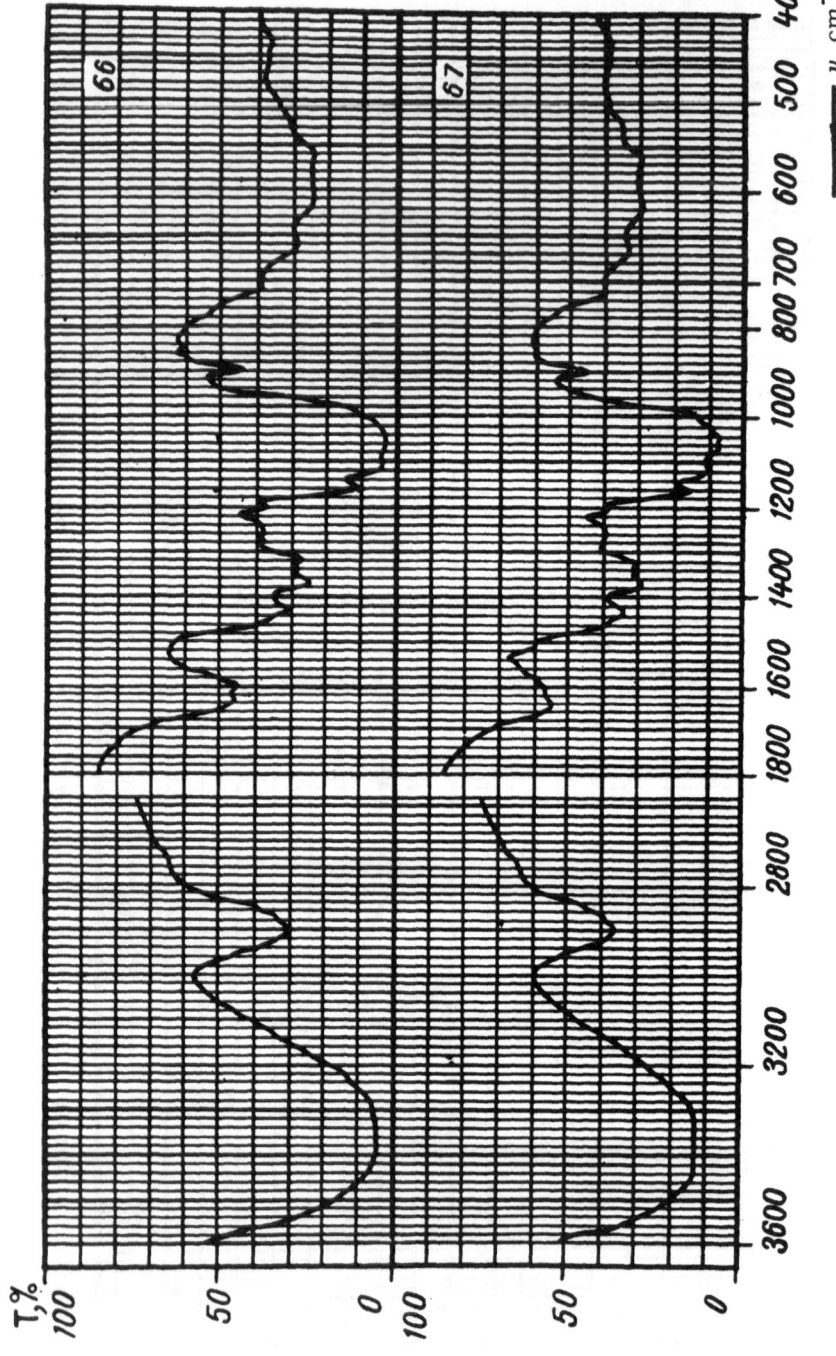

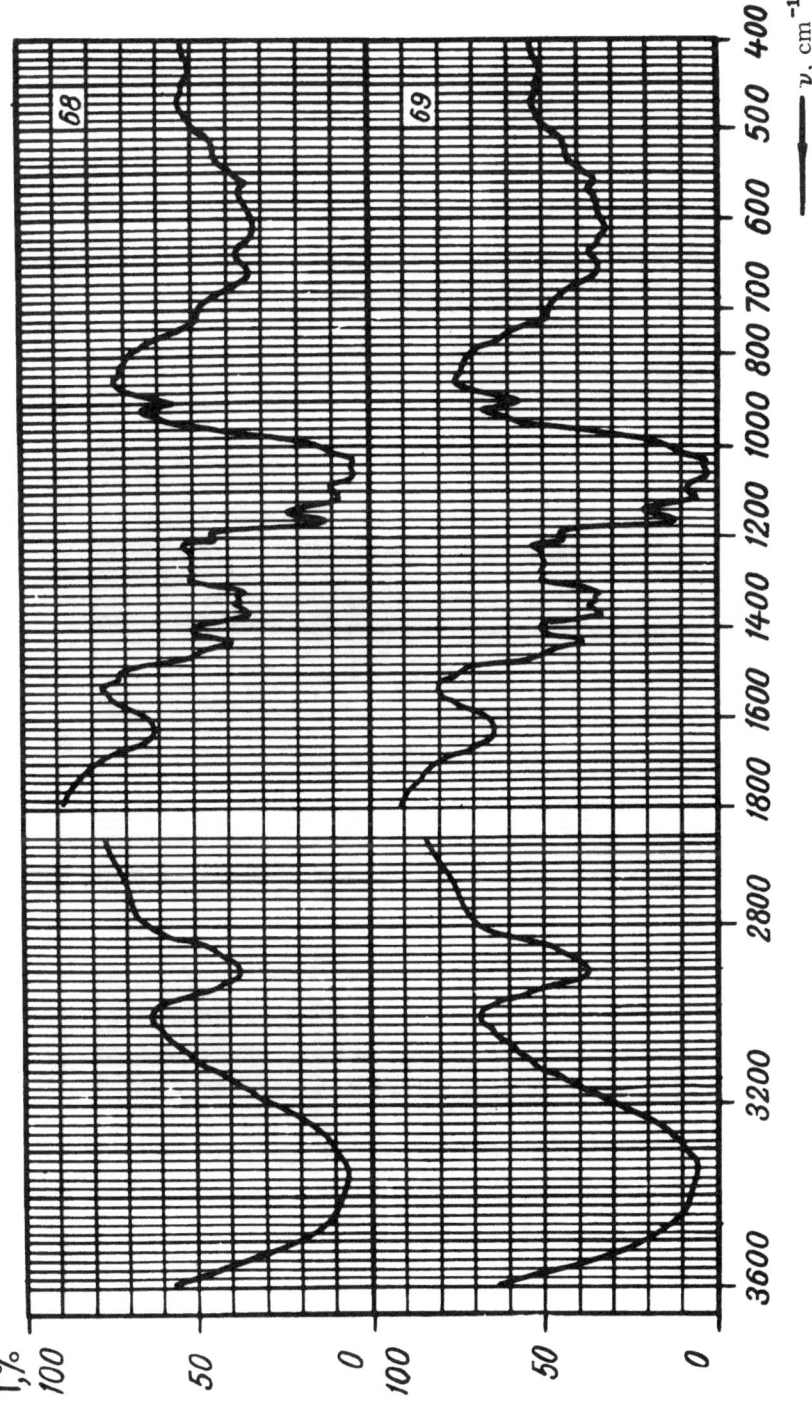

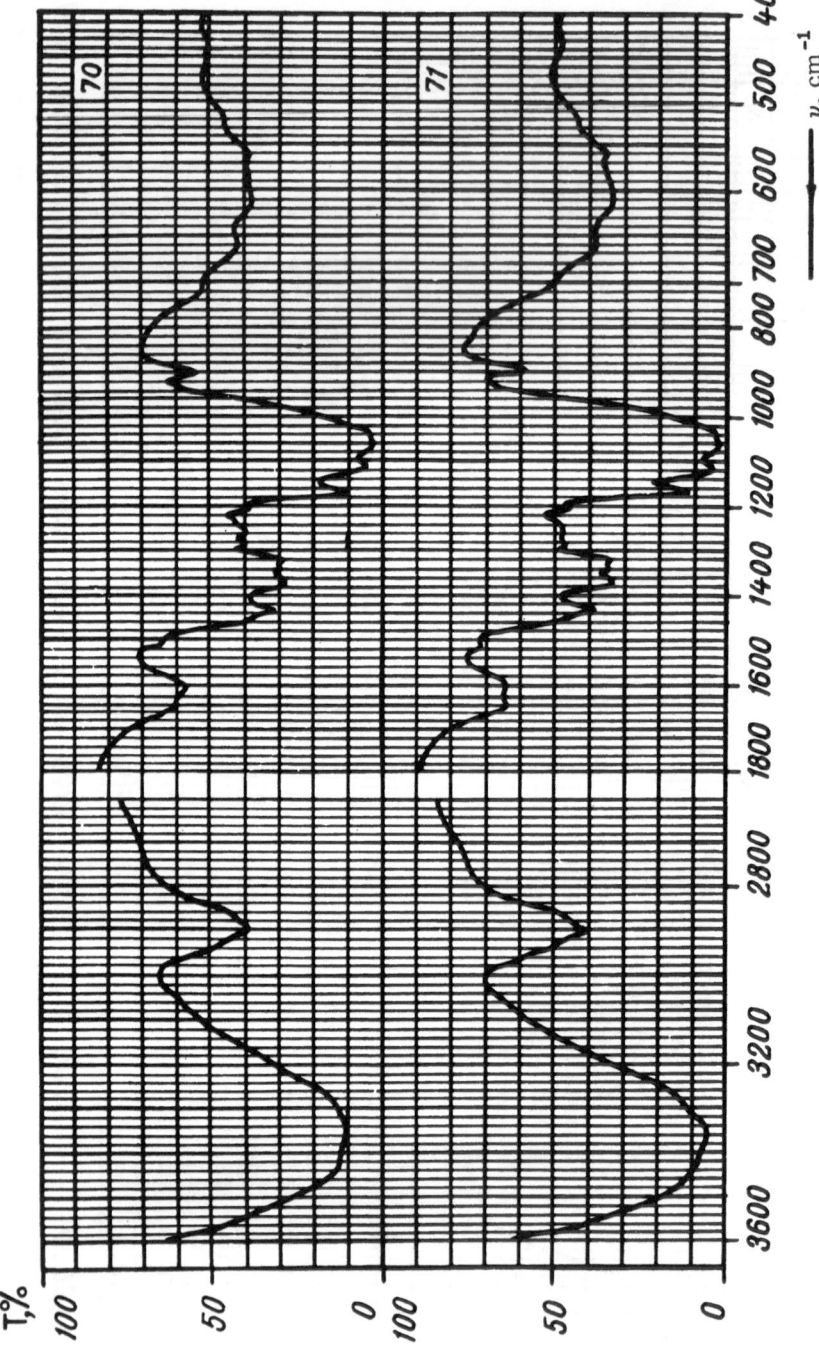

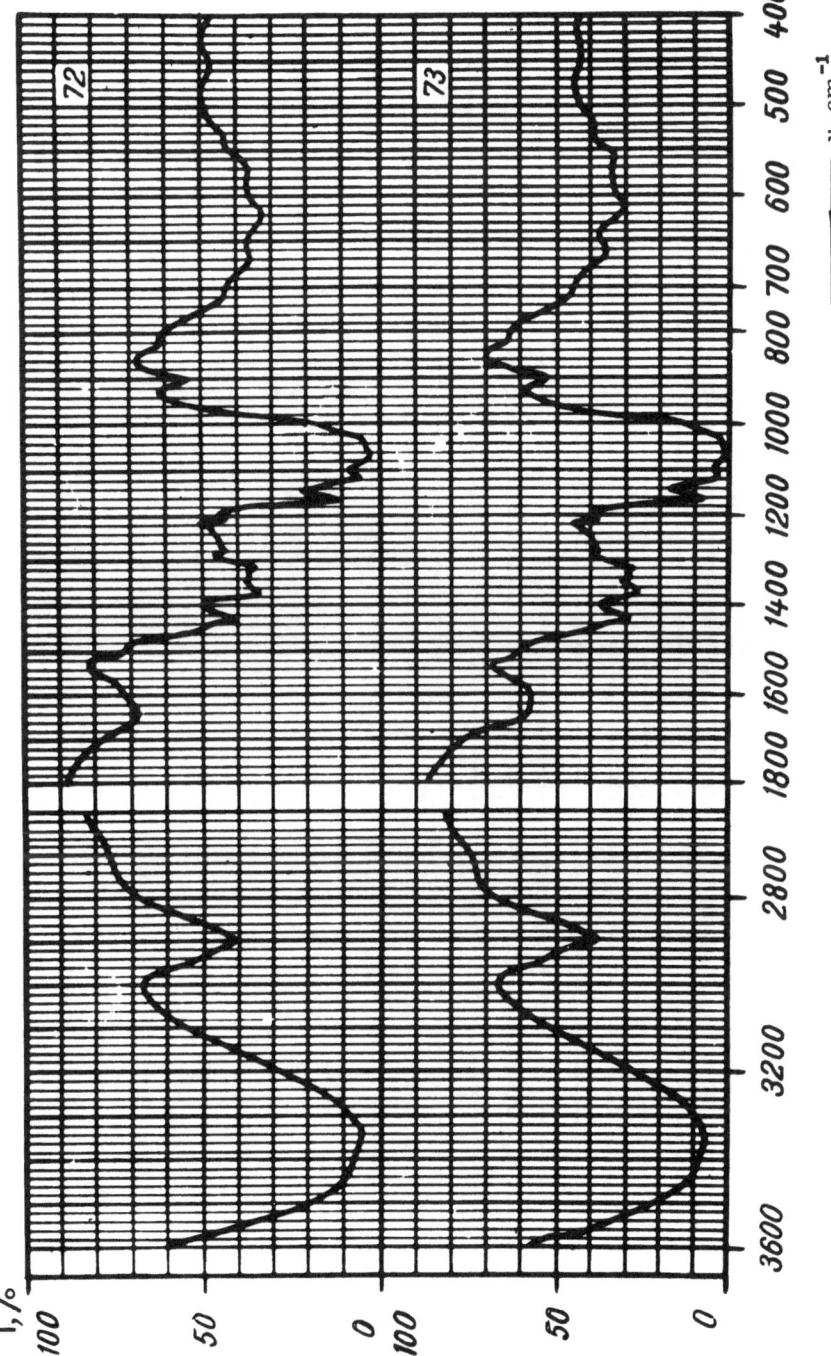

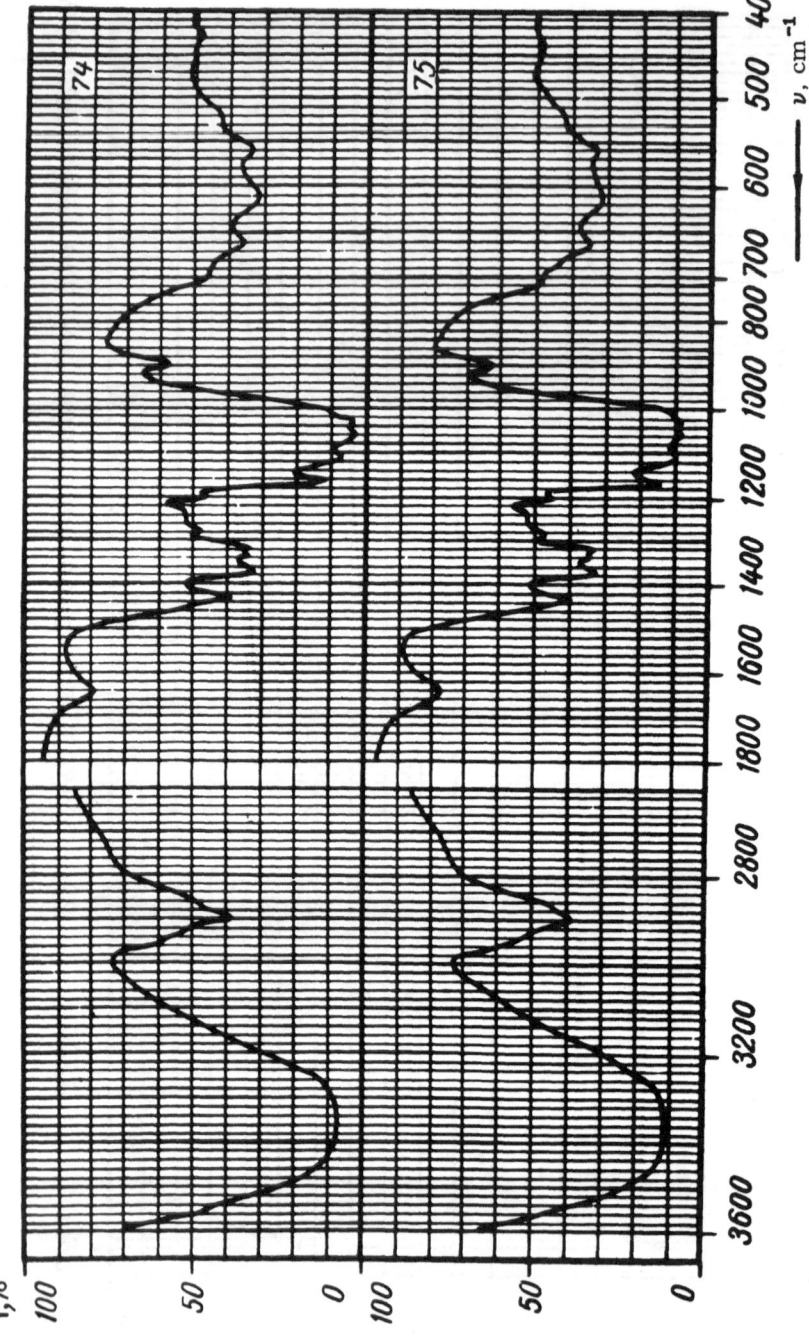

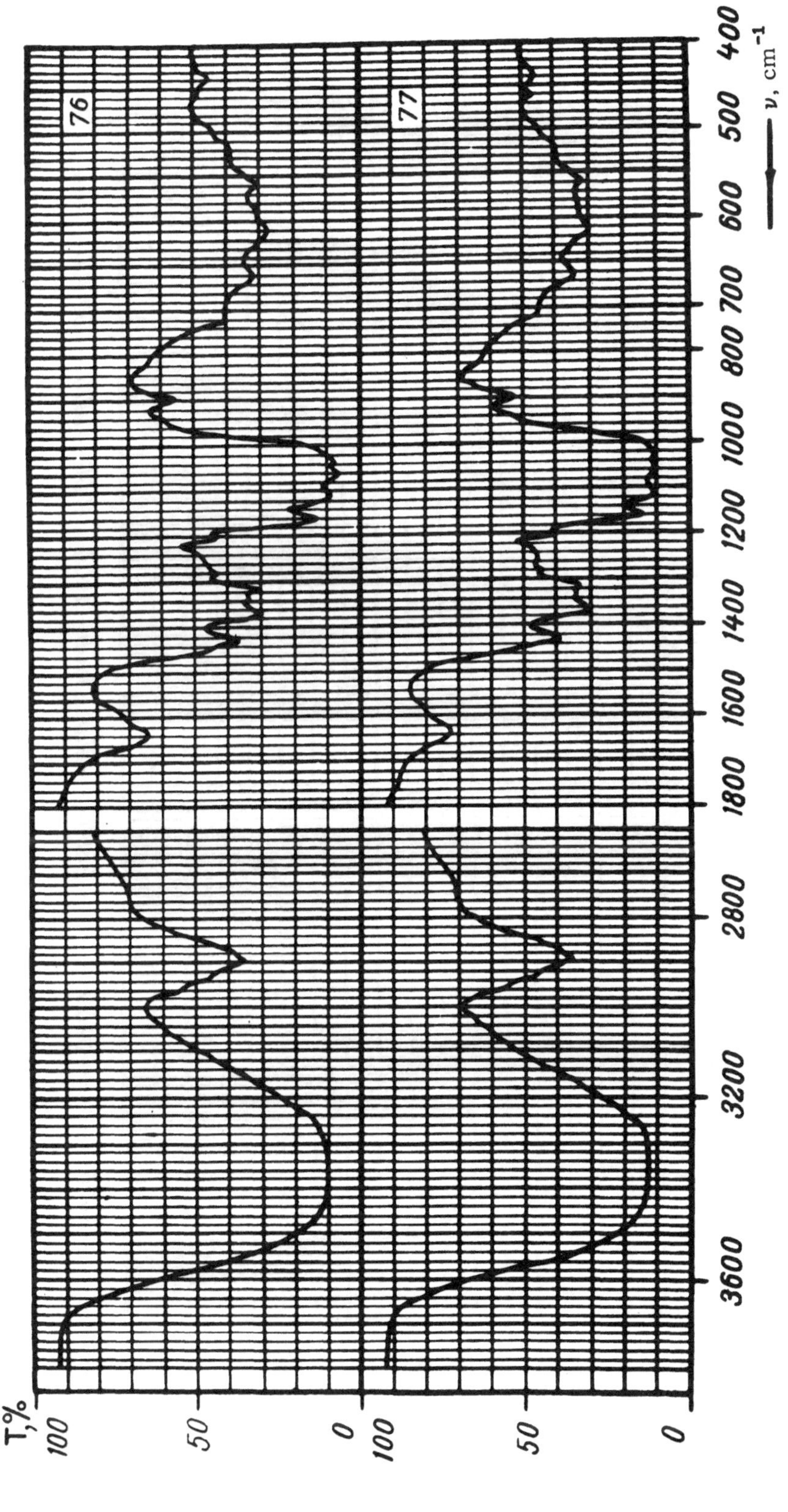
76
77
T,%
100
50
0
100
50
0
ν, cm^{-1}
400
500
600
700
800
1000
1200
1400
1600
1800
2800
3200
3600

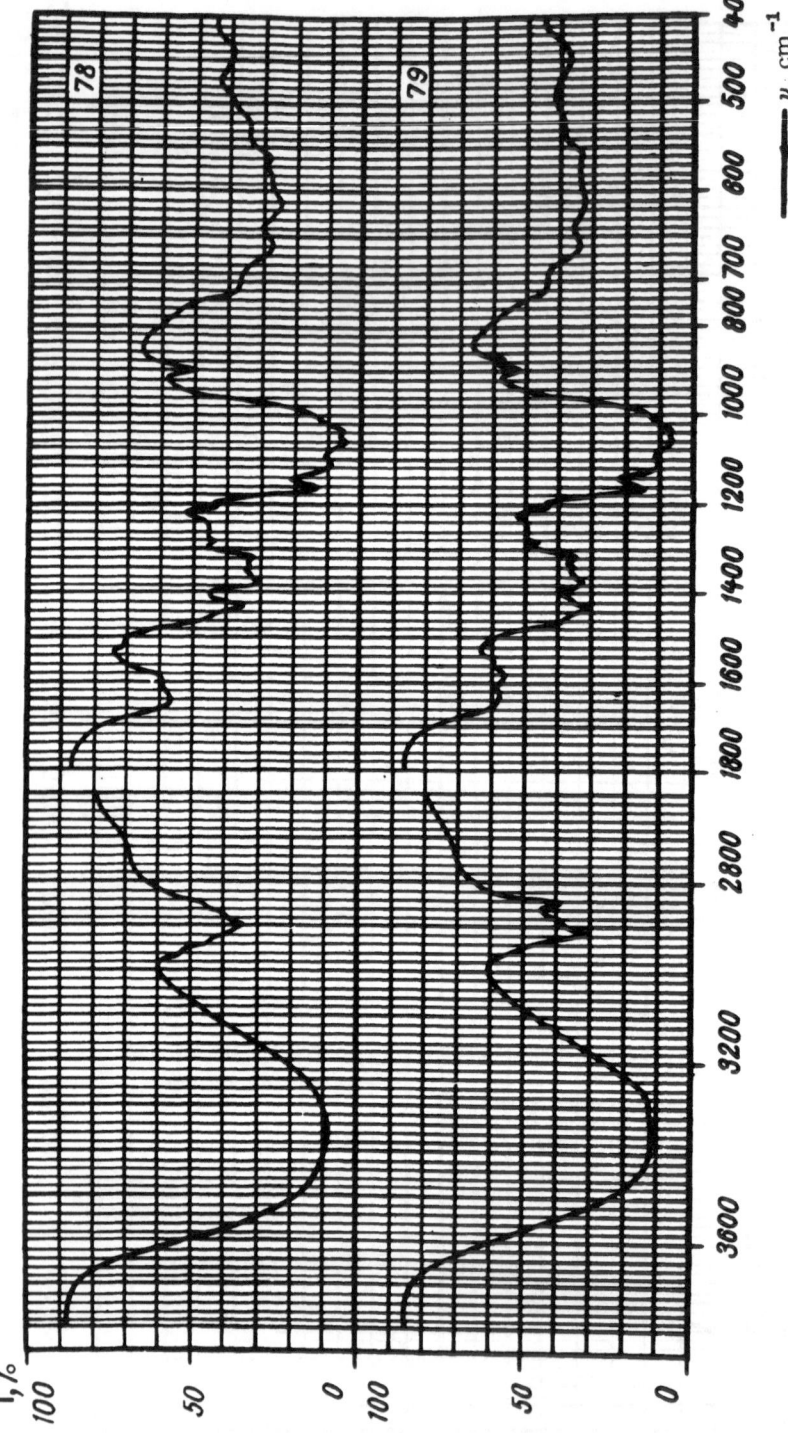

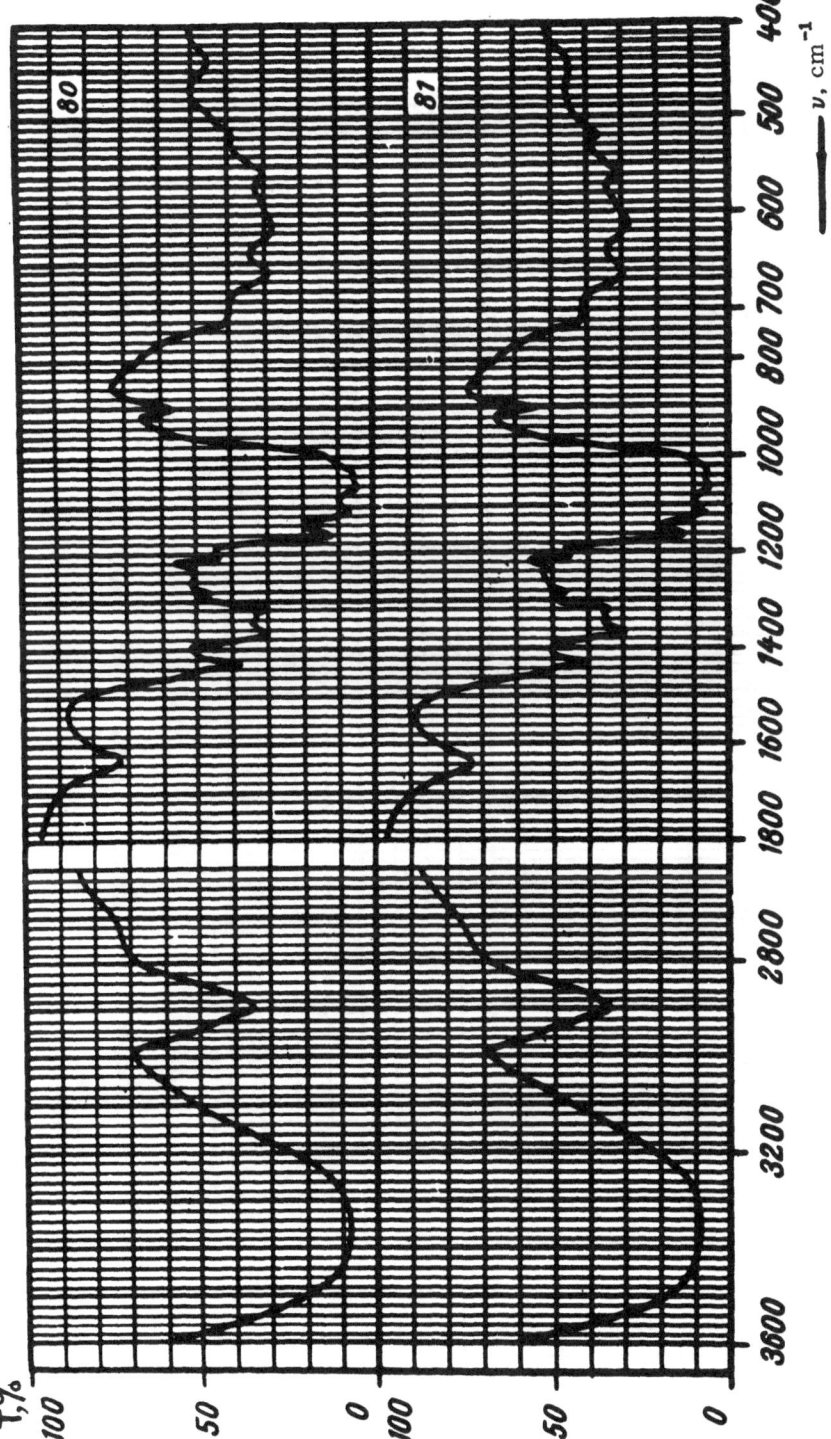

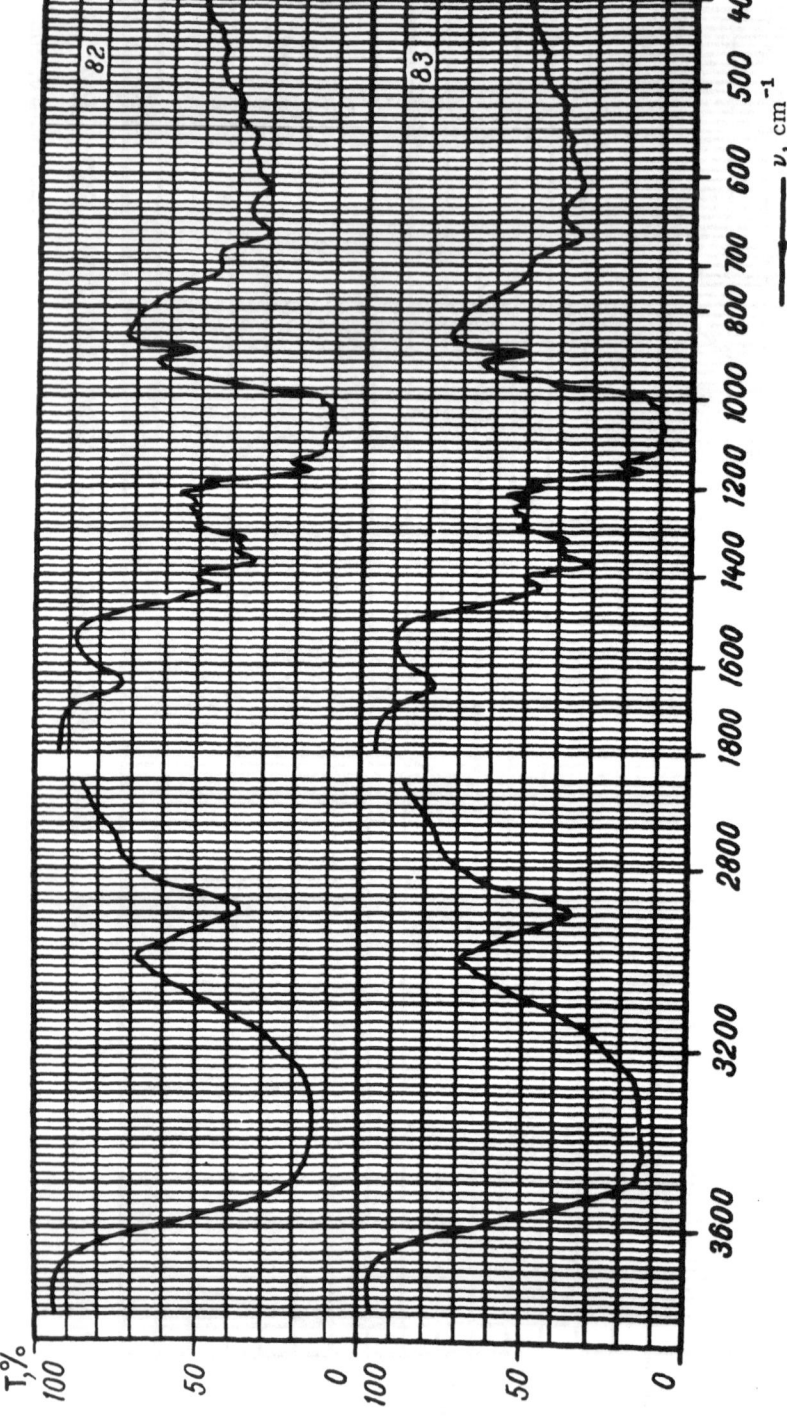

APPENDIX III

SPECTRA OF CELLULOSES REGENERATED AFTER TREATMENT
WITH AQUEOUS ALKALI OF VARIOUS CONCENTRATIONS.
SPECTRA OF VISCOSE SILK AND OF ALKALINE CELLULOSE

84. Wood cellulose regenerated after treatment with 2% NaOH solution (FM)
85. Wood cellulose regenerated after treatment with 5% NaOH solution (FM)
86. Wood cellulose regenerated after treatment with 10% NaOH solution
     (FM)
87. Wood cellulose regenerated after treatment with 12% NaOH solution
     (FM)
88. Wood cellulose regenerated after treatment with 15% NaOH solution
     (FM)
89. Wood cellulose regenerated after treatment with 17.5% NaOH solution
     (FM)
90. Wood cellulose regenerated after treatment with 25% NaOH solution
     (FM)
91. Wood cellulose regenerated after treatment with 35% NaOH solution
     (FM)
92. Cotton cellulose ground for 60 min in a ball mill under HCl vapor (KBr)
93. Cotton cellulose ground for 60 min in a ball mill (KBr)
94. Cotton cellulose ground and recrystallized for 3 h in water at 0°C (KBr)
95. Cotton cellulose ground and recrystallized for 3 h in water at 100°C (KBr)
96. Viscose fiber (FibM)
97. Viscose fiber from cellulose precipitated from 0.5% viscose solution
     (FibM)
98. Viscose cord fiber "VKh" (FibM)
99. Viscose fiber from the firm "America Tyrex, Inc. III" (USA) (FibM).
100. Viscose cord fiber from the firm "Chatillon" (Italy) (FibM)
101-109. Wood cellulose in 0% (101), 2% (102), 5% (103), 10% (104), 12%
     (105), 15% (106), 17.5% (107), 20% (108), and 25% (109) solution of
     NaOD in D_2O (FP)

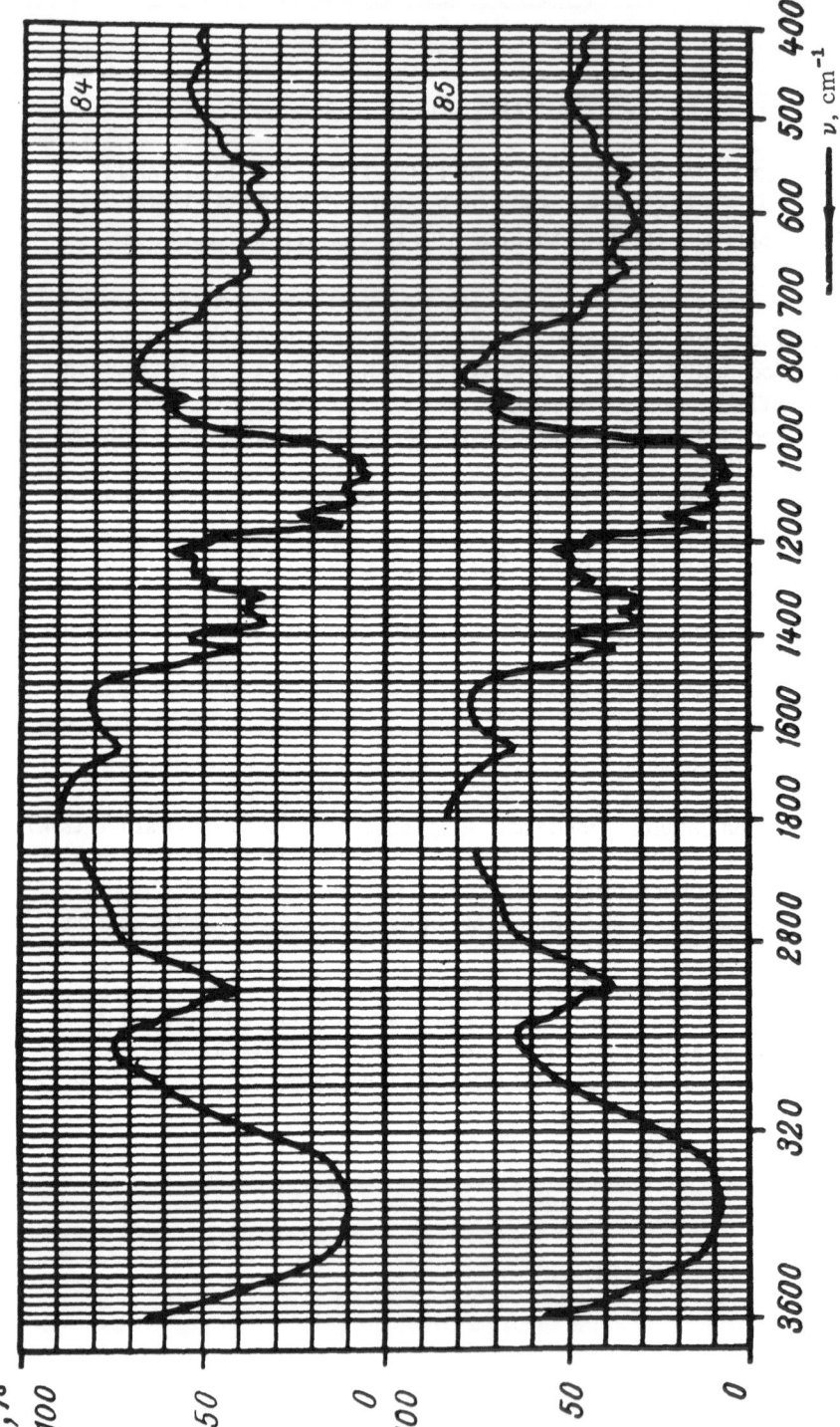

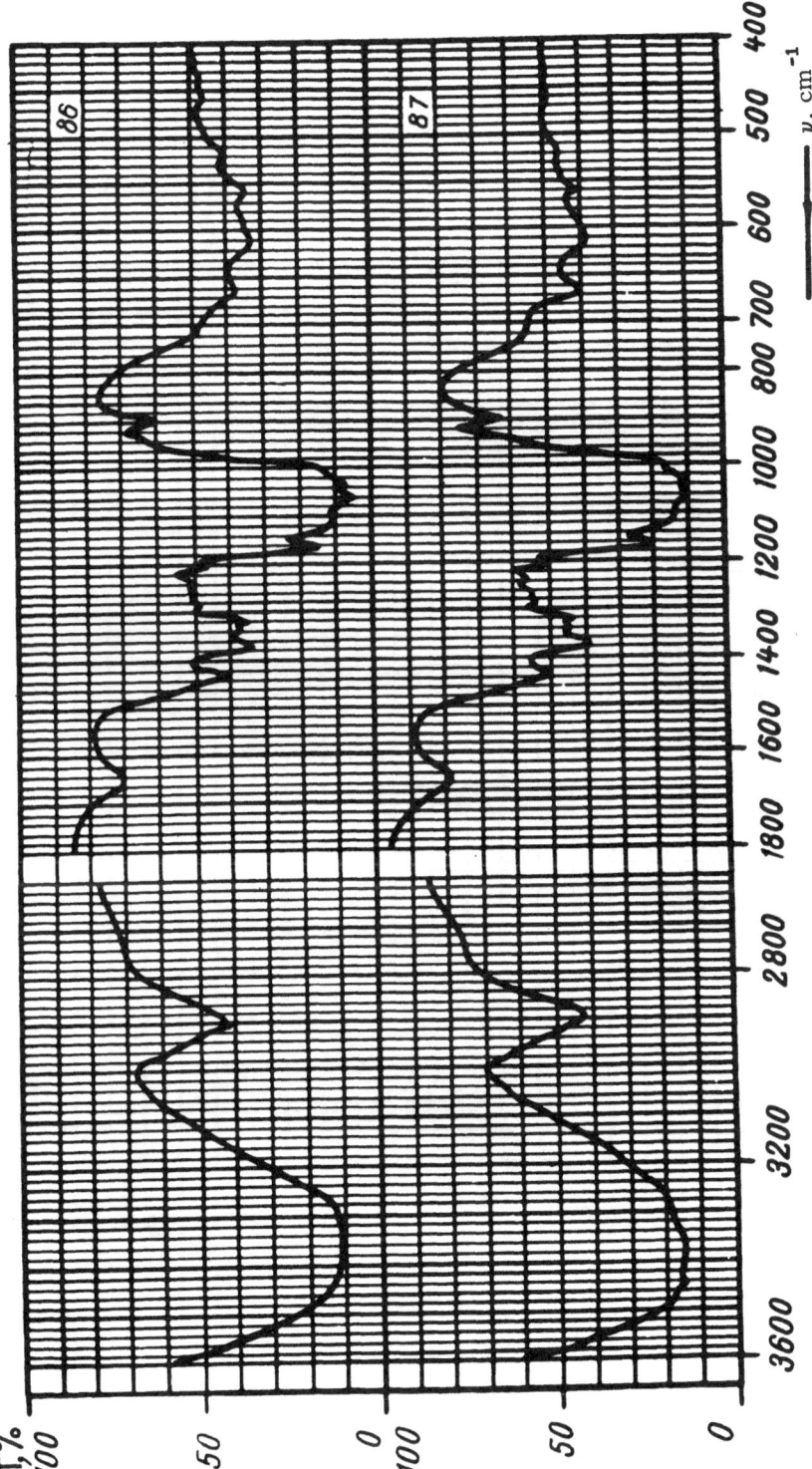

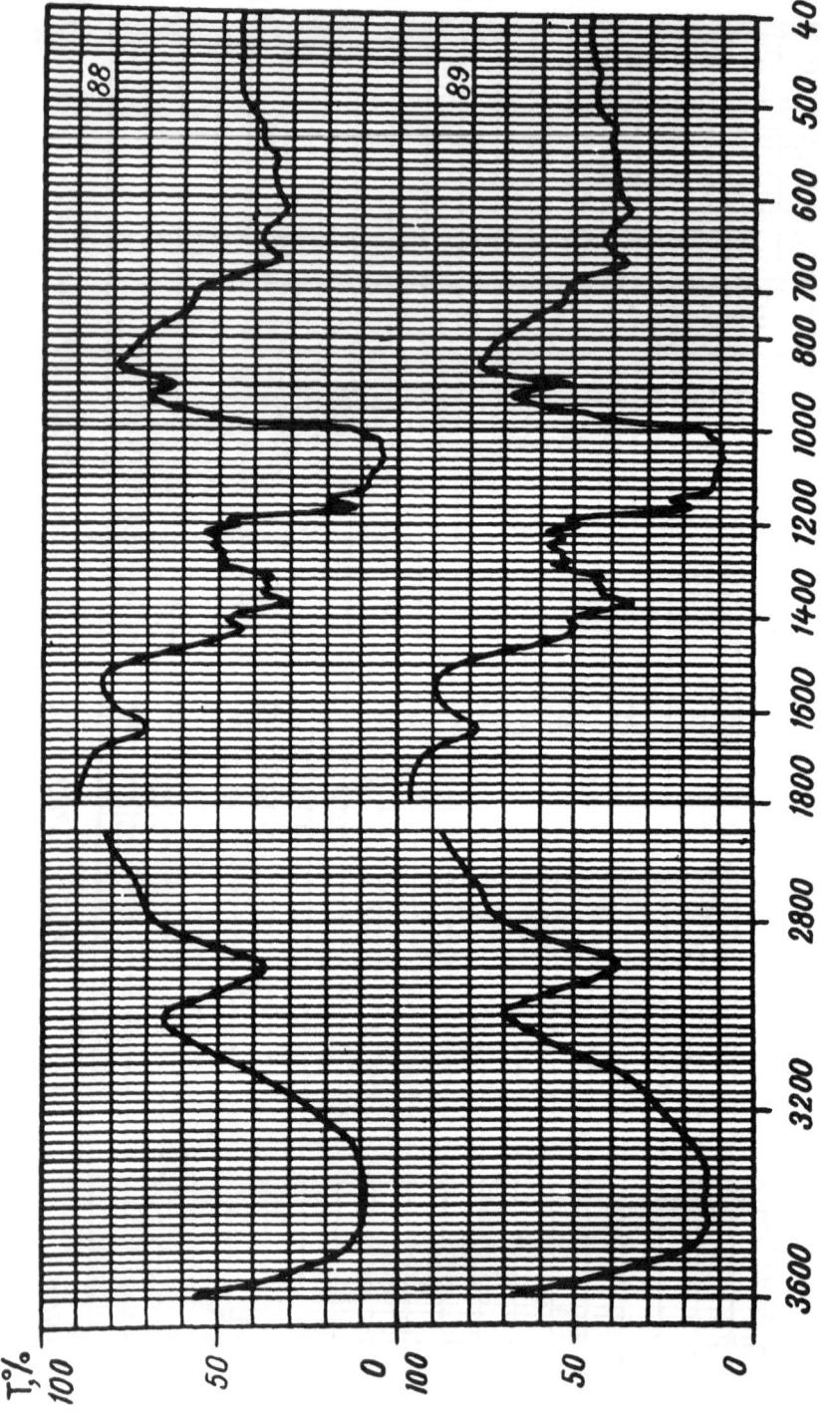

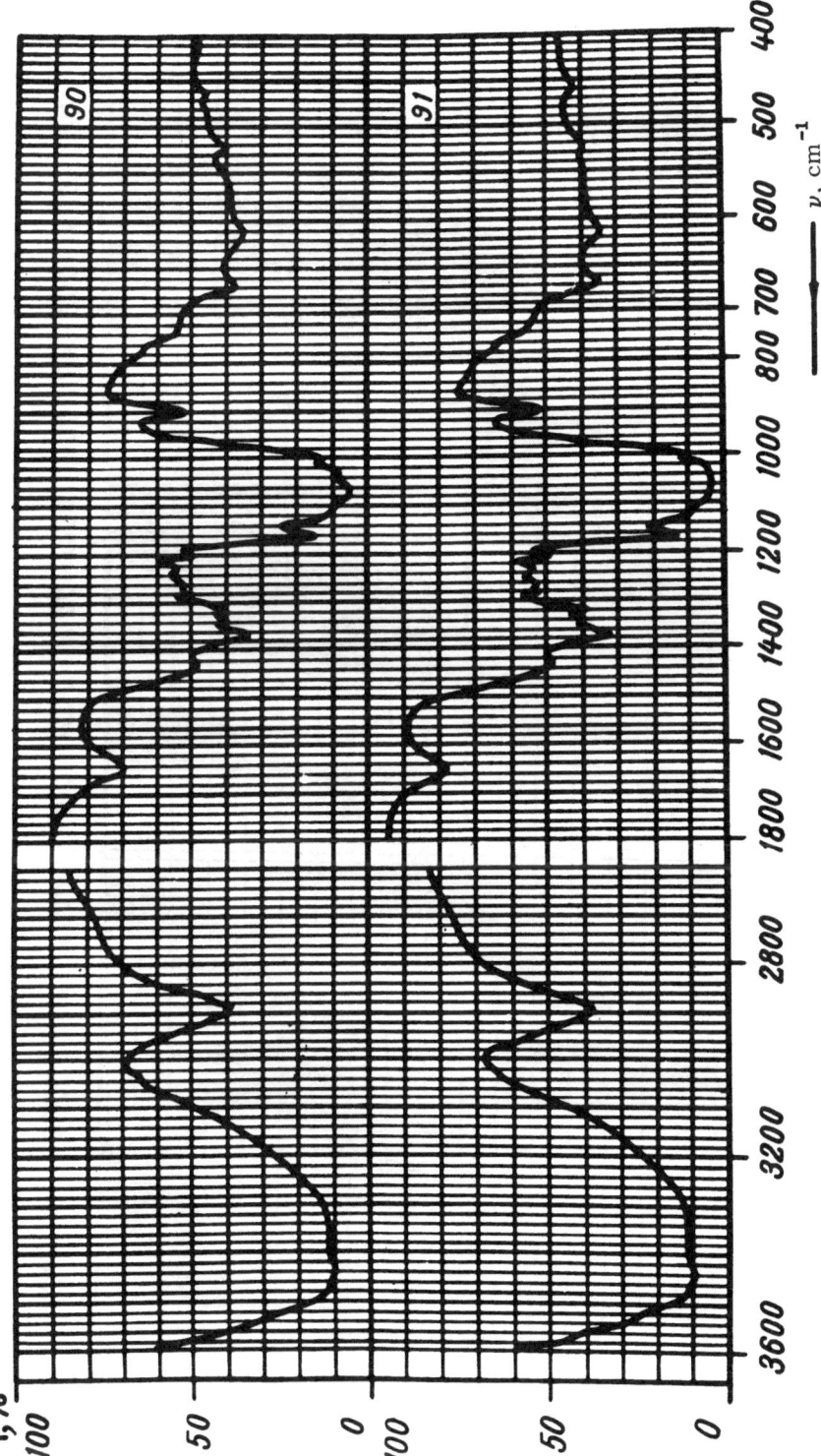

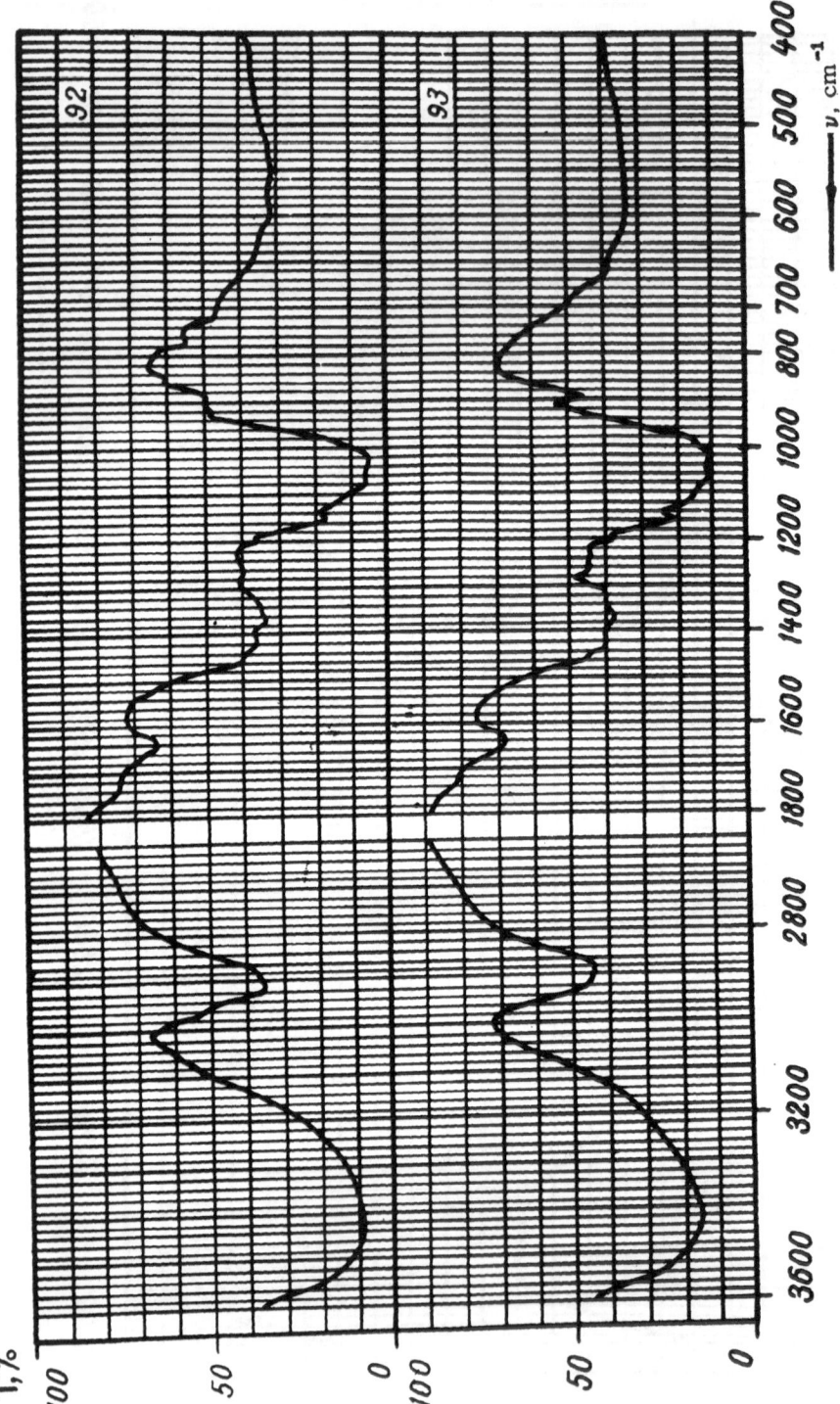

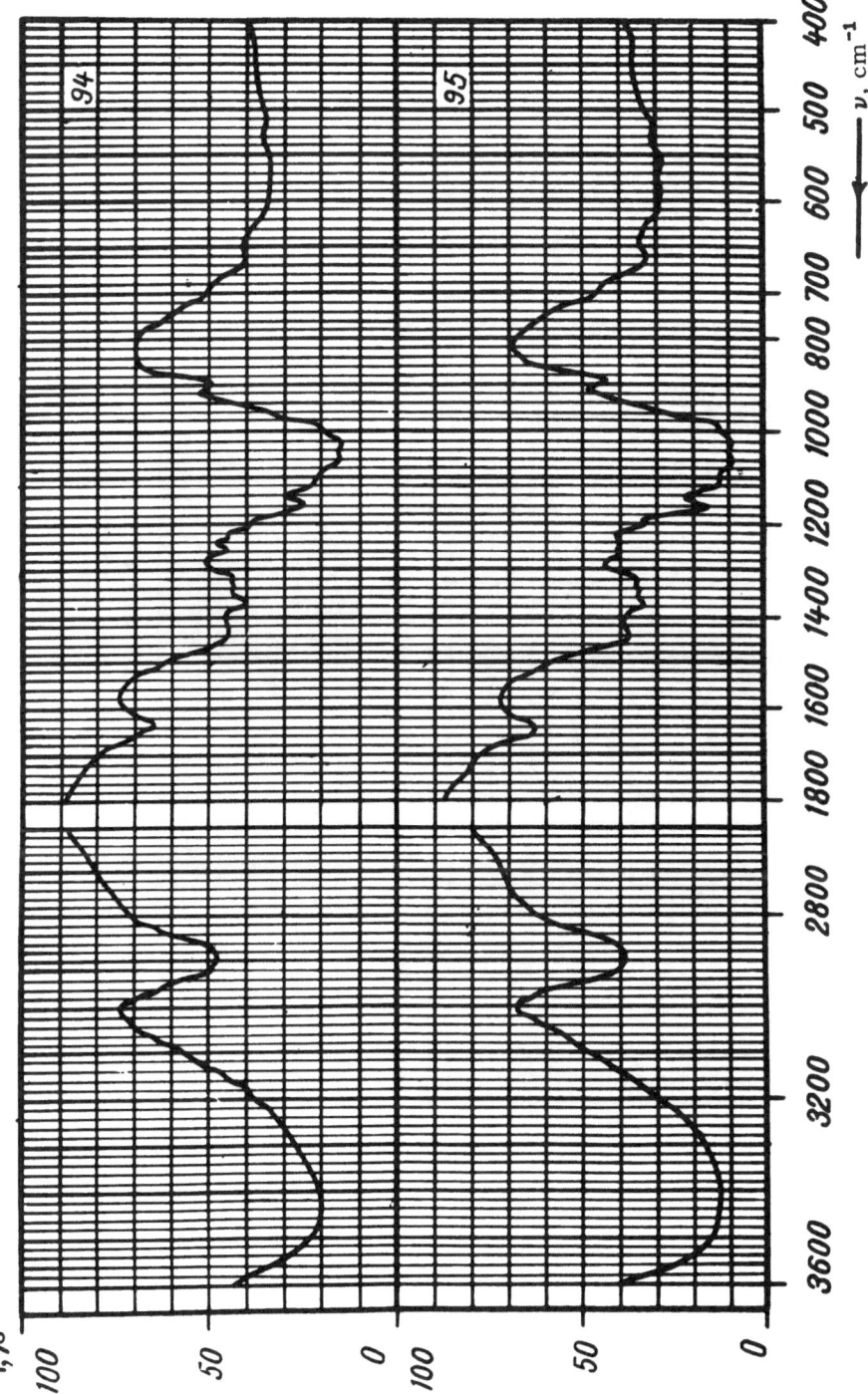

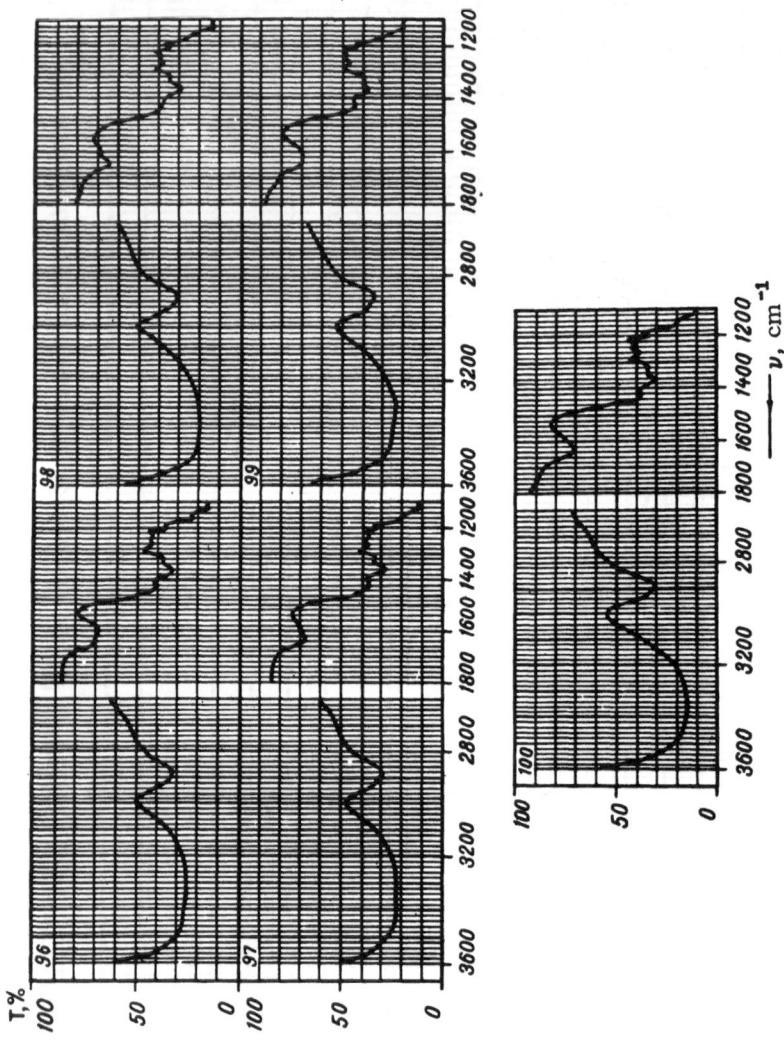

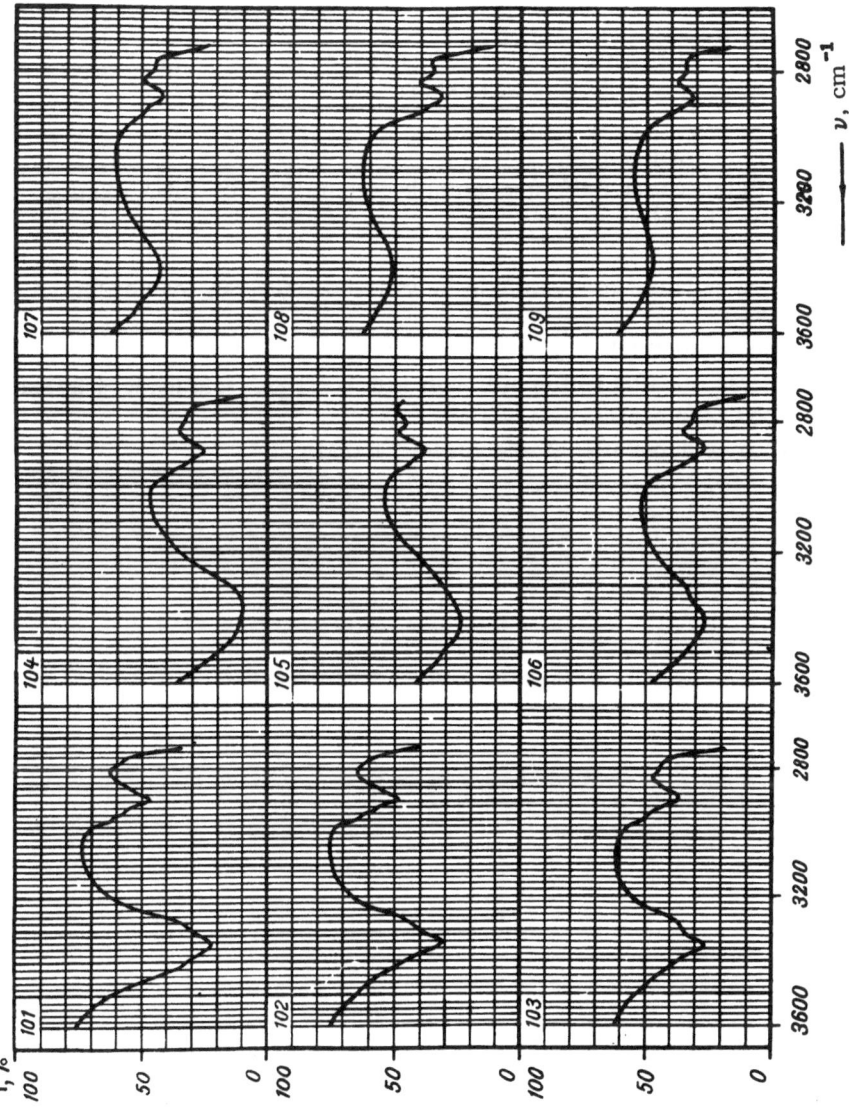

APPENDIX IV

SPECTRA OF PRODUCTS OF PARTIAL HYDROLYSIS
AND ETHANOLYSIS OF CELLULOSE

110. Cellulose cotton fluff.
111. Cellulose cotton fluff after partial hydrolysis with 10% H_2SO_4 solution
for 3 h (FM)
112. Wood cellulose (FM)
113. Wood cellulose after partial hydrolysis with 10% H_2SO_4 solution for 3 h
(FM)
114. Viscose silk (FM)
115. Viscose silk after partial hydrolysis with 10% H_2SO_4 solution for 3 h (FM)
116. Wood cellulose (FM)
117. Wood cellulose after ethanolysis for 8 h (FM)

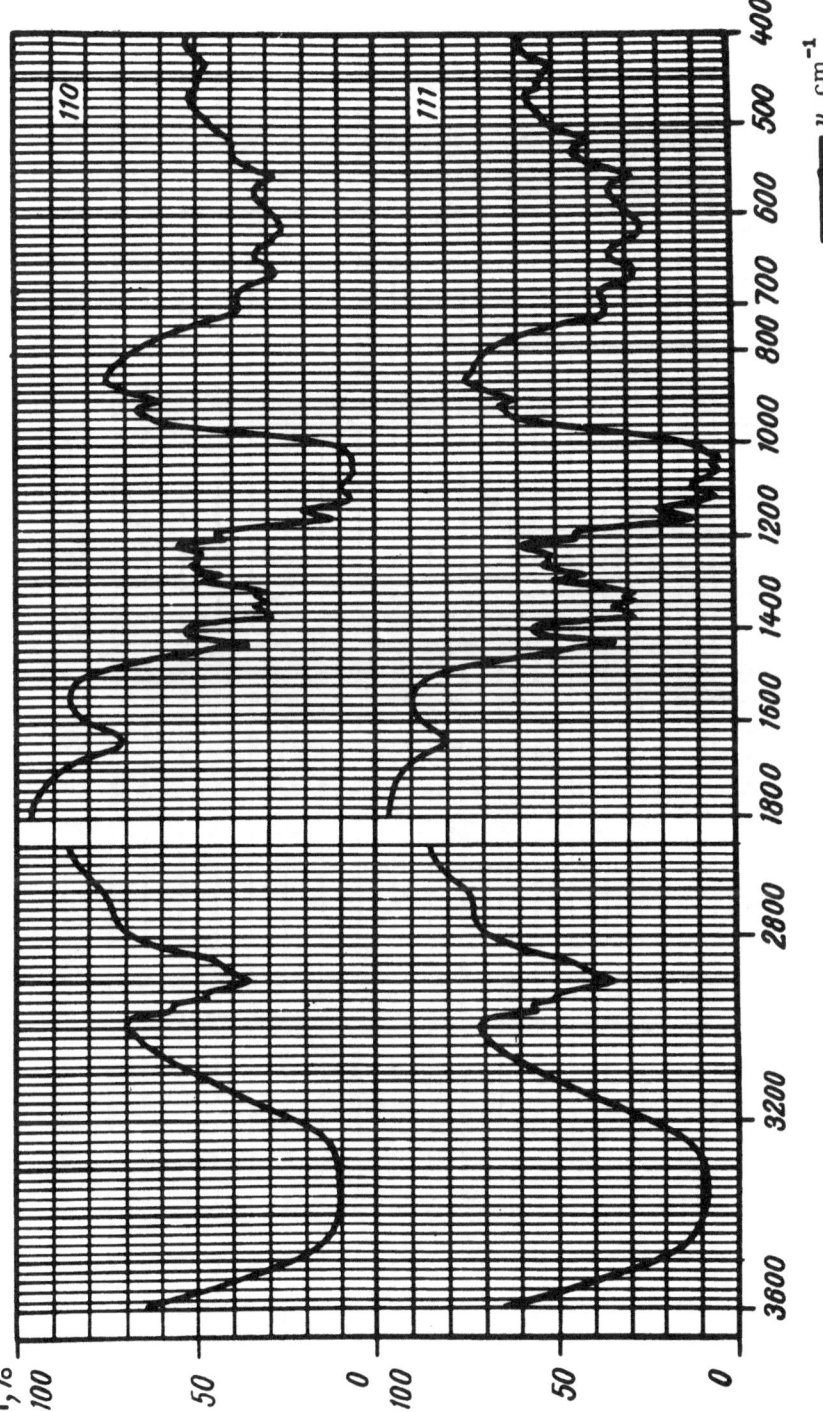

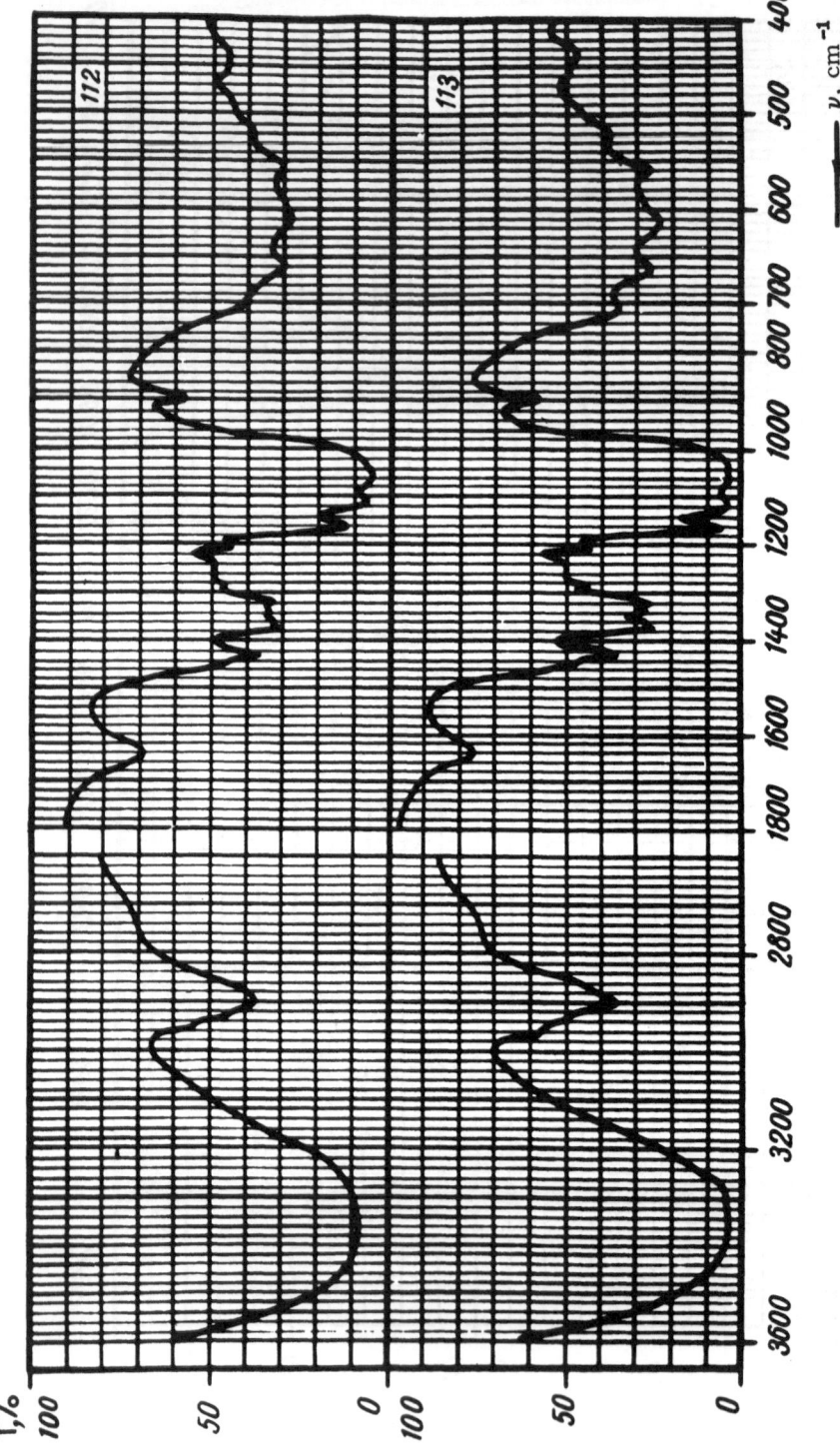

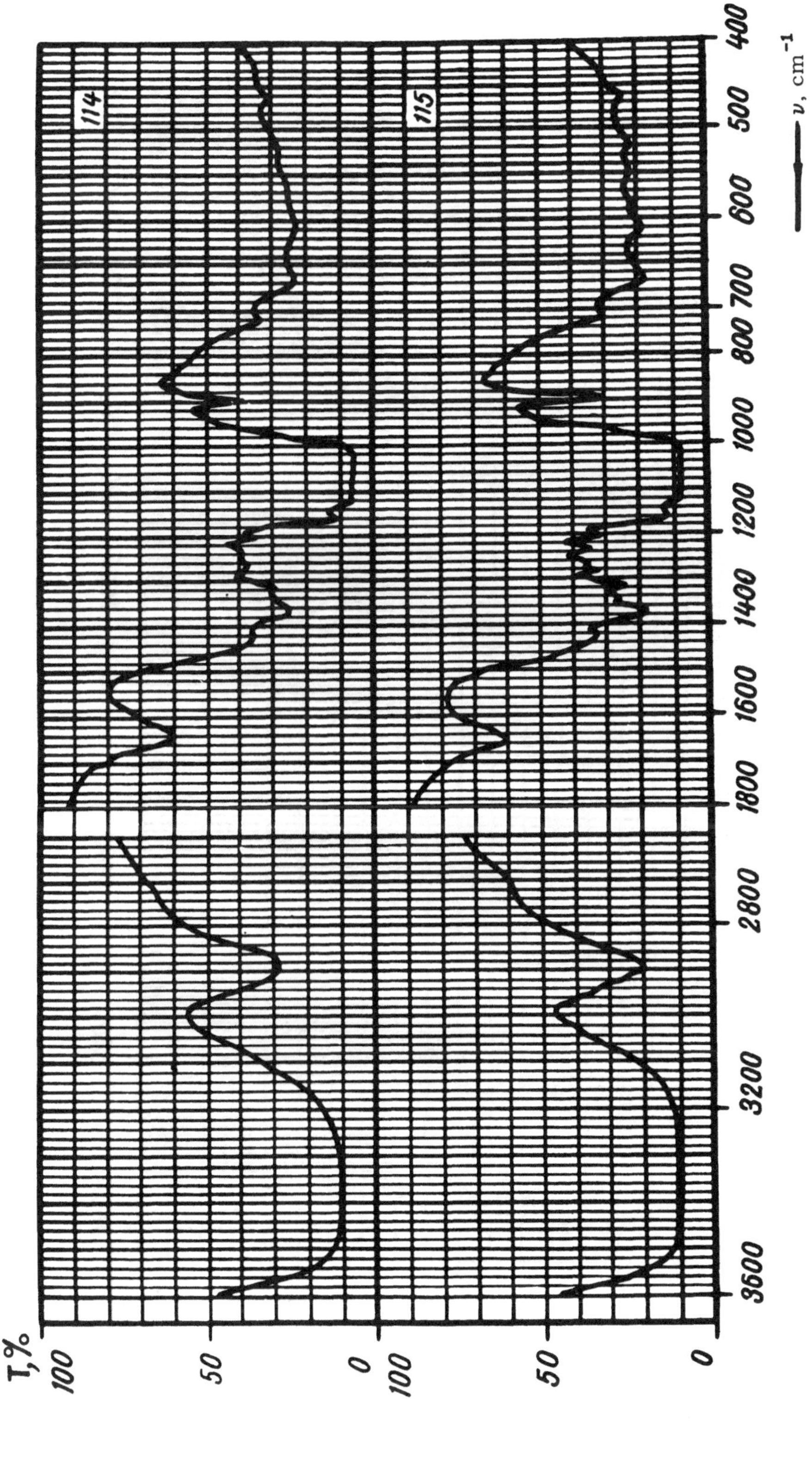
114
115
T,%
100
50
0
100
50
0
3600
3200
2800
1800
1600
1400
1200
1000
800
700
600
500
400
ν, cm^{-1}

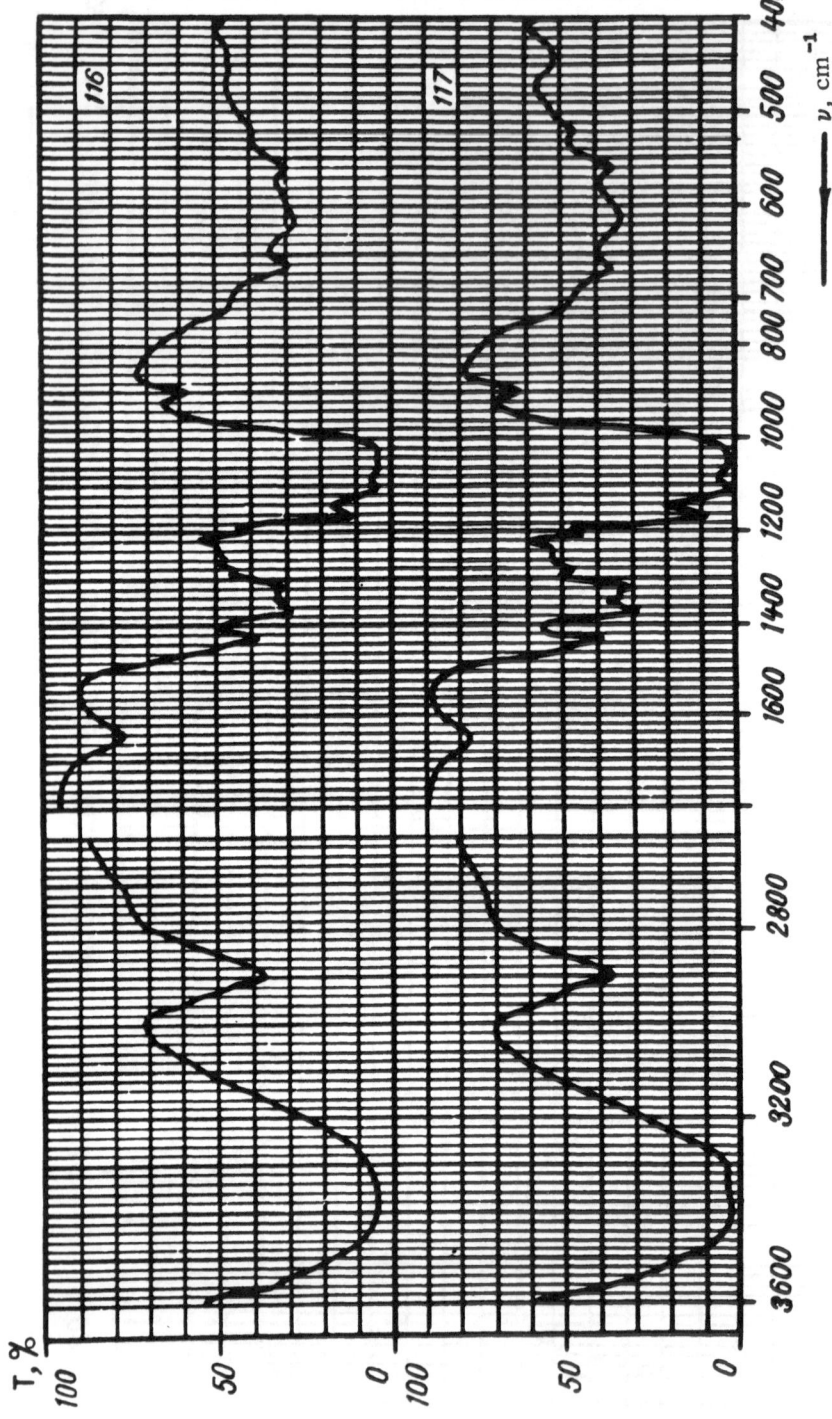

APPENDIX V

SPECTRA OF CELLULOSE ETHERS AND ESTERS

118. Methylcellulose (γ = 280) (KBr)
119. Sodium salt of carboxymethylcellulose (γ = 60) (KBr)
120. Ethylcellulose (γ = 230) (F)
121. Nonylcellulose (γ = 190) (F)
122. Hydroxyethylcellulose (substitution coefficient 0.521) (FM)
123. Hydroxyethylcellulose (substitution coefficient 1.384) (KBr)
124. Tertiary-butyl ether of cellulose (γ = 30) (F)
125. Isopropyl ether of cellulose (γ = 30) (F)
126. Benzylcellulose (γ = 200) (KBr)
127. Trityl ether of cellulose (γ = 101) (F)
128. Ditosylcellulose (13.4% S) (KBr)
129. Tosyliodocellulose (5.13% S, 26.23% I) (KBr)
130. Cyanoethylcellulose (γ = 250) (F)
131. Cellulose formate (γ = 65) (FM)
132. Cellulose formate (γ = 69) (FM)
133. Cellulose acetate (γ = 300) (FM)
134. Cellulose acetobutyrate (γ_{total} = 300) (FM)
135. Cellulose acetate, produced under heterogeneous conditions of acetylation and partial saponification (γ = 280) (F)
136. Cellulose acetate, produced under homogeneous conditions of acetylation and partial saponification (γ = 280) (F)
137. Cellulose acetoenanthate (γ_{total} = 300, γ_{ac} = 240) (FP)
138. Cellulose acetopropionate (γ_{total} = 300, γ_{ac} = 240) (FP)
139. Cellulose stearate (γ = 300) (F)
140. Cellulose ester of 2-dimethylpropionic acid (γ = 210) (F)
141. Cellulose nitrate (γ = 200) (FM)
142. Cellulose nitrate (γ = 250) (FM)

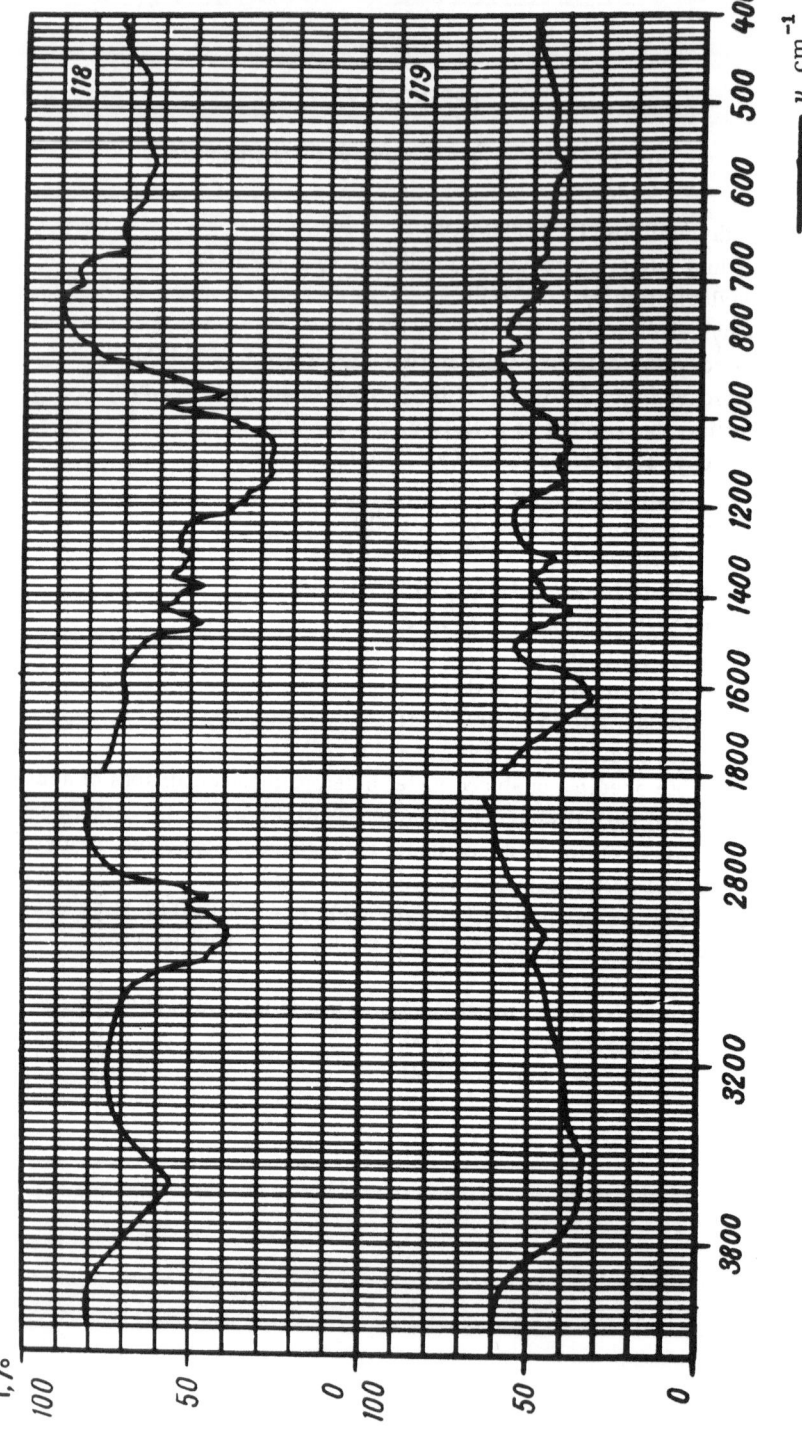

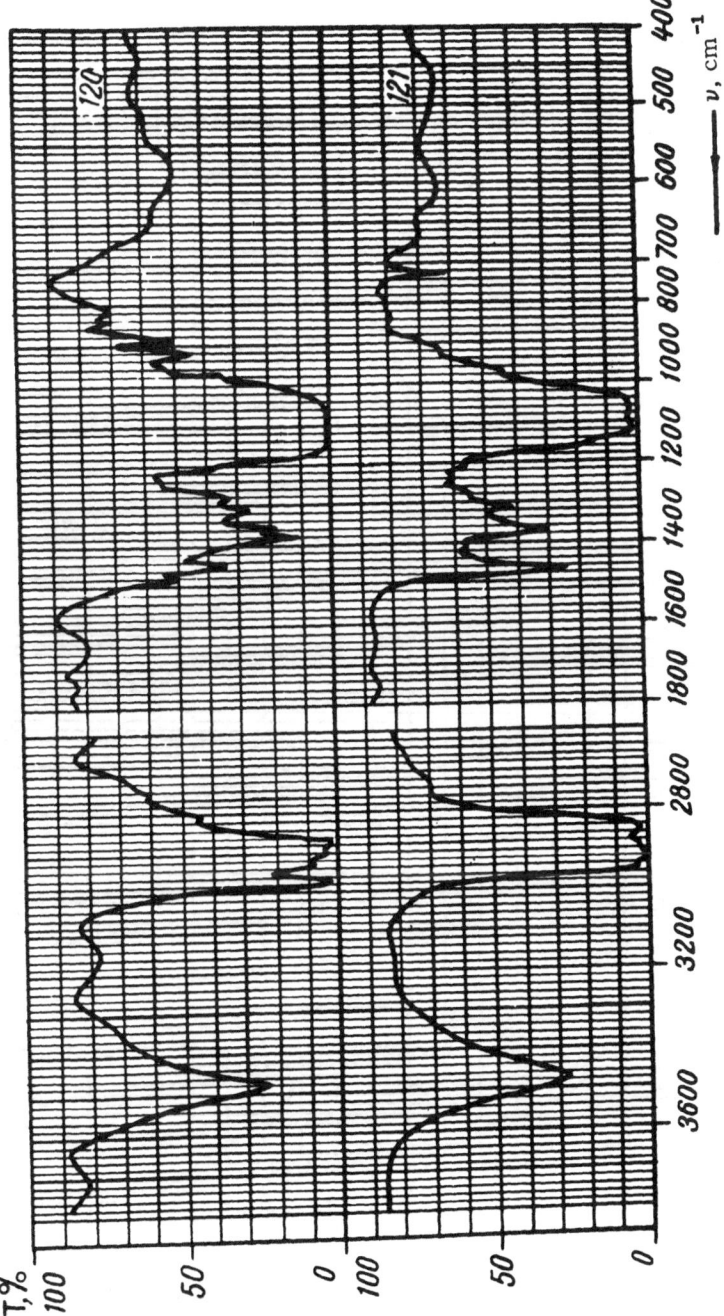

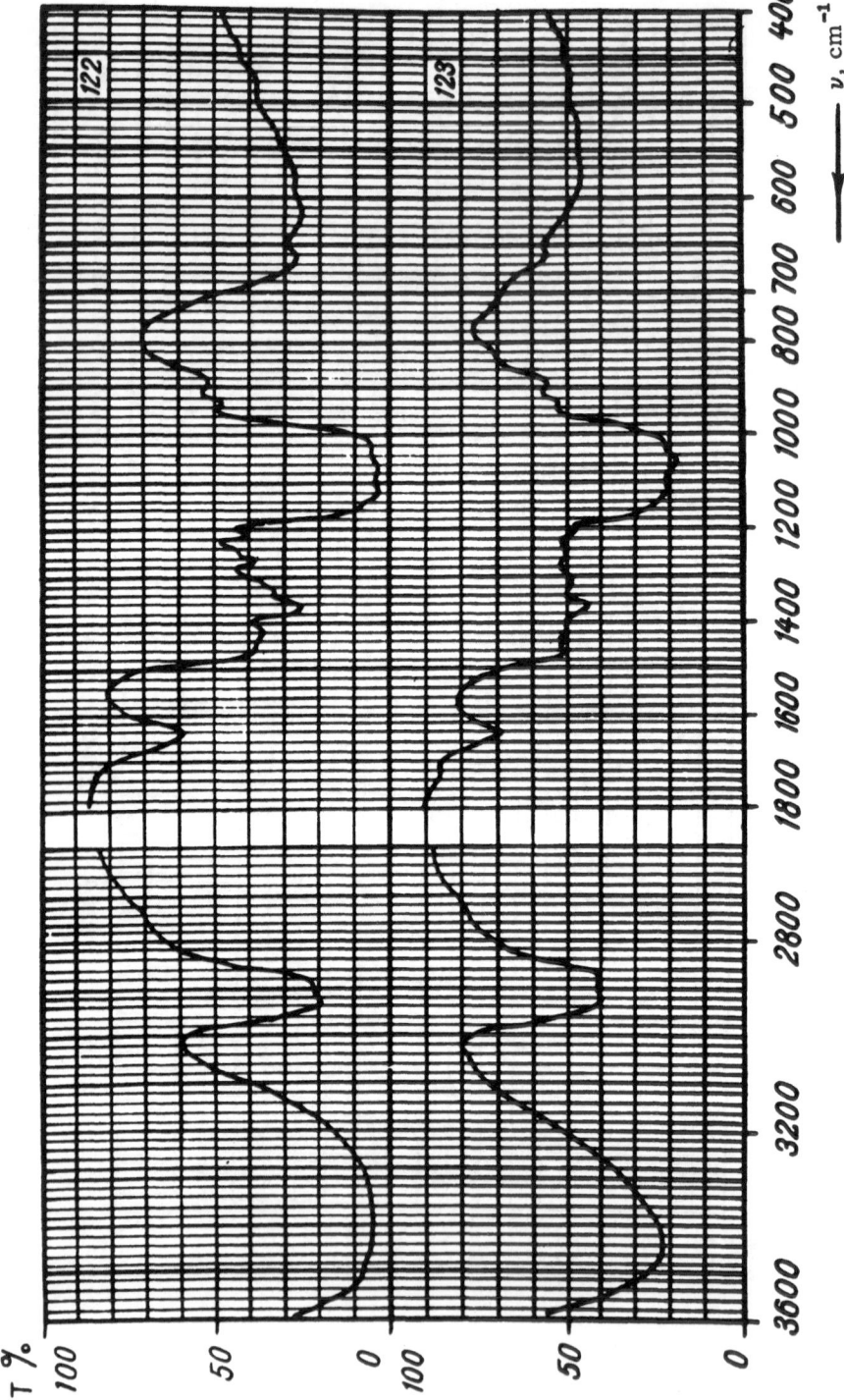

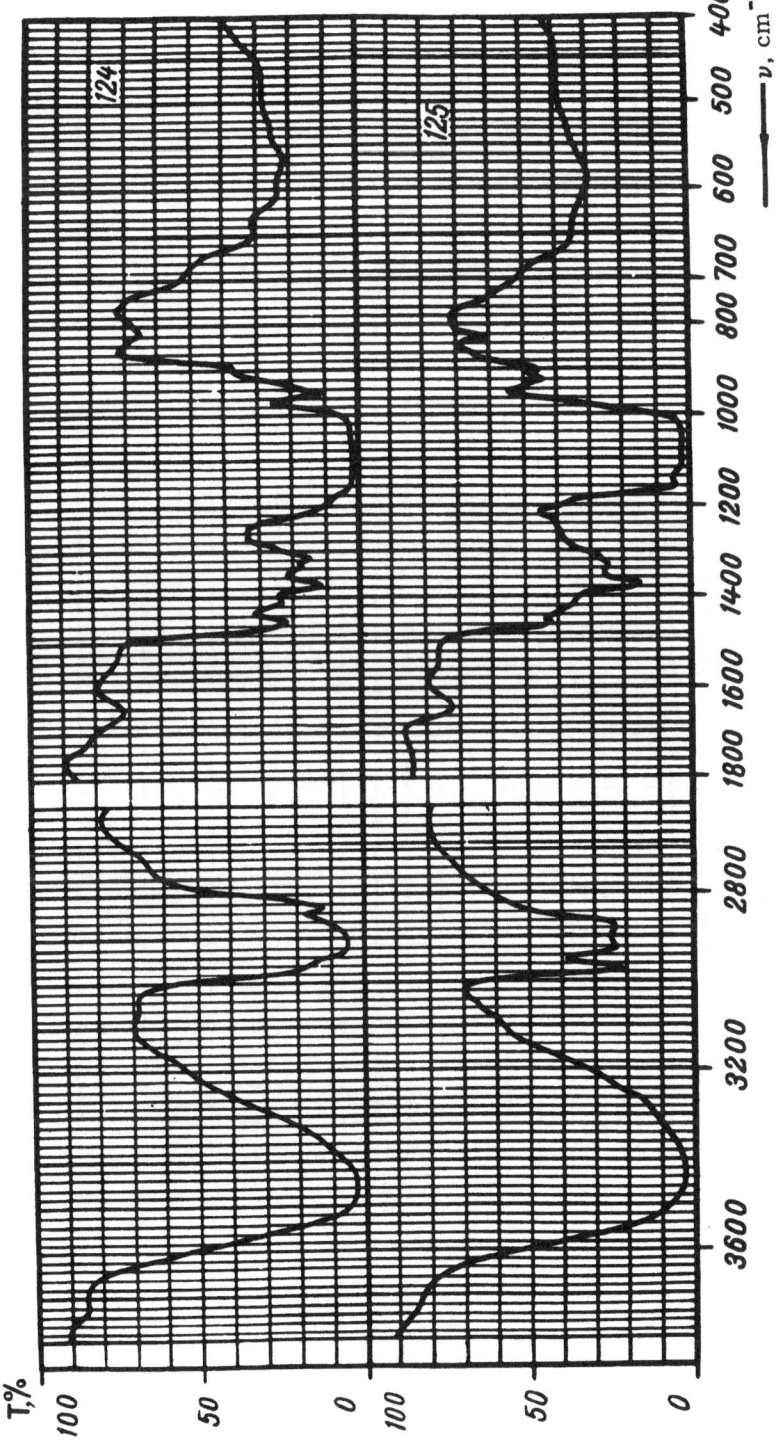

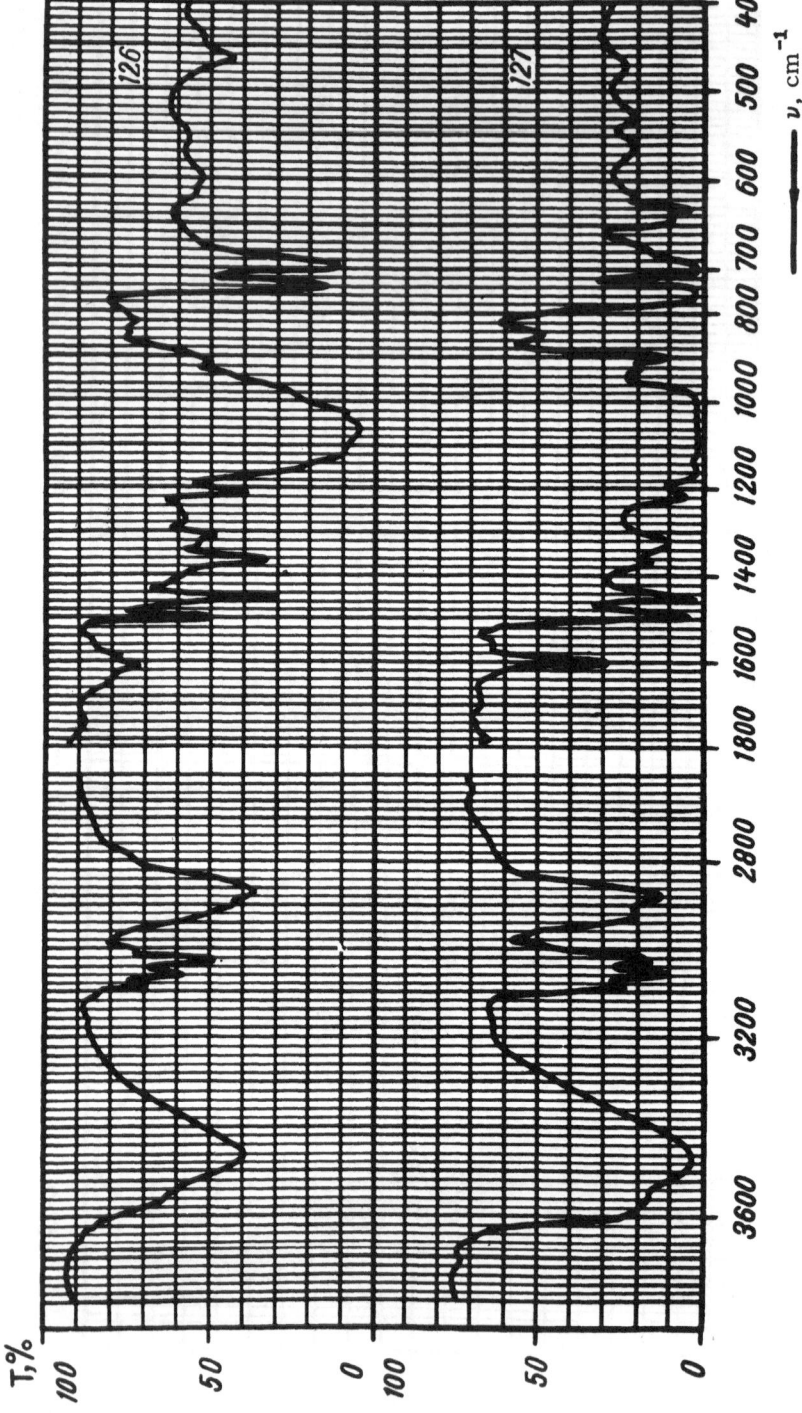

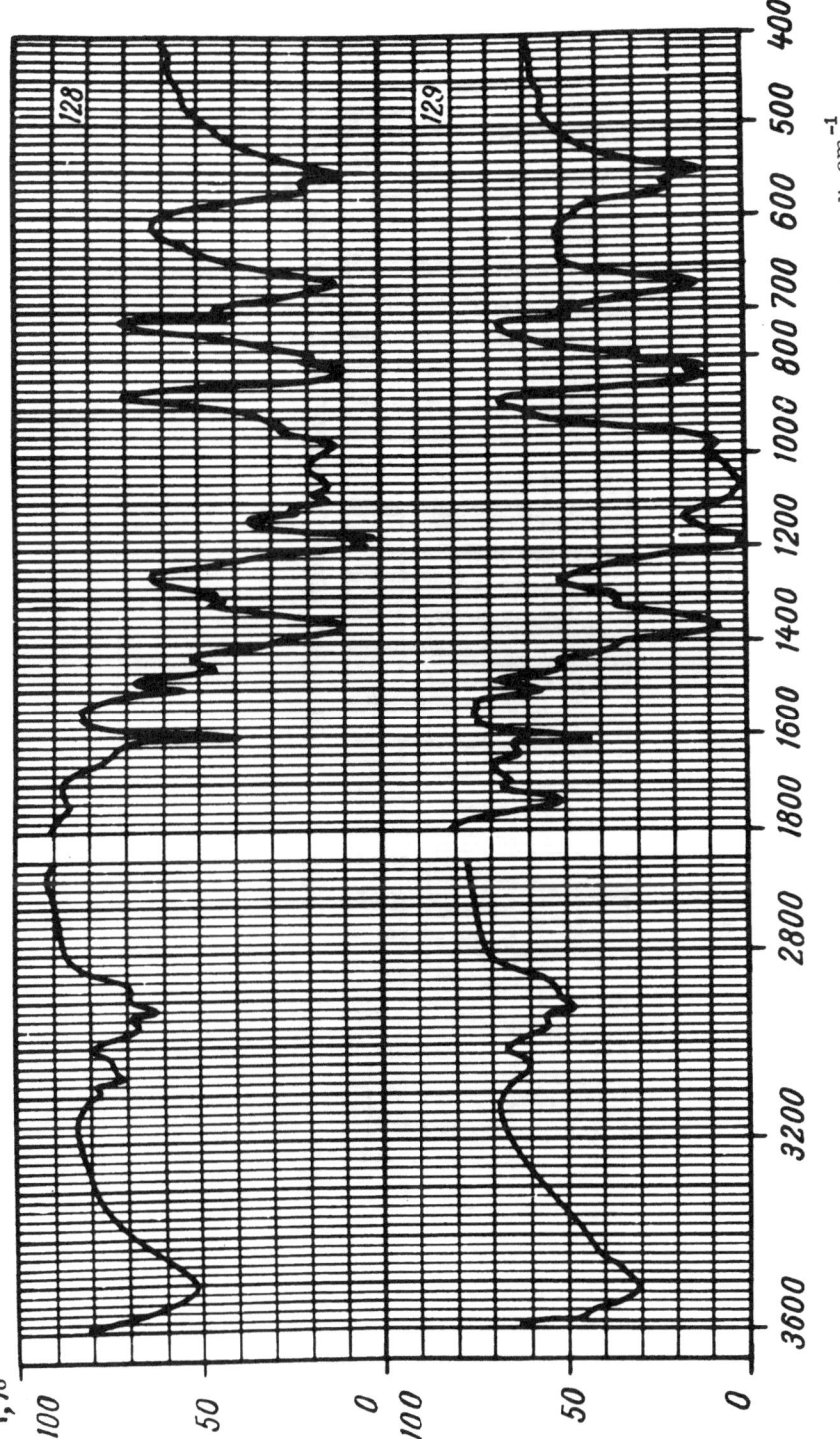

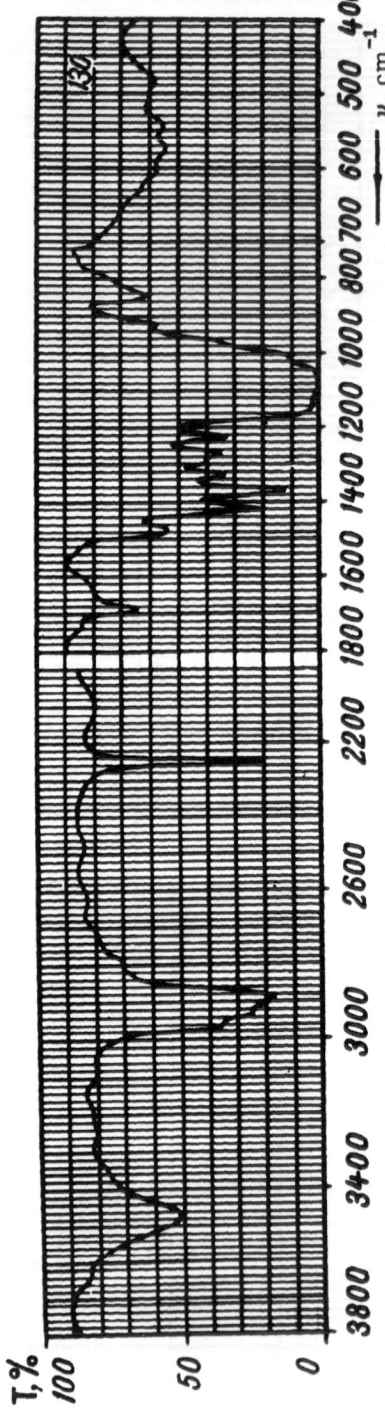

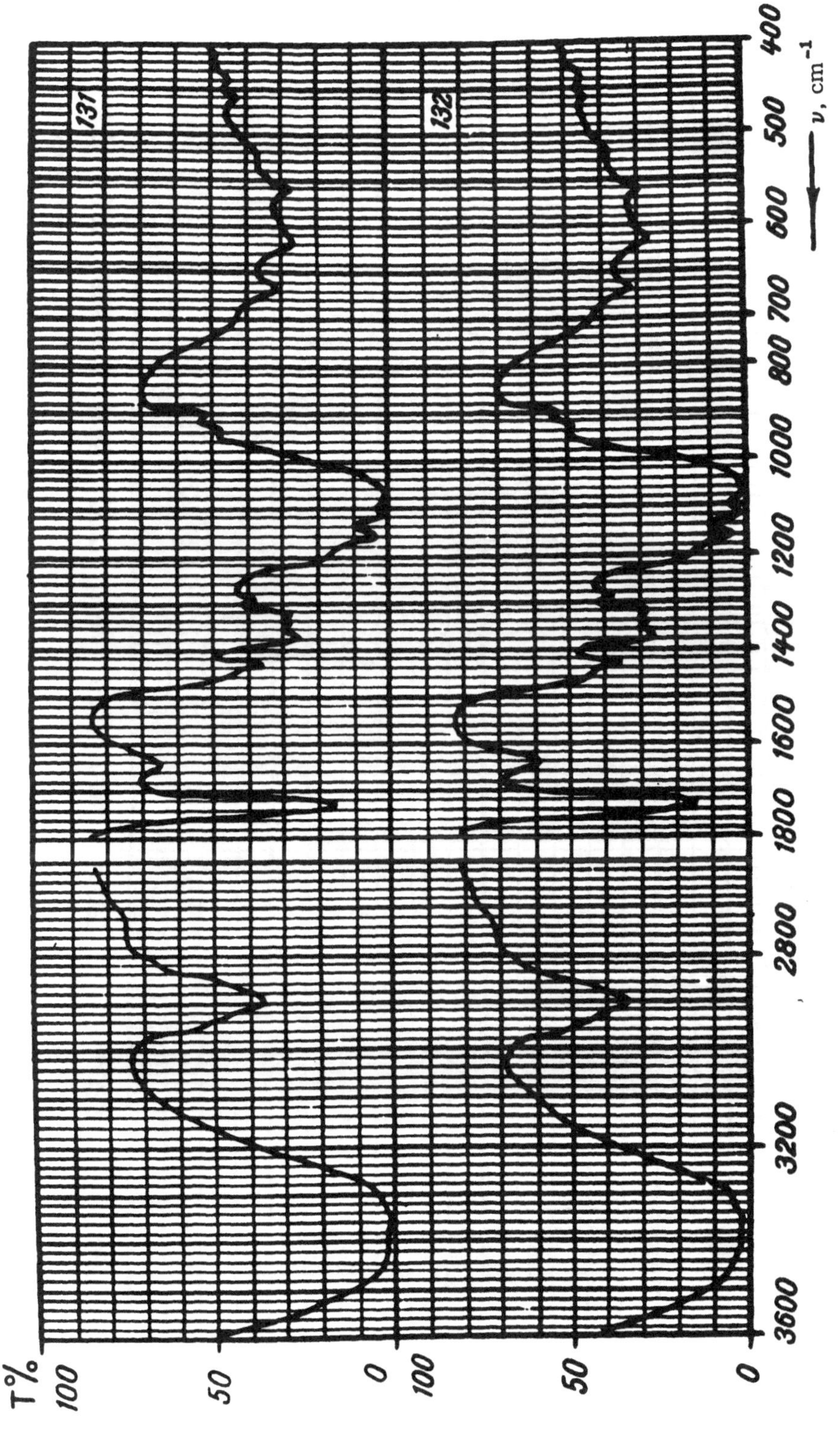
131
132
400
500
600
700
800
1000
1200
1400
1600
1800
2800
3200
3600
ν, cm^{-1}
T%
100
50
0
100
50
0

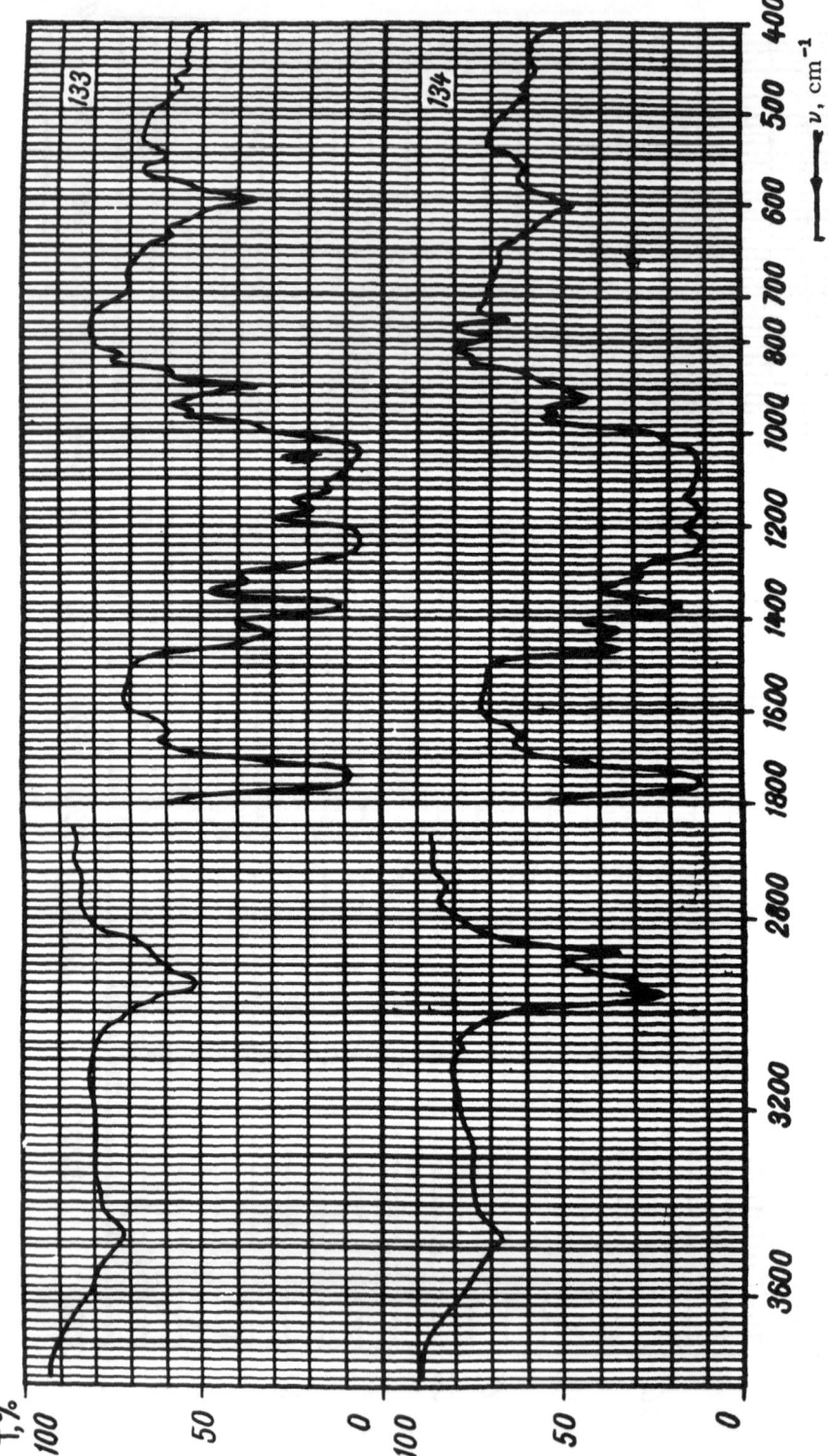

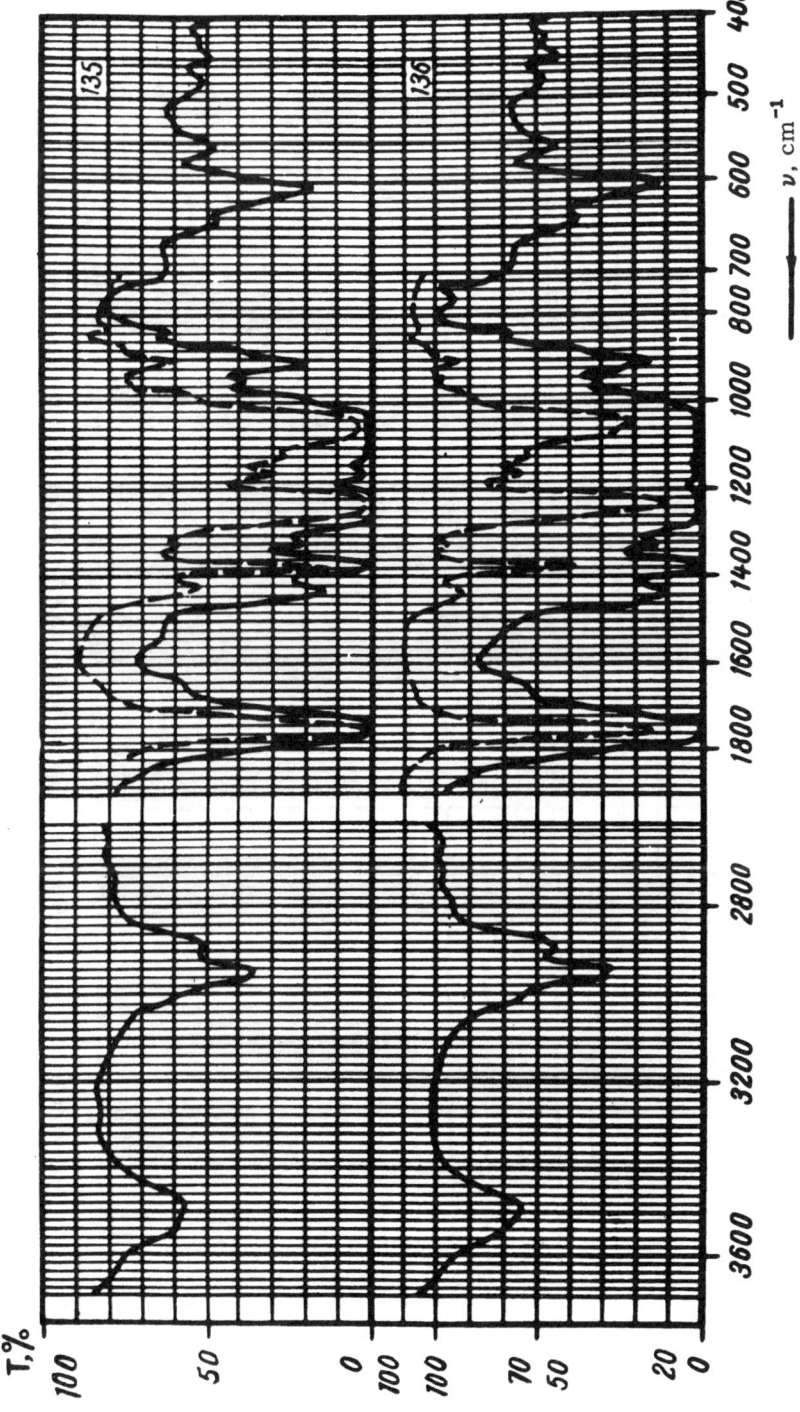

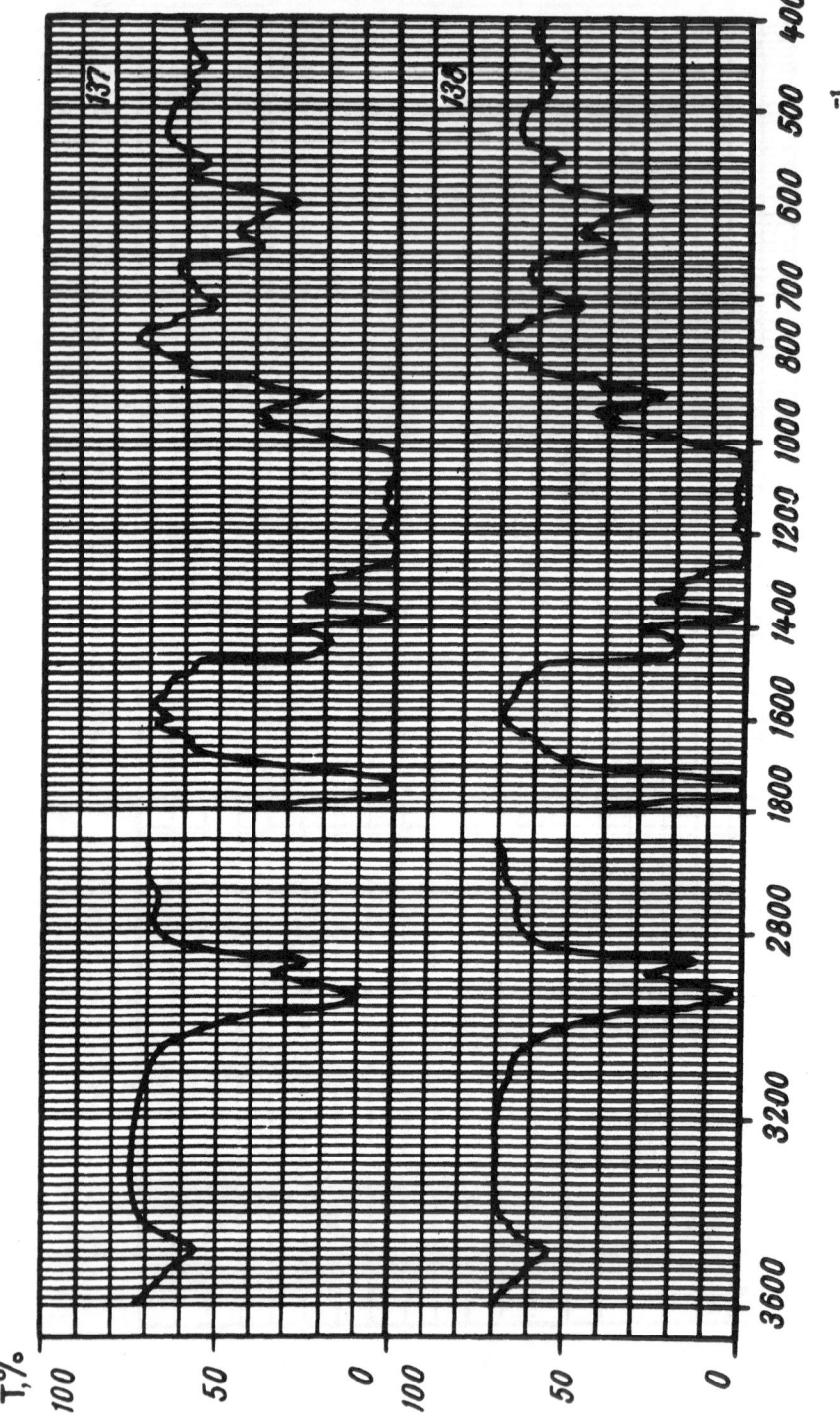

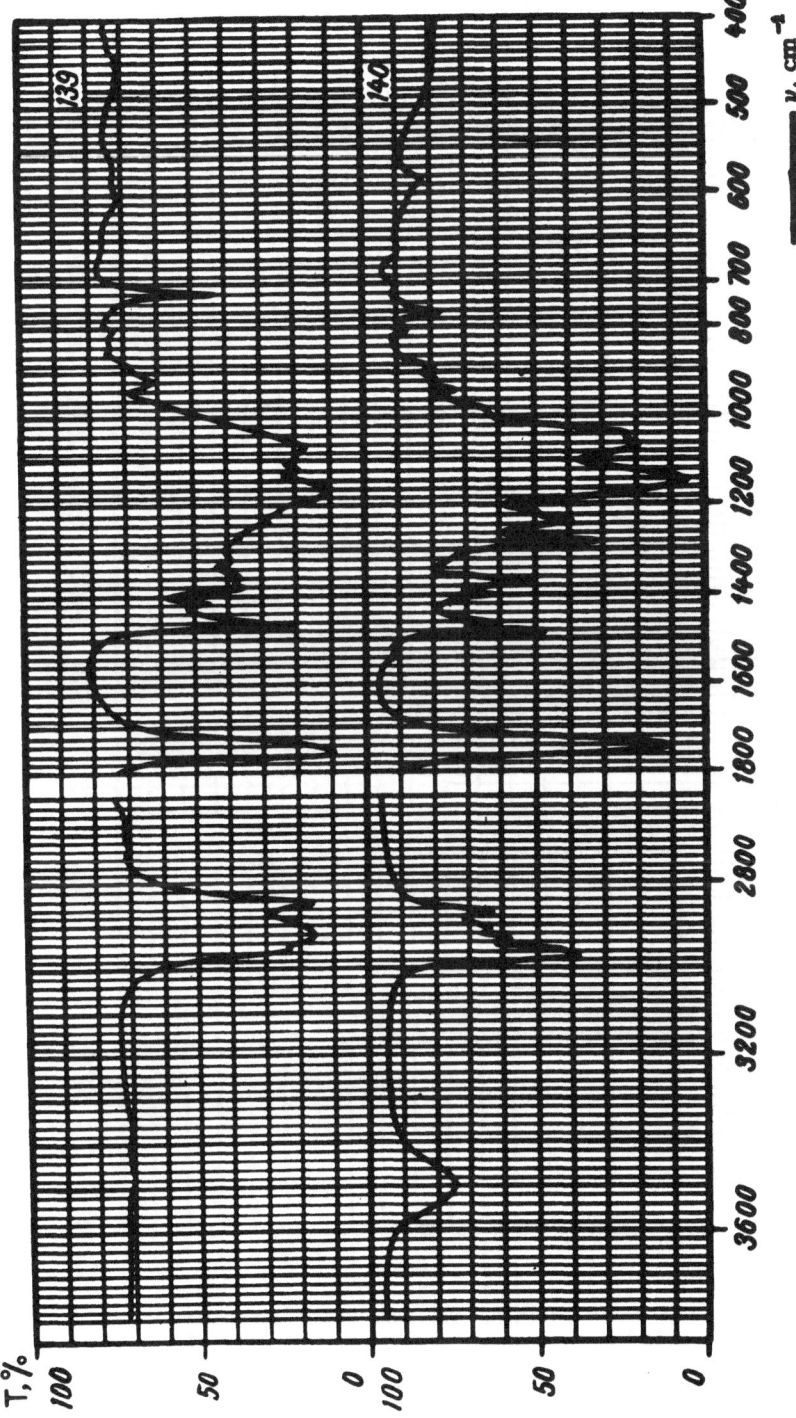

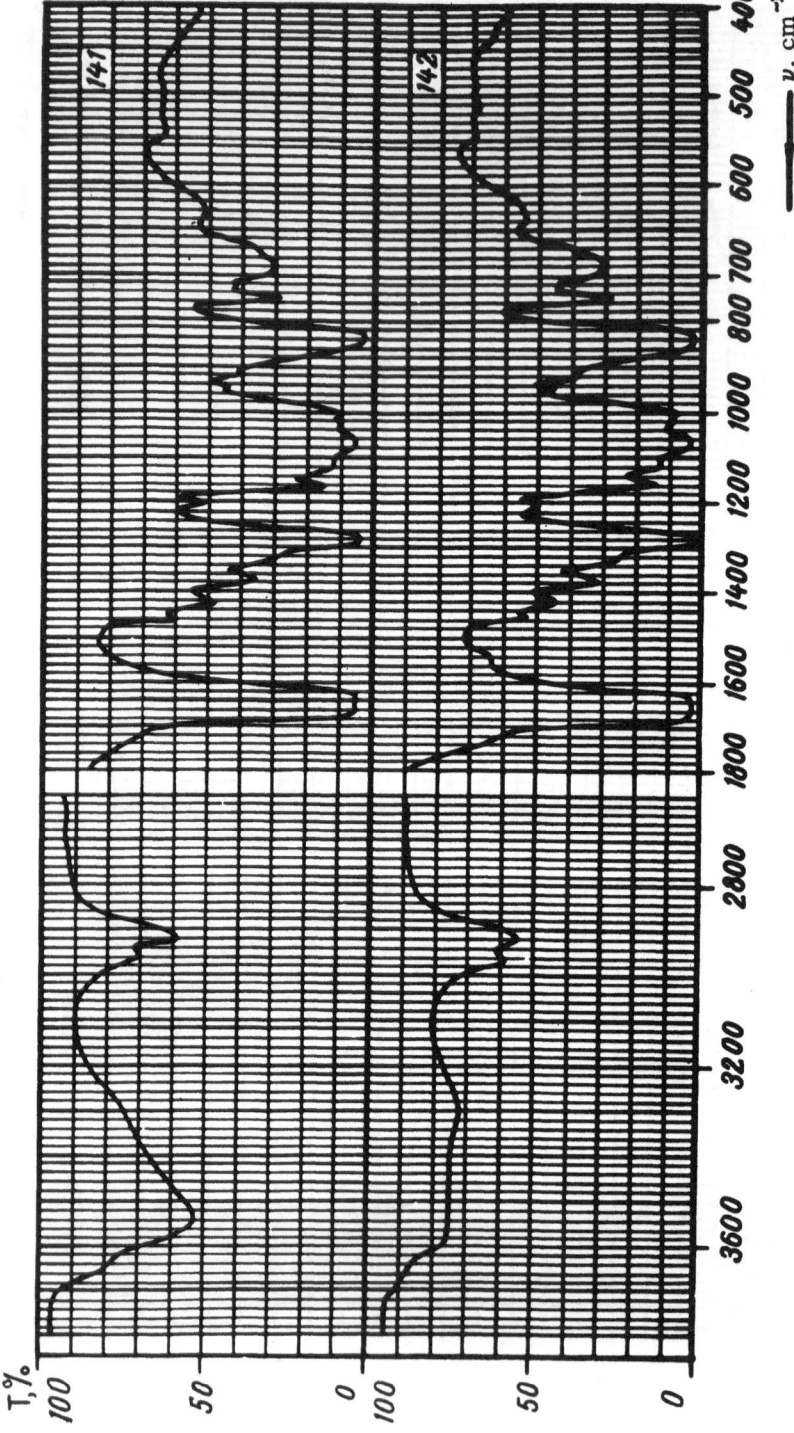

APPENDIX VI

SPECTRA OF OXIDATION PRODUCTS OF CELLULOSE

143. Cotton cellulose, oxidized by nitrogen oxides to monocarboxycellulose (6% COOH, 0.3% N) (FM)
144. Cotton cellulose, oxidized by nitrogen oxides (12.3% COOH, 1.08% N) (FM)
145. Staple fiber (FM)
146. Staple fiber, oxidized by nitrogen oxides (20% COOH) (FM)
147. Cellulose cotton fluff (FM)
148. Cellulose cotton fluff, oxidized by iodic acid to dialdehydocellulose (23.1% CHO) (FM)
149. Cotton cellulose, oxidized successively with iodic acid and sodium chlorite to dicarboxycellulose (0.4% COOH) (FM)
150. Cotton cellulose, oxidized successively with iodic acid and sodium chlorite to dicarboxycellulose (2% COOH) (FM)
151. Cotton cellulose, oxidized successively with iodic acid (7.2% CHO) and nitrogen oxides (23.2% COOH) (KBr)
152. Cotton cellulose, oxidized successively with iodic acid (34% CHO) and nitrogen oxides (42.4% COOH) (KBr)
153. Cotton cellulose, oxidized with iodic acid (7.62% CHO) and then reduced with sodium borohydride to cellulose dialcohol (FM)
154. Cotton cellulose, oxidized with iodic acid (25.8% CHO) and then reduced with sodium borohydride to cellulose dialcohol (FM)
155. Cotton cellulose, oxidized with chromic acid (4.92% CO, 2.5% COOH) (FM)
156. Cotton cellulose, oxidized with nitric acid (1.93% CO, 1.06% COOH) (KBr)
157. Ethylcellulose, oxidized by atmospheric oxygen at 130°C for 1 h, with a rate of passage of oxygen of 3 liters/h (F)
158. Ethylcellulose, unoxidized (F)
159. Staple fiber after irradiation with a dose of $0.28 \cdot 10^7$ roentgens (FM)
160. Staple fiber after irradiation with a dose of $2.8 \cdot 10^7$ roentgens (FM)

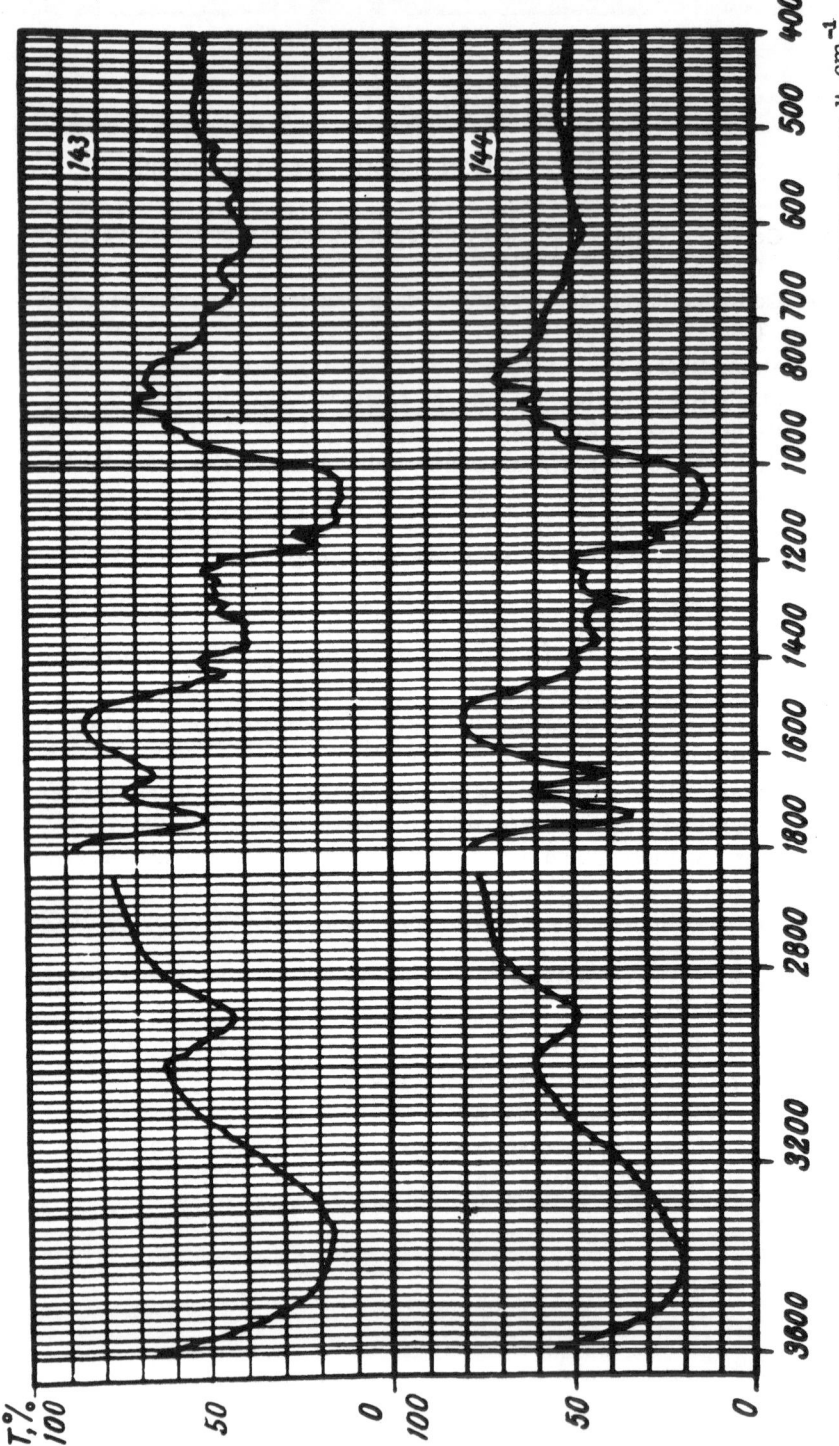

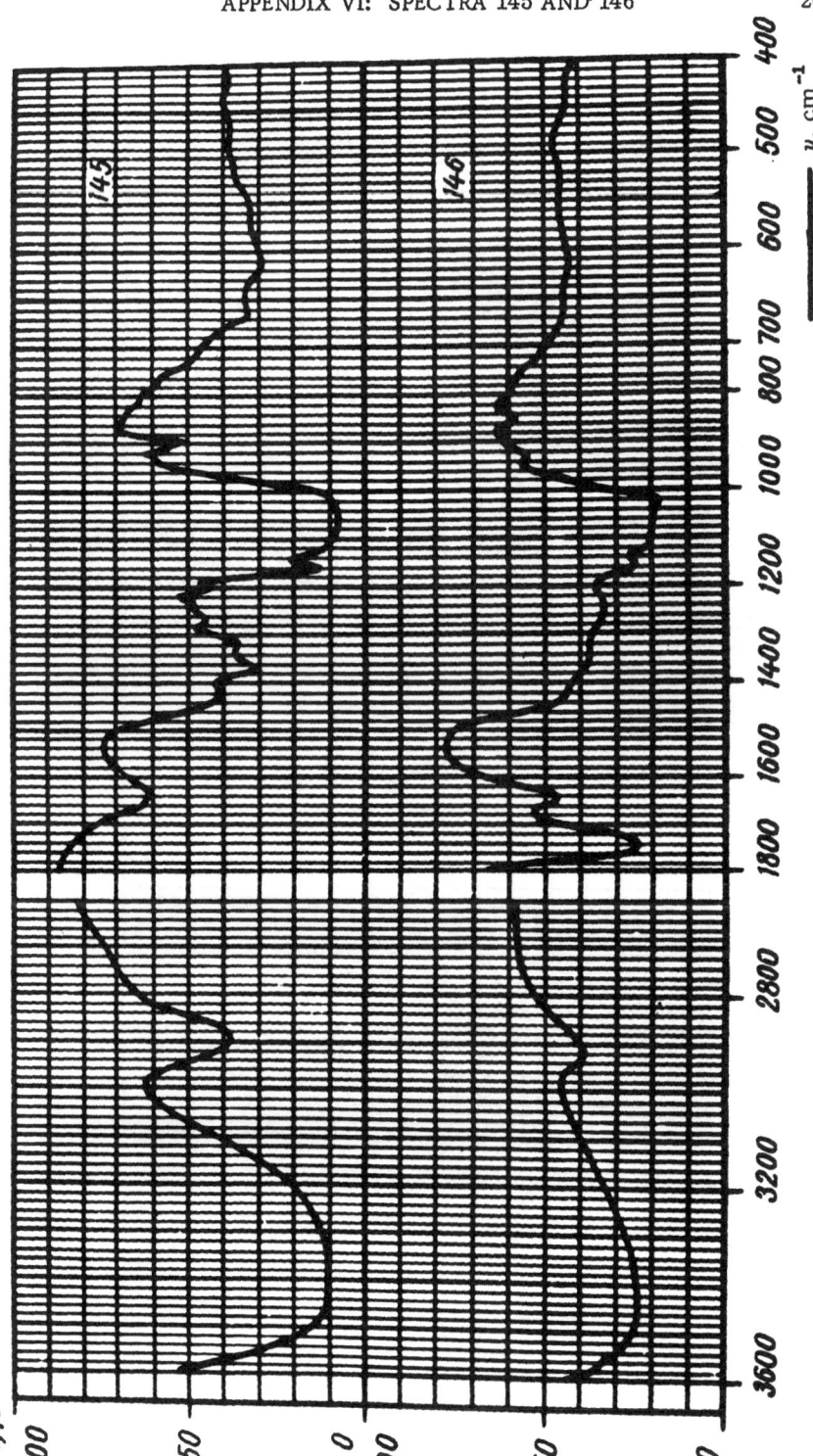

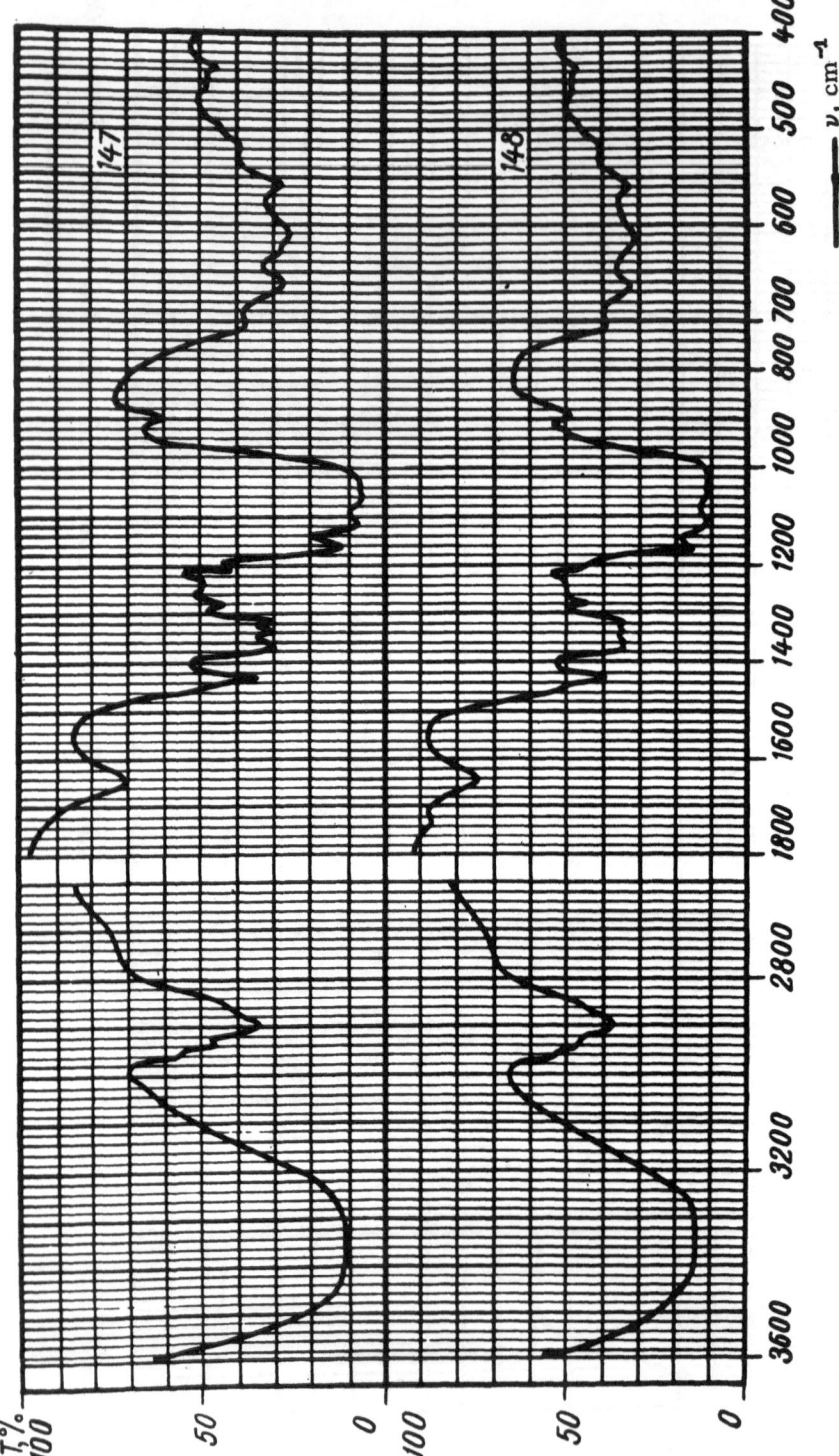

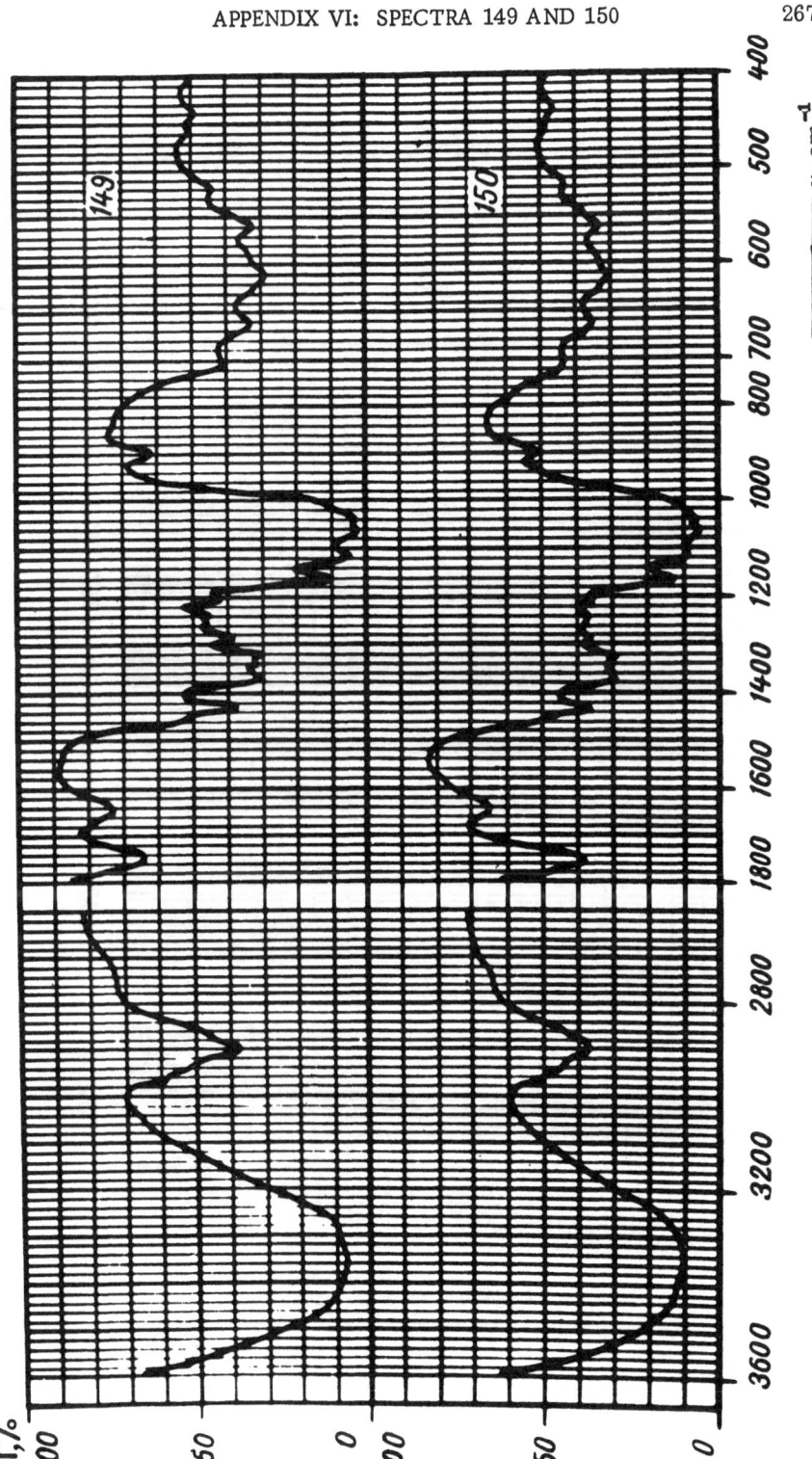

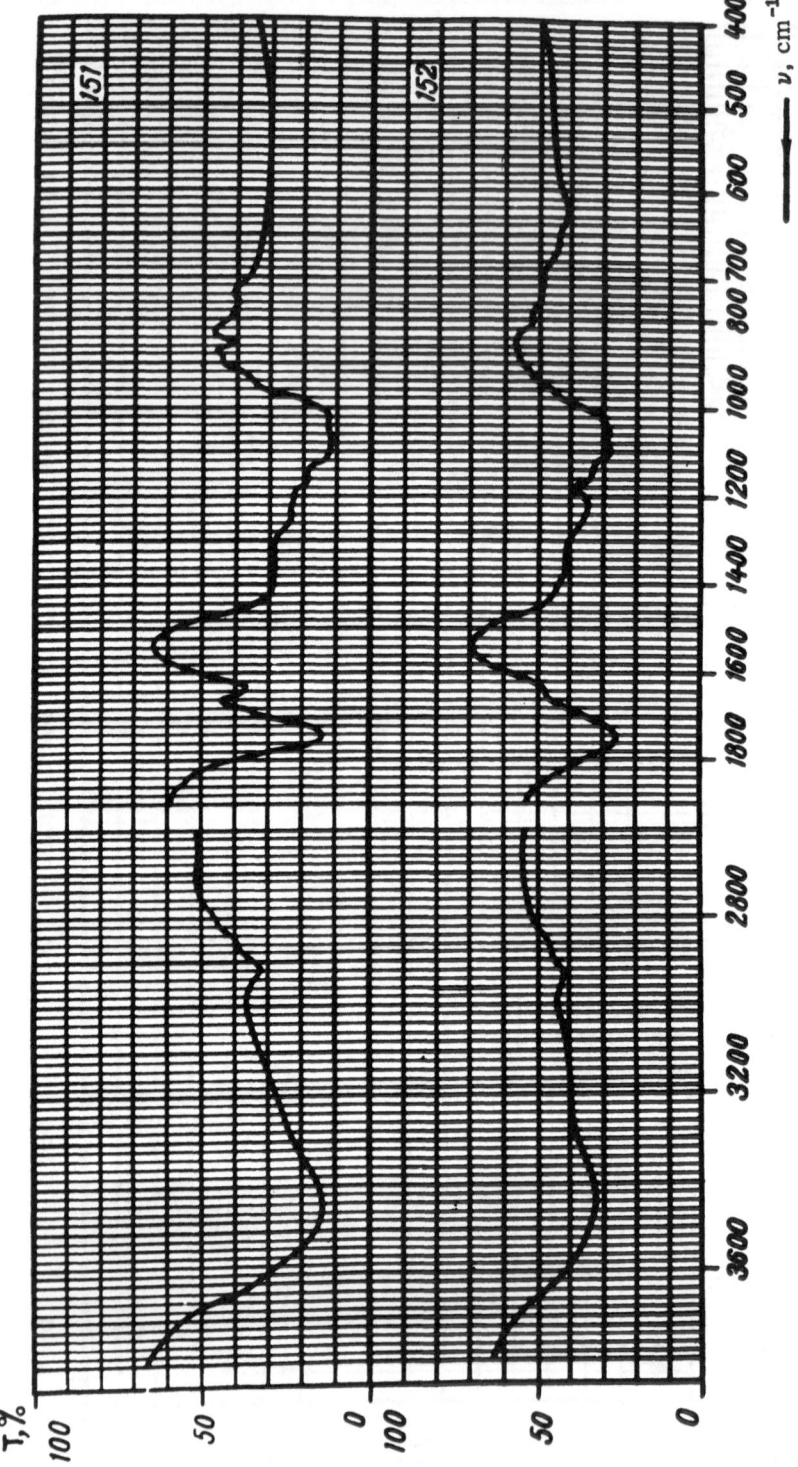

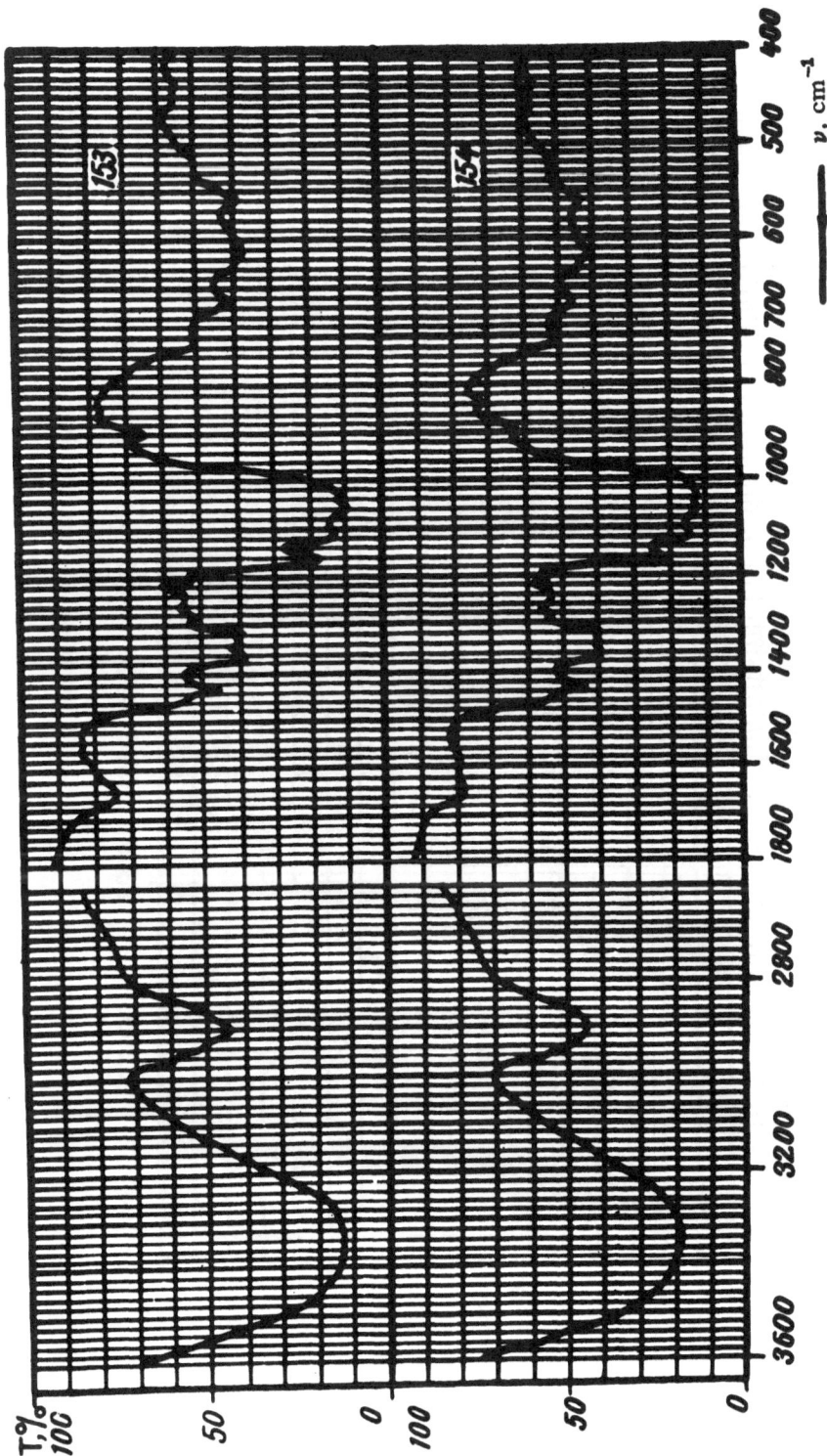

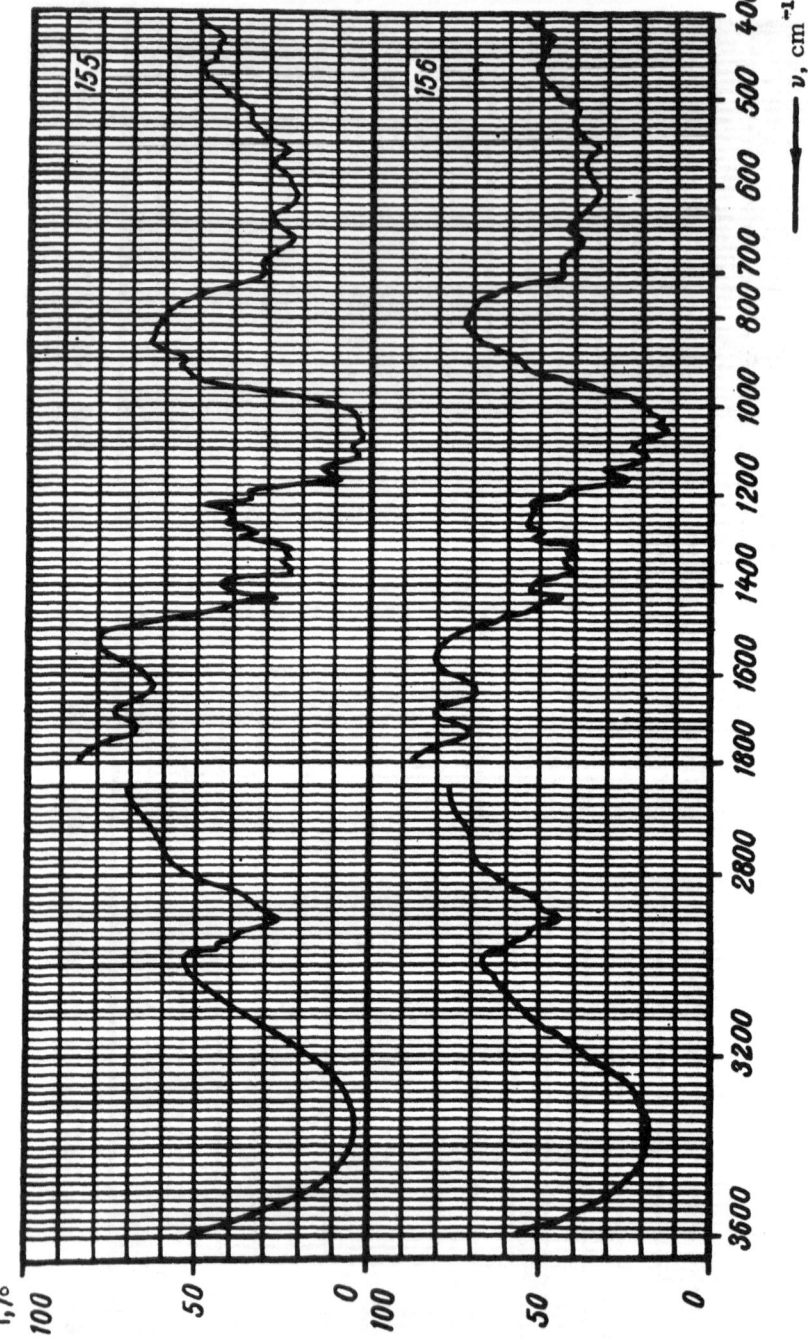

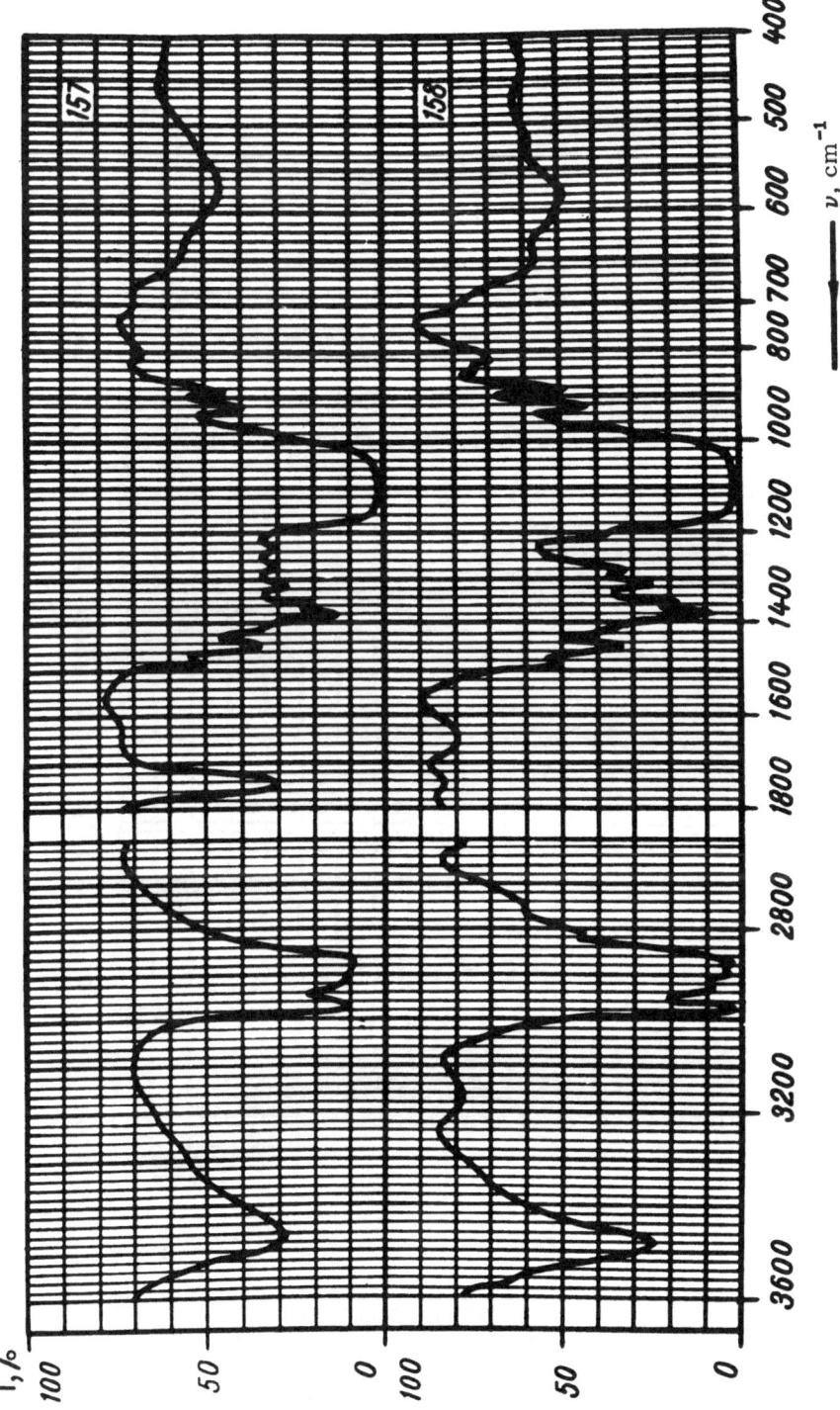

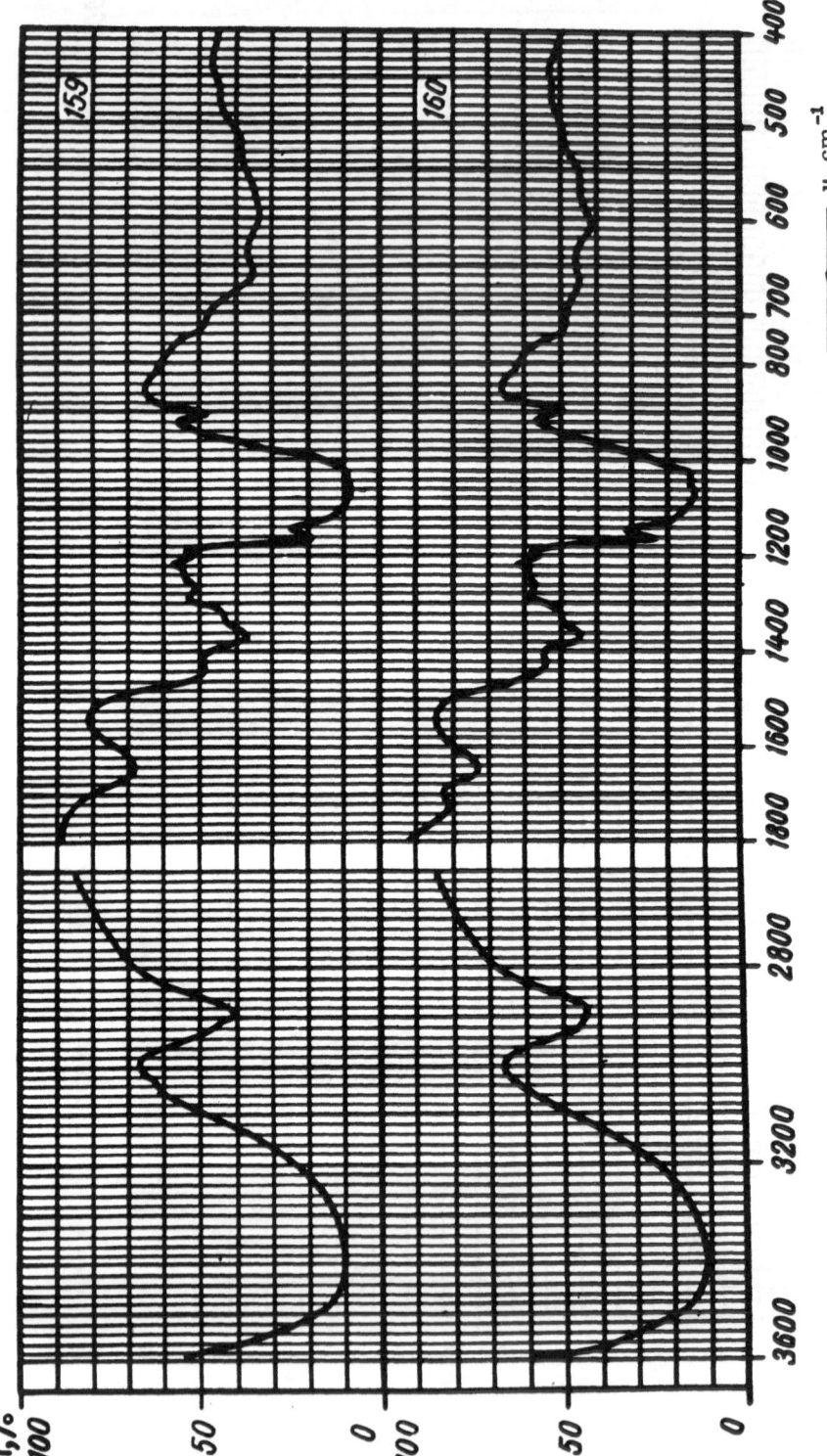

APPENDIX VII

SPECTRA OF SALTS OF CELLULOSE OXIDATION PRODUCTS

161. Cotton cellulose, oxidized with nitrogen oxides (9.6% COOH, 0.4% N) (FM)

162. Cotton cellulose, oxidized with nitrogen oxides (9.6% COOH, 0.4% N) and then treated with an 0.05 M solution of aluminum acetate in water (FM)

163. Cotton cellulose, oxidized with nitrogen oxides (9.6% COOH, 0.4% N) (FM)

164. Cotton cellulose, oxidized with nitrogen oxides (9.6% COOH, 0.4% N) and then treated with an 0.05 M solution of barium acetate in water (FM)

165. Cotton cellulose, oxidized with nitrogen oxides (9.6% COOH, 0.4% N) and then treated with an 0.05 M solution of magnesium acetate in water (FM)

166. Cotton cellulose, oxidized with nitrogen oxides (9.6% COOH, 0.4% N) and then treated with an 0.05 M solution of lead acetate in water (FM)

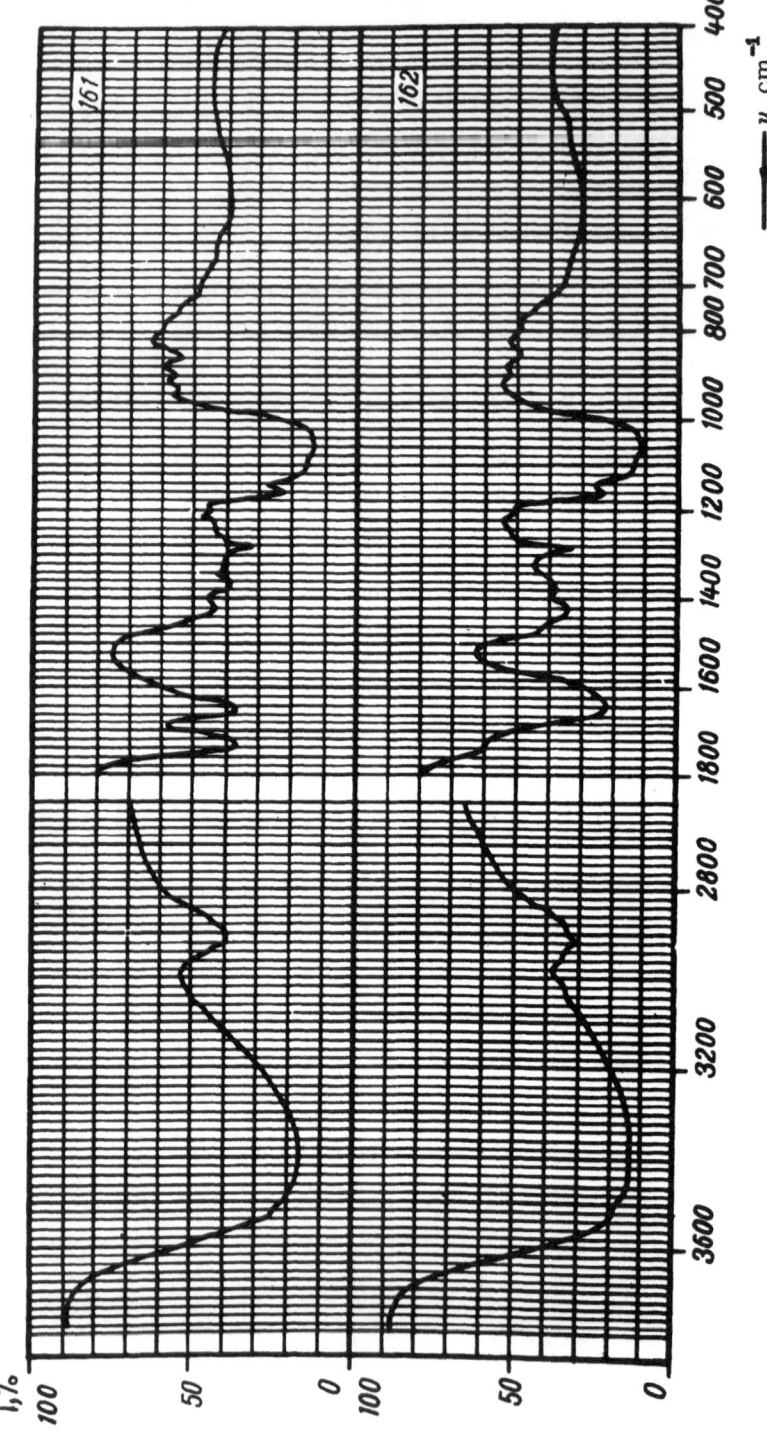

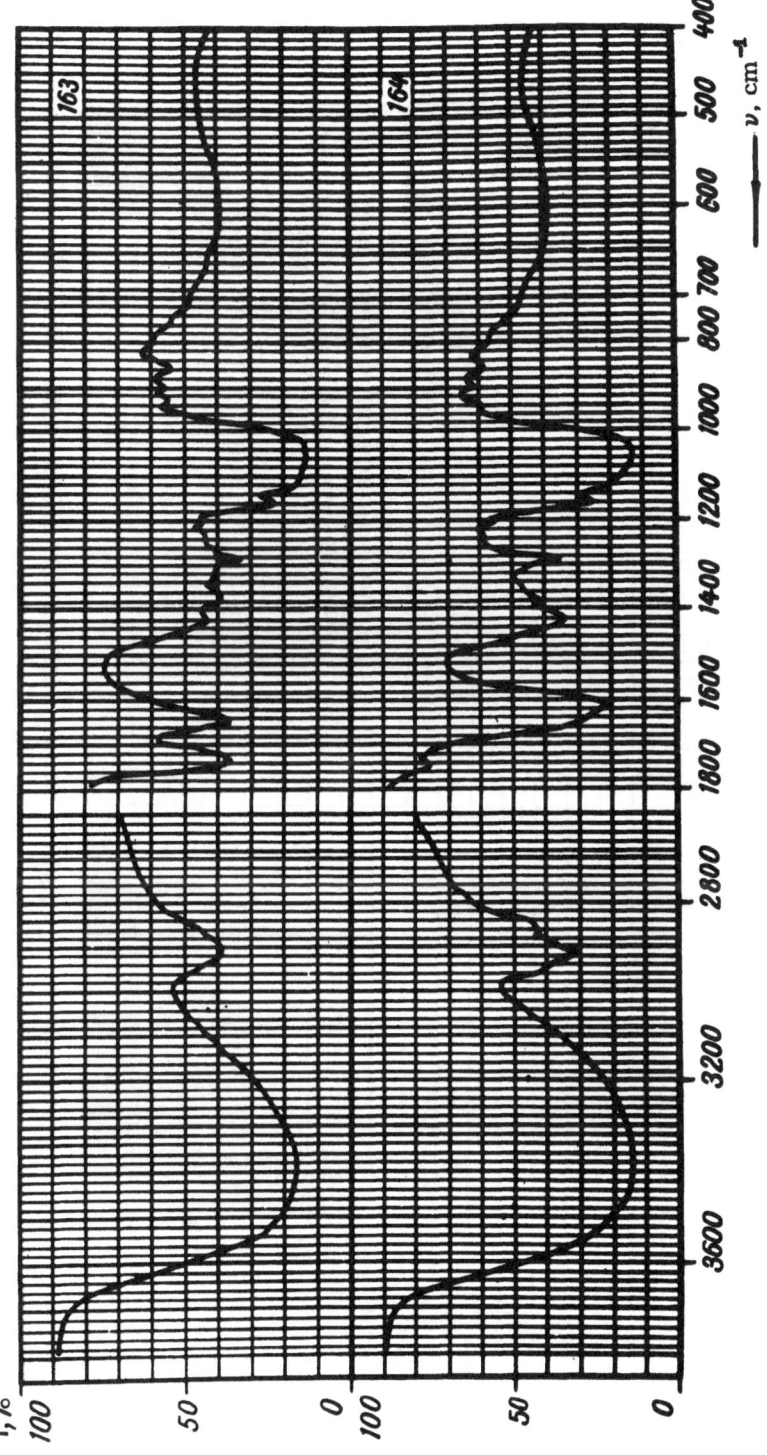

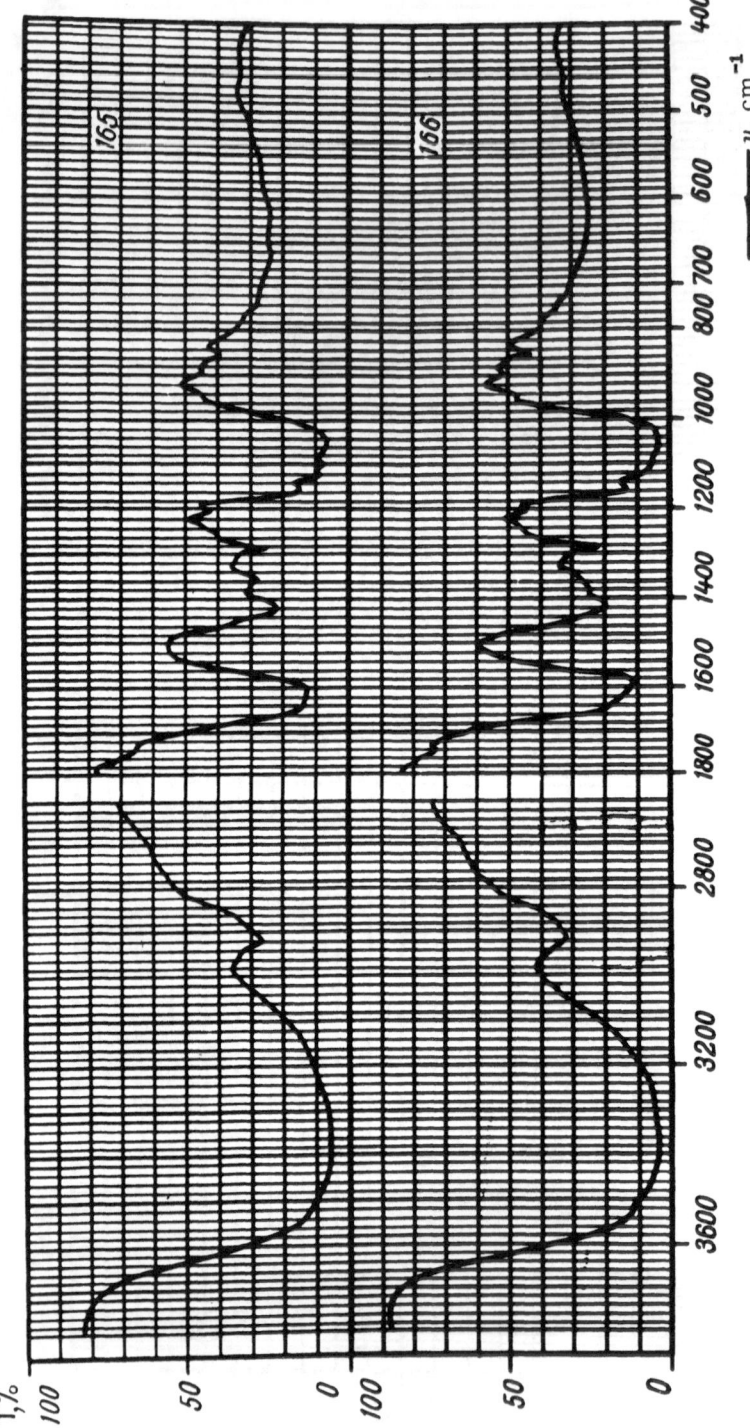

APPENDIX VIII

SPECTRA OF NEW TYPES OF CELLULOSE DERIVATIVES AND OF MODIFIED CELLULOSES

167. Cellulose methylphosphonate (2% P) (KBr)
168. Cellulose methylphosphonate (8% P) (KBr)

$$(cell.-O)_2 P(O)CH_3$$

169. Cellulose methyl phosphate (KBr)

cell. $\diamondsuit P(O)OCH_3$ (O, O)

170. Cellulose phenylphosphonate (KBr)

$$(cell.-O)_2 P(O)C_6H_5$$

171. Cellulose phenyl phosphate (KBr)

$$cell.-O-P(O)(OH)OC_6H_5$$

172. Cellulose diphenyl phosphate (KBr)

$$cell.-O-P(O)(OC_6H_5)_2$$

173. Cellulose phenyl phosphate after treatment with NH_4OH (KBr)

$$cell.-O-P(O)(OC_6H_5)ONH_4$$

174. Cellulose chloroethyl phosphate, obtained by the re-esterification method (KBr)

$$cell.-O-P(O)(OC_2H_4Cl)_2$$

175. Cellulose chloroethyl phosphate (KBr)

$$(cell.-O)_2P(O)OC_2H_4Cl$$

176. Cellulose chloroethyl phosphate after oxidation with H_2O_2 (KBr)
177. Cellulose chlorovalerate ($\gamma = 280$) (KBr)

cell.$-O-C\overset{O}{\underset{(CH_2)_4Cl}{\diagup}}$

178. Cellulose mixed ester of acetic and chlorovaleric acids [$\gamma_{OCO(CH_2)_4Cl}$ = 280, γ_{OCOCH_3} = 20] (KBr)

179. Cellulose methylxanthate (γ = 30) (KBr)

$$\text{cell.}-O-C\overset{S}{\underset{SCH_3}{\diagup}}$$

180. Cellulose methylxanthate (γ = 160) (KBr)

181. Cellulose p-nitrophenylxanthate (γ = 25) (KBr)

$$\text{cell.}-O-C\overset{S}{\underset{SC_6H_4NO_2}{\diagup}}$$

182. Cellulose diethylacetamidoxanthate (γ = 30) (KBr)

$$\text{cell.}-O-C\overset{S}{\underset{SCH_2CON(C_2H_5)_2}{\diagup}}$$

183. Cellulose thiourethane, obtained with the aid of $CdSO_4$ (γ = 25) (KBr)

$$\text{cell.}-O-C\overset{S}{\underset{NH(CH_2)_6NHC}{\diagup}}\overset{S}{\underset{O-\text{'cell.}}{\diagup}}$$

184. Cellulose thiourethane, obtained with the aid of $ZnSO_4$ (γ = 22) (KBr)

185. Cellulose methylxanthate (γ = 160), before thermal treatment (KBr)

186. Cellulose methylxanthate (γ = 160), after thermal treatment at 200°C for 20 h (KBr)

187. Cellulose ether of 3-hydroxyethylsulfonylmethoxybenzonitrile

$$\text{cell}-O-CH_2CH_2SO_2\diagup\!\!\!\bigcirc\overset{SCN}{\diagdown OCH_3}$$

188. Cellulose ether of 4-β-hydroxyethylsulfonyl-2-aminoanisole (γ = 25) (FM)

$$\text{cell.}-O-CH_2CH_2SO_2\diagup\!\!\!\bigcirc\overset{NH_2}{\diagdown OCH_3}$$

189. Graft copolymer of cellulose with polyacrylonitrile (FM)

$$\text{cell.}-O-CH_2CH_2SO_2\diagup\!\!\!\bigcirc\overset{\overset{O}{\underset{||}{}}HN-(CH_2-CHCN-)_n}{\diagdown OCH_3}$$

190. Graft copolymer of cellulose with poly-2-methyl-5-vinylpyridine (FM)

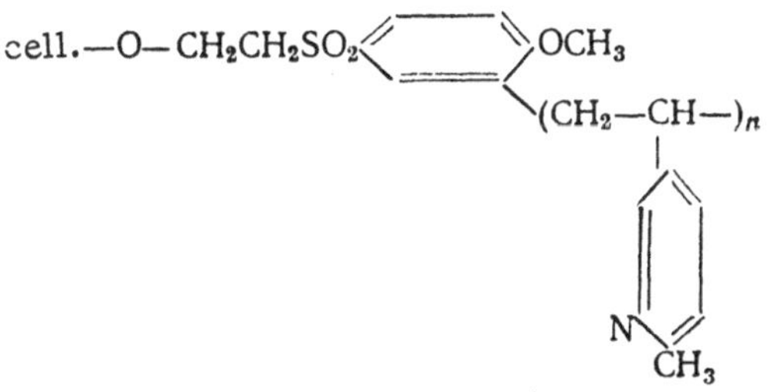

$$cell.—O—CH_2CH_2SO_2\diagup\diagdown OCH_3$$
$$(CH_2—CH—)_n$$
$$N\diagdown CH_3$$

191. Graft copolymer of cellulose with polyacrylhydroxamic acid (3.35% N) (FM)

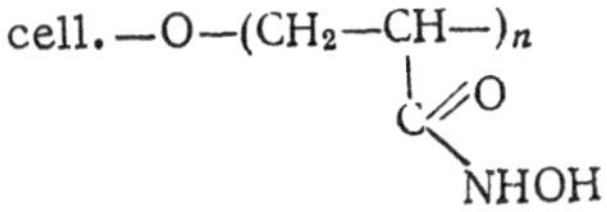

$$cell.—O—(CH_2—CH—)_n$$
$$C\diagdown^{O}$$
$$NHOH$$

192. Graft copolymer of cellulose with polyacrylhydroxamic acid (2.9% N) (FM)
193. Product of ion exchange of compound 191 with iron cations (FM)
194. Product of ion exchange of compound 191 with copper cations (FM)

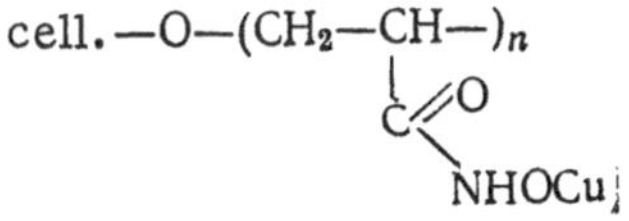

$$cell.—O—(CH_2—CH—)_n$$
$$C\diagdown^{O}$$
$$NHOCu \downarrow$$

195. Graft copolymer of cellulose and the polymer of the diethylaminoethyl ester of methacrylic acid (FM)

$$cell.—O—CH_2CH_2SO_2\diagup\diagdown OCH_3$$
$$CH_3$$
$$(CH_2—C—)_n$$
$$O^{\diagup C}\diagdown OCH_2CH_2N(C_2H_5)_2$$

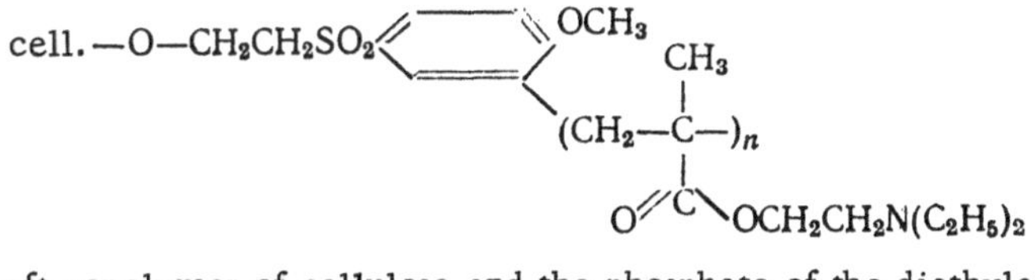

196. Graft copolymer of cellulose and the phosphate of the diethylamino-ethyl ester of methacrylic acid (FM)

$$cell.—O—CH_2CH_2SO_2\diagup\diagdown OCH_3$$
$$CH_3$$
$$(CH_2—C—)_n$$
$$O^{\diagup C}\diagdown OCH_2CH_2N(C_2H_5)H_3PO_4$$

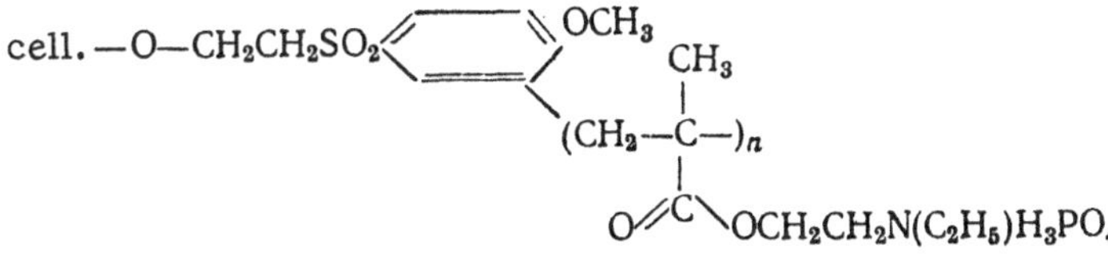

197. Graft copolymer of cellulose with polyisobutene

$$\text{cell.} - \text{O} - \text{CH}_2\text{CH}_2\text{SO}_2 \overset{}{\underset{}{\bigcirc}} - (\text{CH}_2 - \overset{\overset{\text{CH}_3}{|}}{\text{C}} = \text{CH} -)_n$$

198. Graft copolymer of cellulose and polyacrylic acid (13.6% COOH) (FM)
199. Graft copolymer of cellulose and polyvinylidene chloride (13% Cl) (FM)

$$\text{cell.} - \text{O} - (\text{CH}_2\text{CCl}_2 -)\text{n}$$

200. Graft copolymer of cellulose and polyvinylidene chloride (21% Cl) (FM)
201. Cellulose mixed ester of acetic and levulinic acids [γOCOCH_3 = 150, $\gamma\text{OCO(CH}_2)_2\text{COCH}_3$ = 100] (KBr)

$$\text{cell.} \left]\begin{array}{l} -\text{O}-\text{COCH}_3 \\ -\text{O}-\text{CO(CH}_2)_2\text{COCH}_3 \end{array}\right.$$

202. Cellulose mixed ester of toluenesulfonic and levulinic acids
[$\gamma\text{OSO}_2\text{C}_6\text{H}_4\text{CH}_3$ = 15, $\gamma\text{OCO(CH}_2)_2\text{COCH}_3$ = 75) (KBr)

$$\text{cell.} \left]\begin{array}{l} -\text{O}-\text{SO}_2\text{C}_6\text{H}_4\text{CH}_3 \\ -\text{O}-\text{CO(CH}_2)_2\text{COCH}_3 \end{array}\right.$$

203. Phenylhydrazinedeoxycellulose ($\gamma\text{OSO}_2\text{C}_6\text{H}_4\text{CH}_3$ = 100, $\gamma\text{NHNHC}_6\text{H}_5$ = 100) (KBr)

$$\text{cell.} \left]\begin{array}{l} -\text{O}-\text{SO}_2\text{C}_6\text{H}_4\text{CH}_3 \\ -\text{NHNHC}_6\text{H}_5 \end{array}\right.$$

204. Cellulose pyridinium sulfate (γ = 262, 15.7% S) (KBr)

$$\text{cell.} -\text{OSO}_2\text{O}-\text{C}_5\text{H}_5\text{N}^+\text{H}$$

205. Cellulose diacetate (γ = 230) (KBr)
206. Cellulose mixed ester with acetic and hexafluoroisobutyric acids (γ = 230, 14% F) (KBr)

$$\text{cell.} \left]\begin{array}{l} -\text{O}-\text{COCH}_3 \\ -\text{O}-\text{COCH(CF}_3)_2 \end{array}\right.$$

207. Trimethylsilylcellulose (3.4% Si) (KBr)

$$\text{cell.} - \text{O} - \text{Si(CH}_3)_3$$

208. Ether of cellulose and 1-hydroxyundecamethylpentasilane (10.77% Si) (KBr)

$$\text{cell.}-O-[Si(CH_3)_2]_4Si(CH_3)_3$$

209. 3,6-Anhydrocellulose (γ = 20) (FM)

210. 3,6-Anhydrocellulose (γ = 70, $\gamma_{OSO_2C_6H_5}$ = 7) (FM)

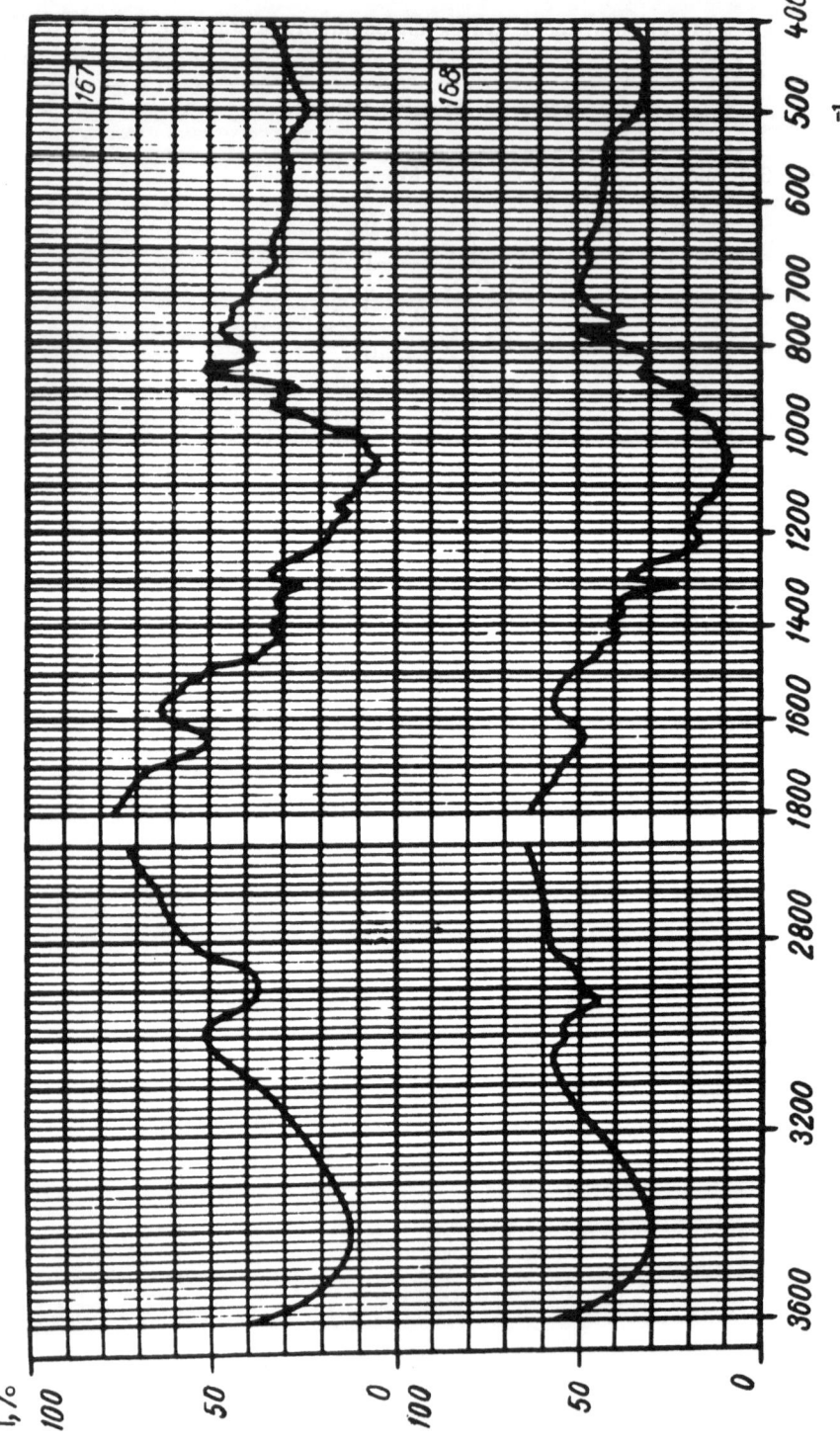

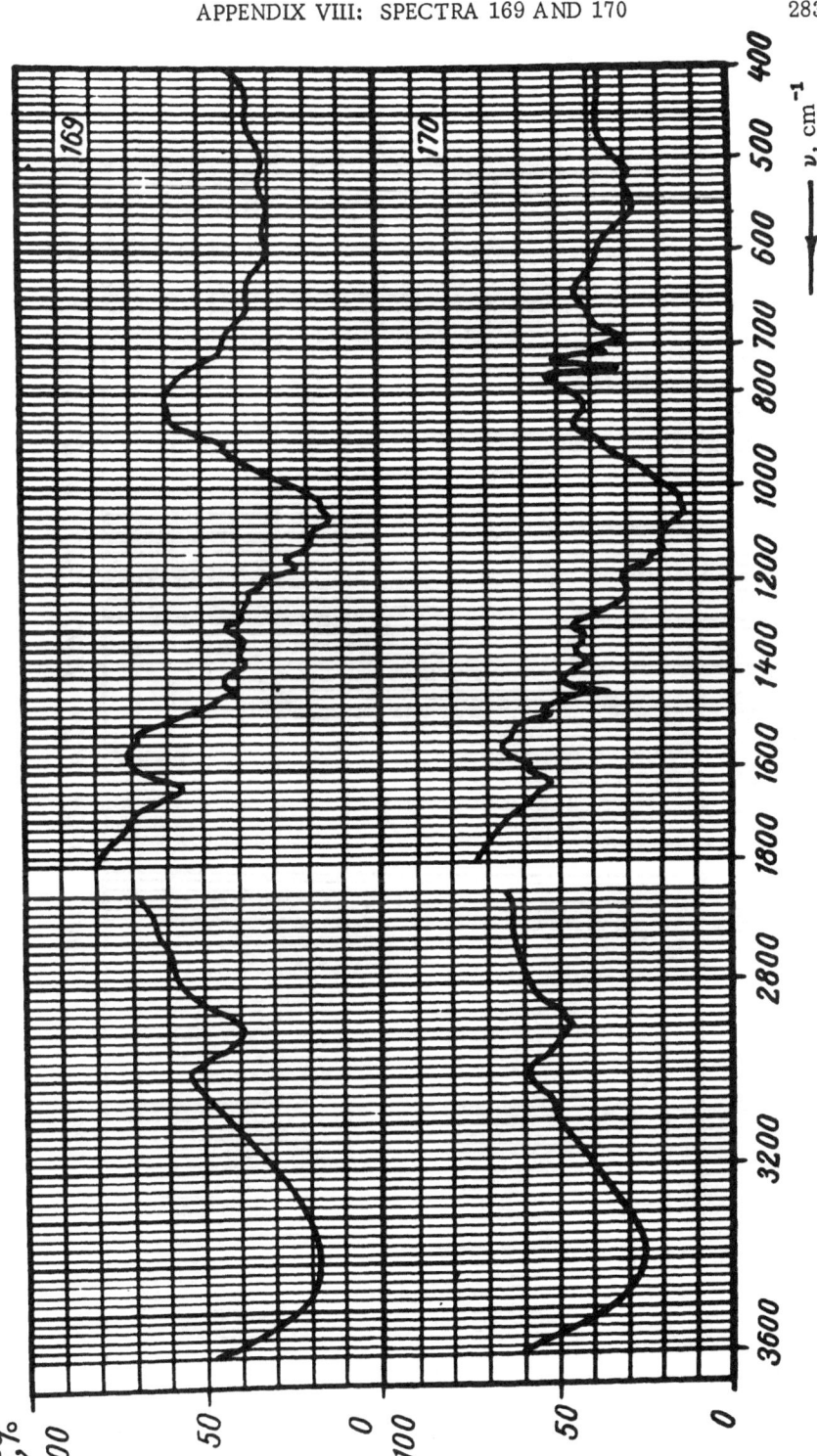

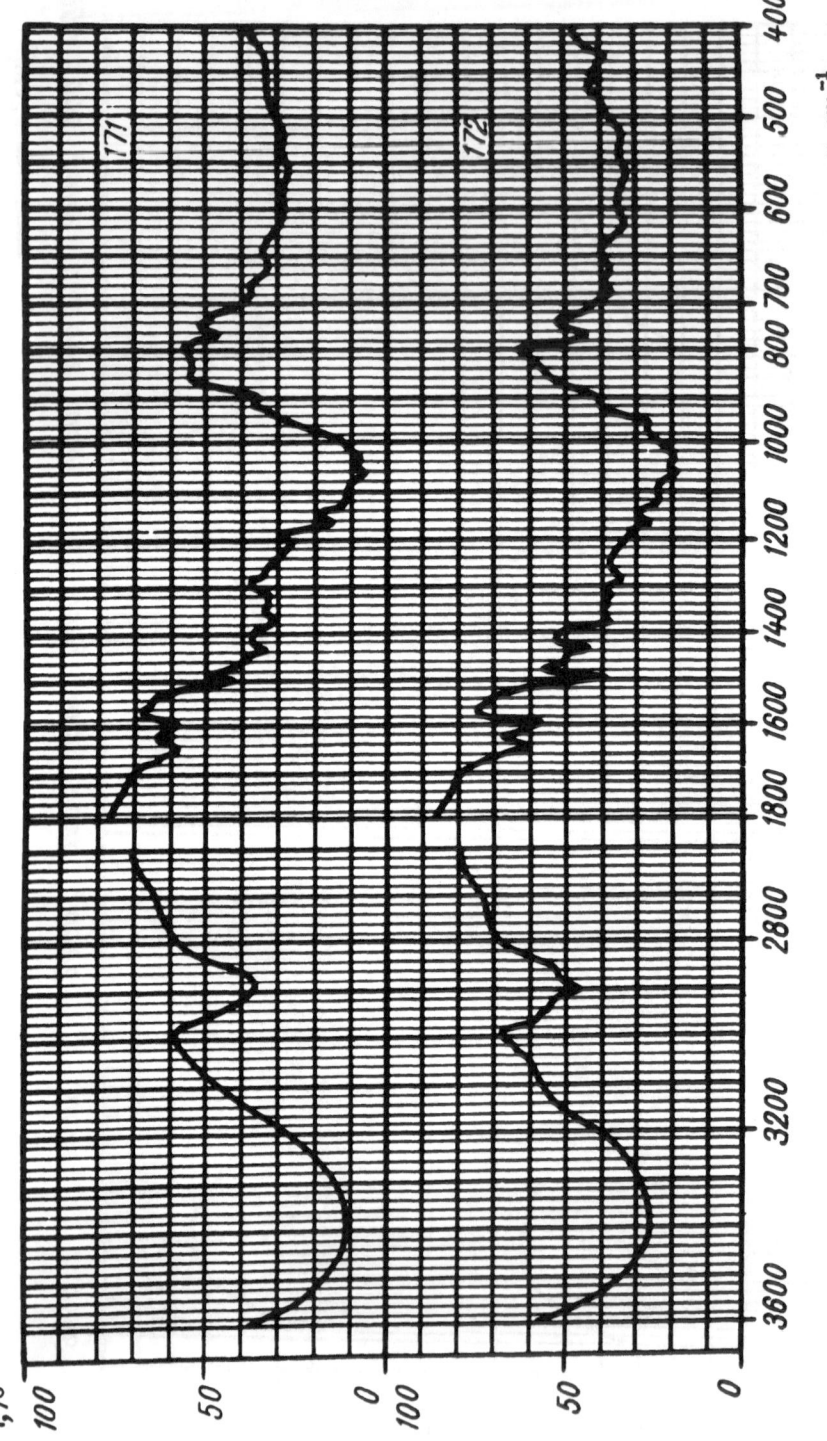

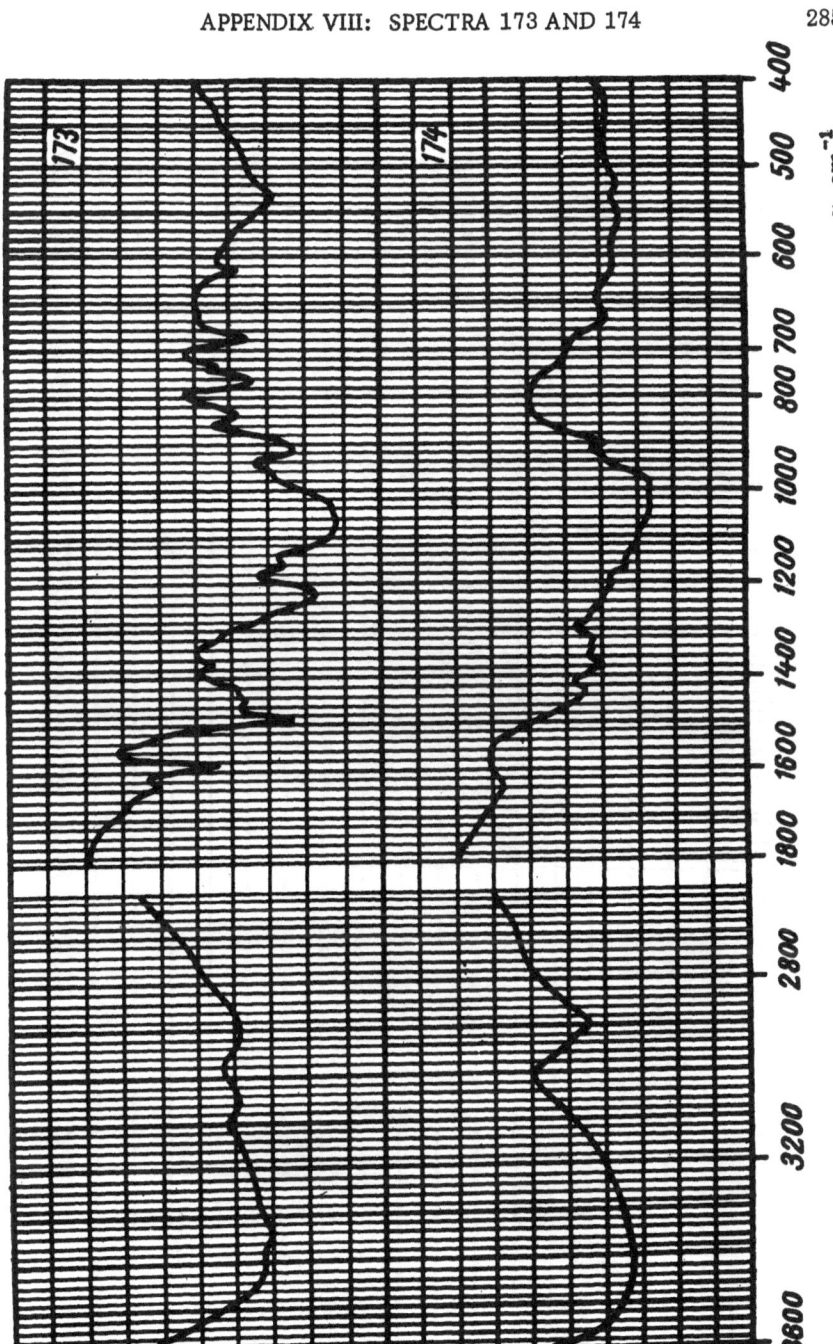

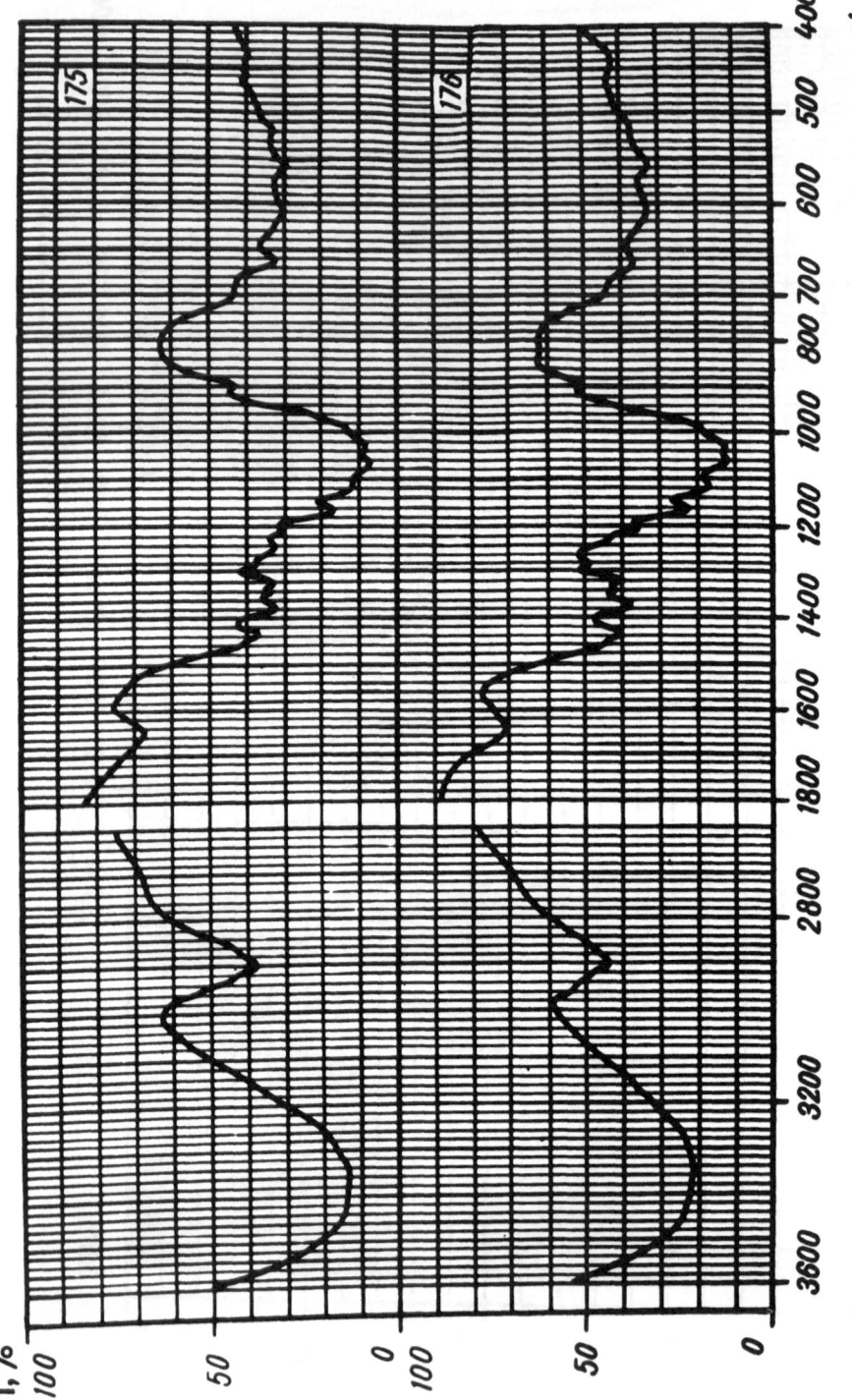

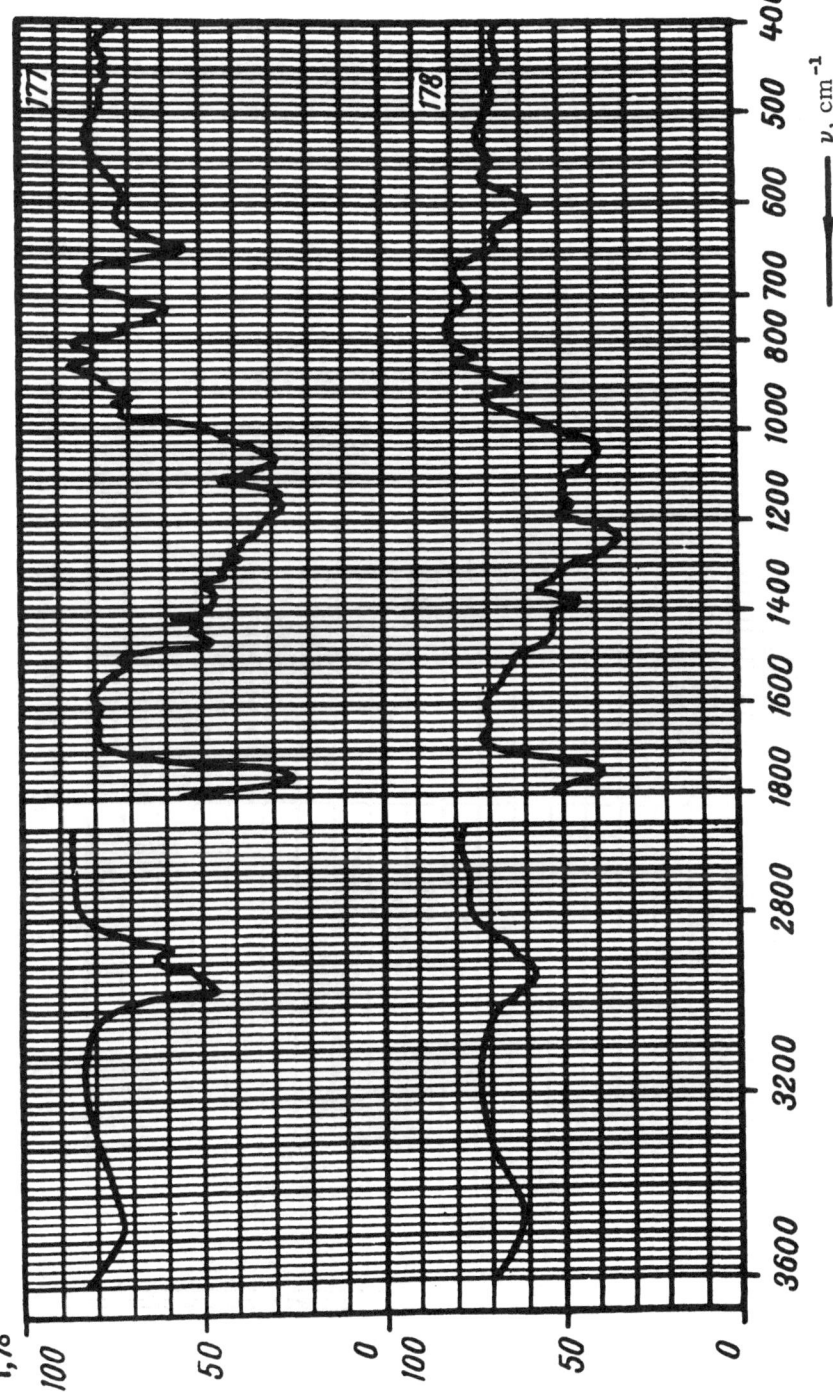

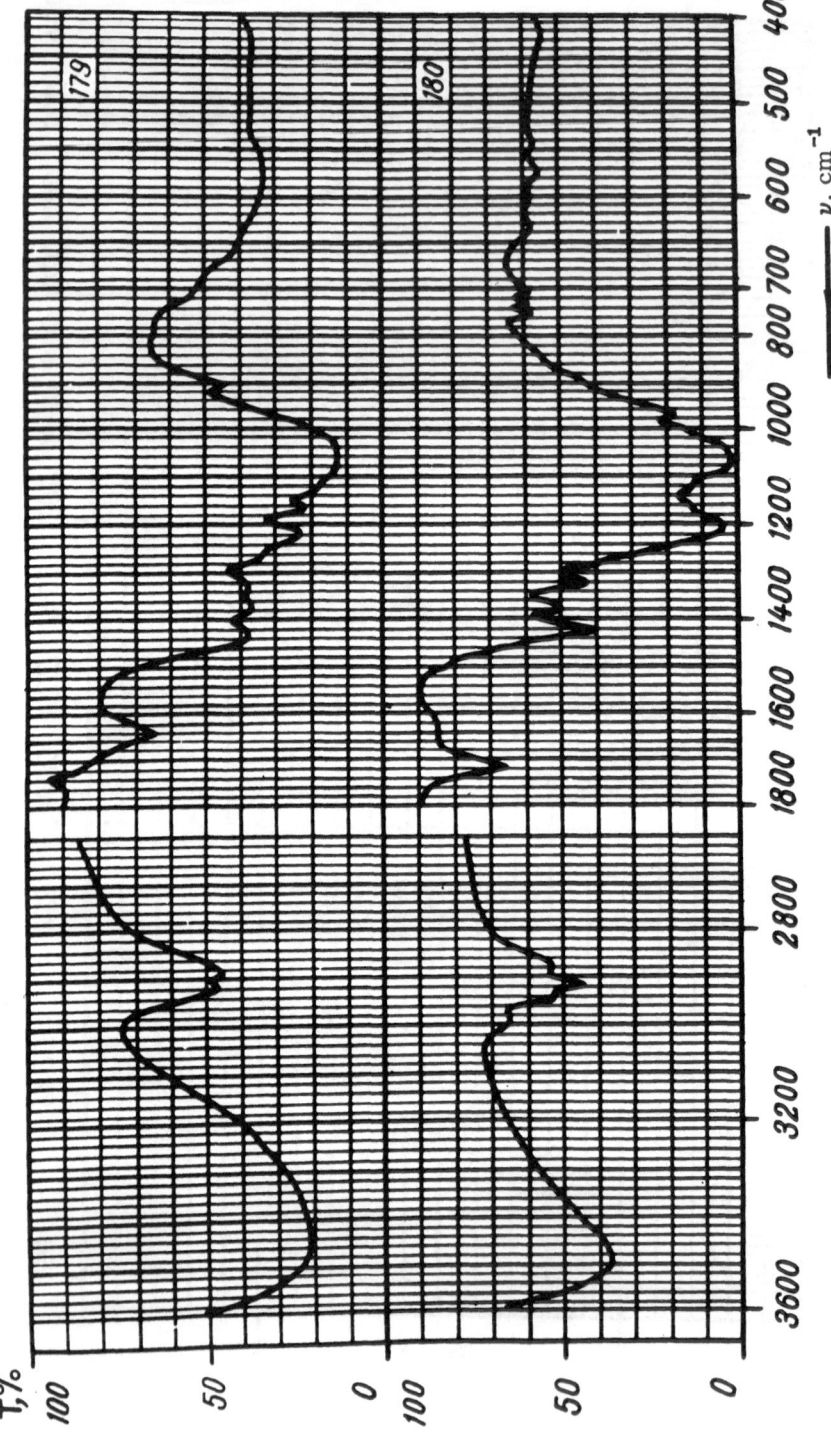

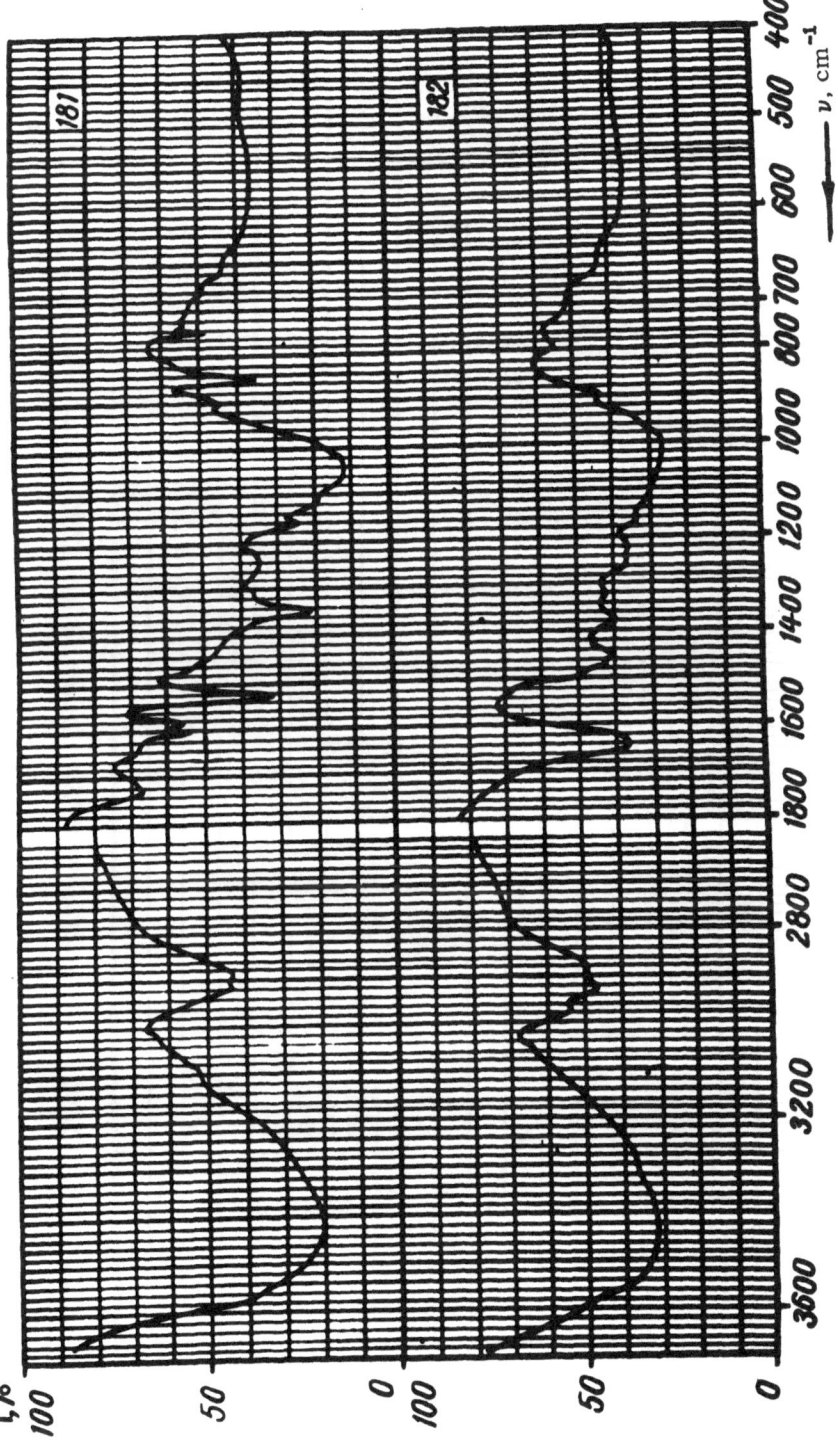

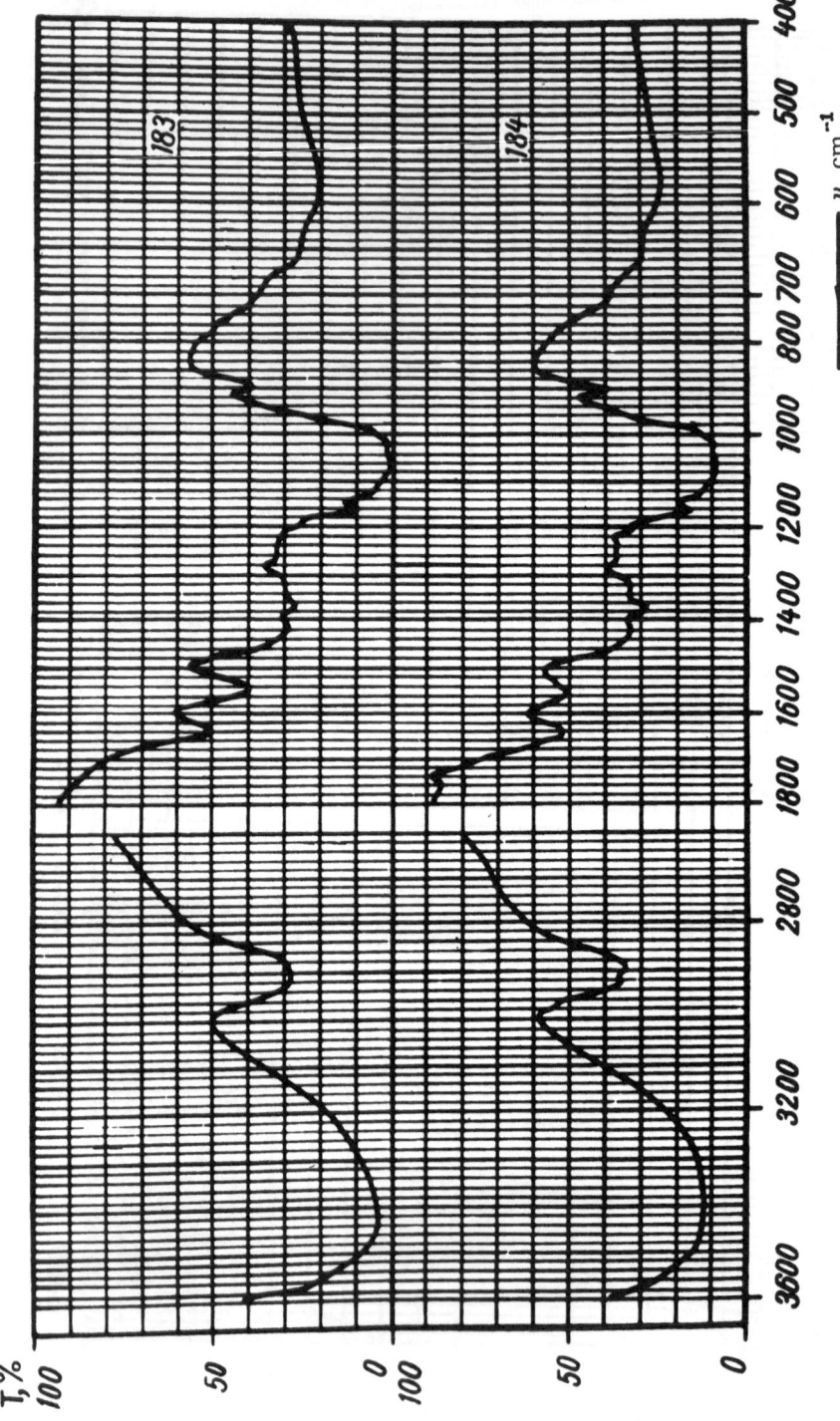

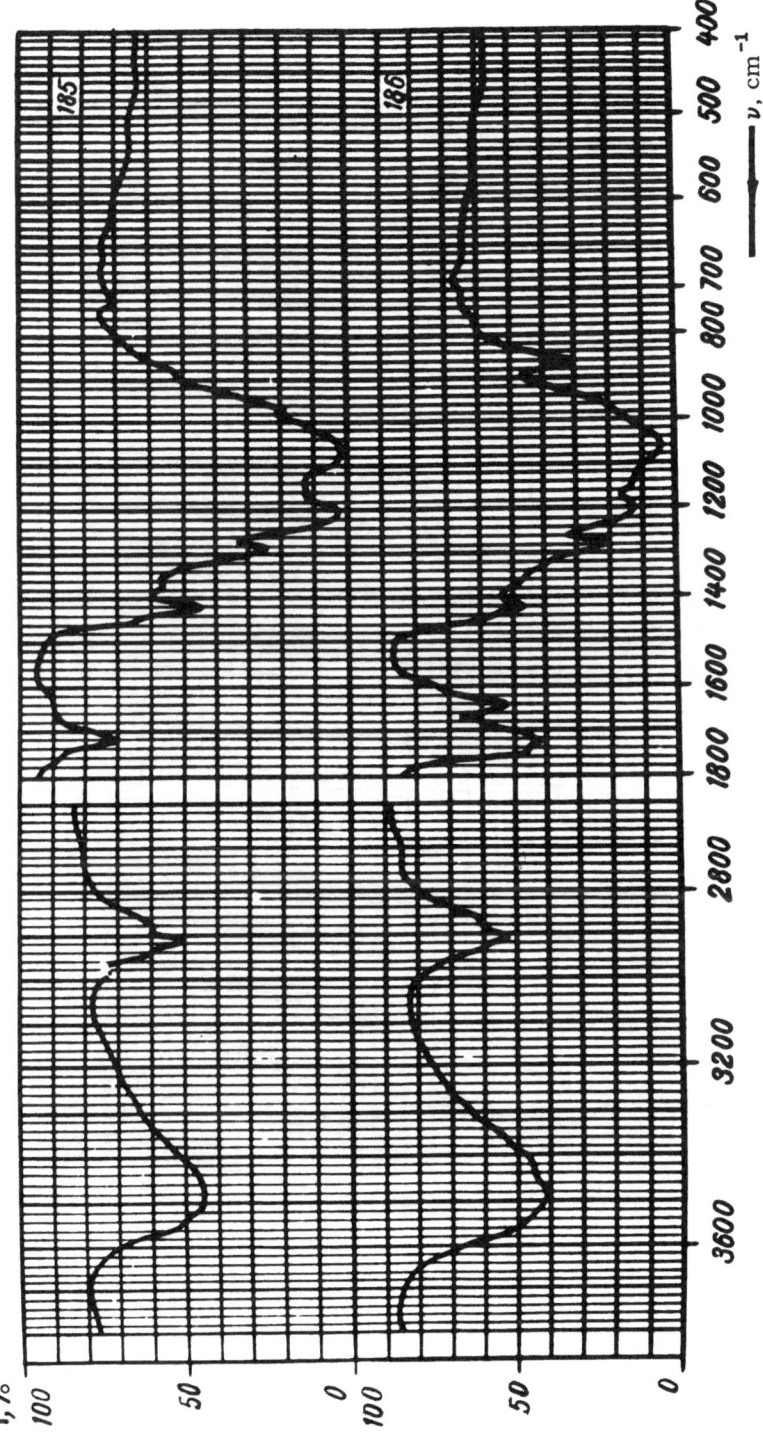

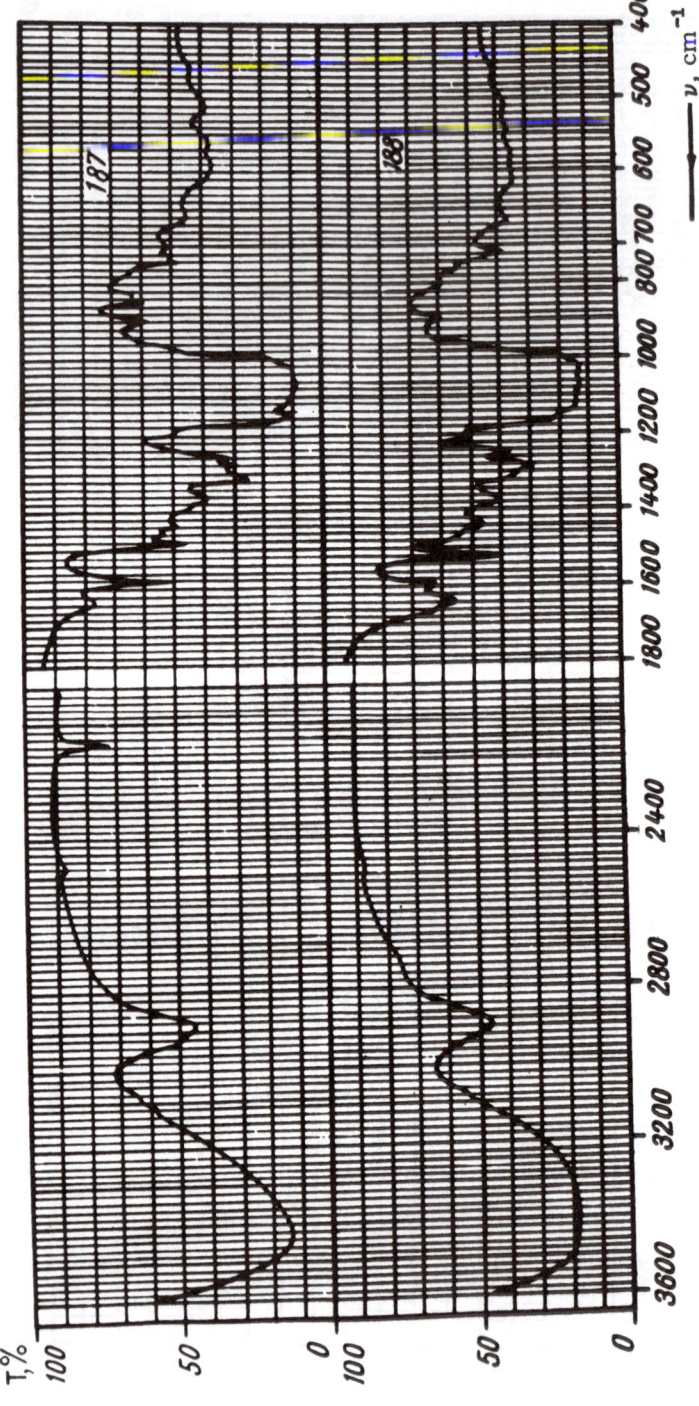

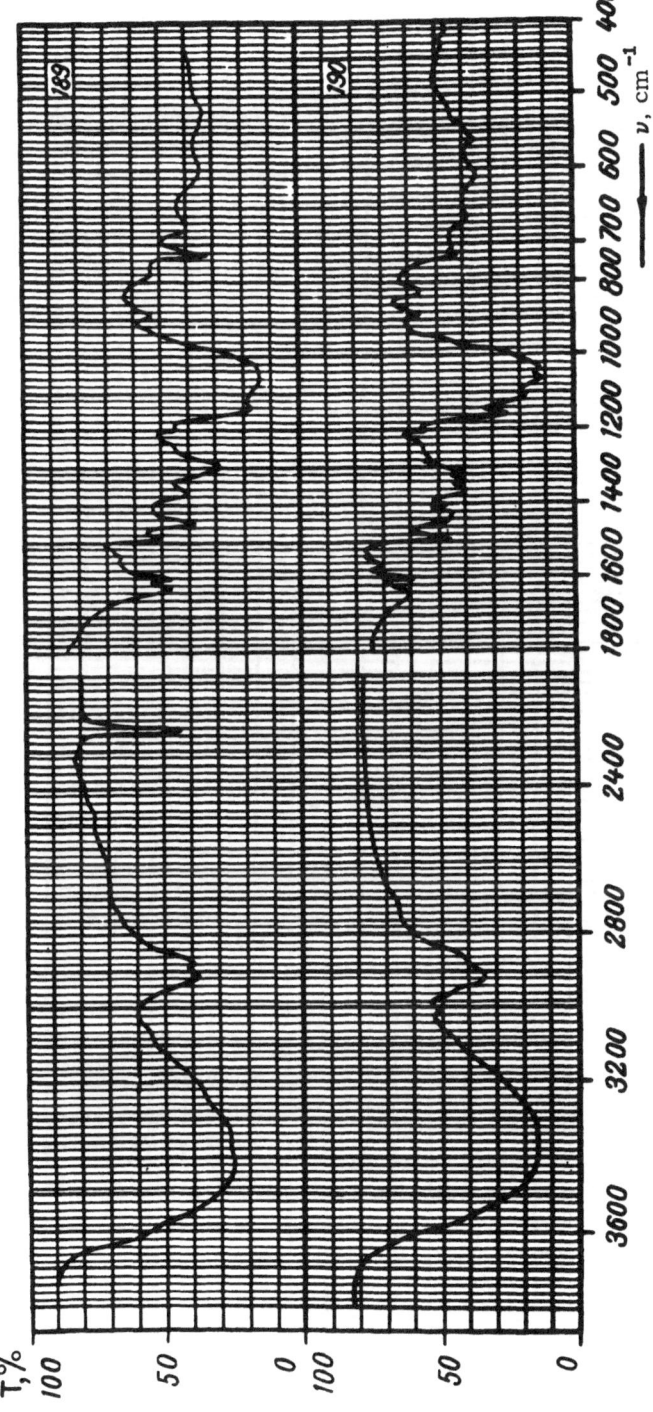

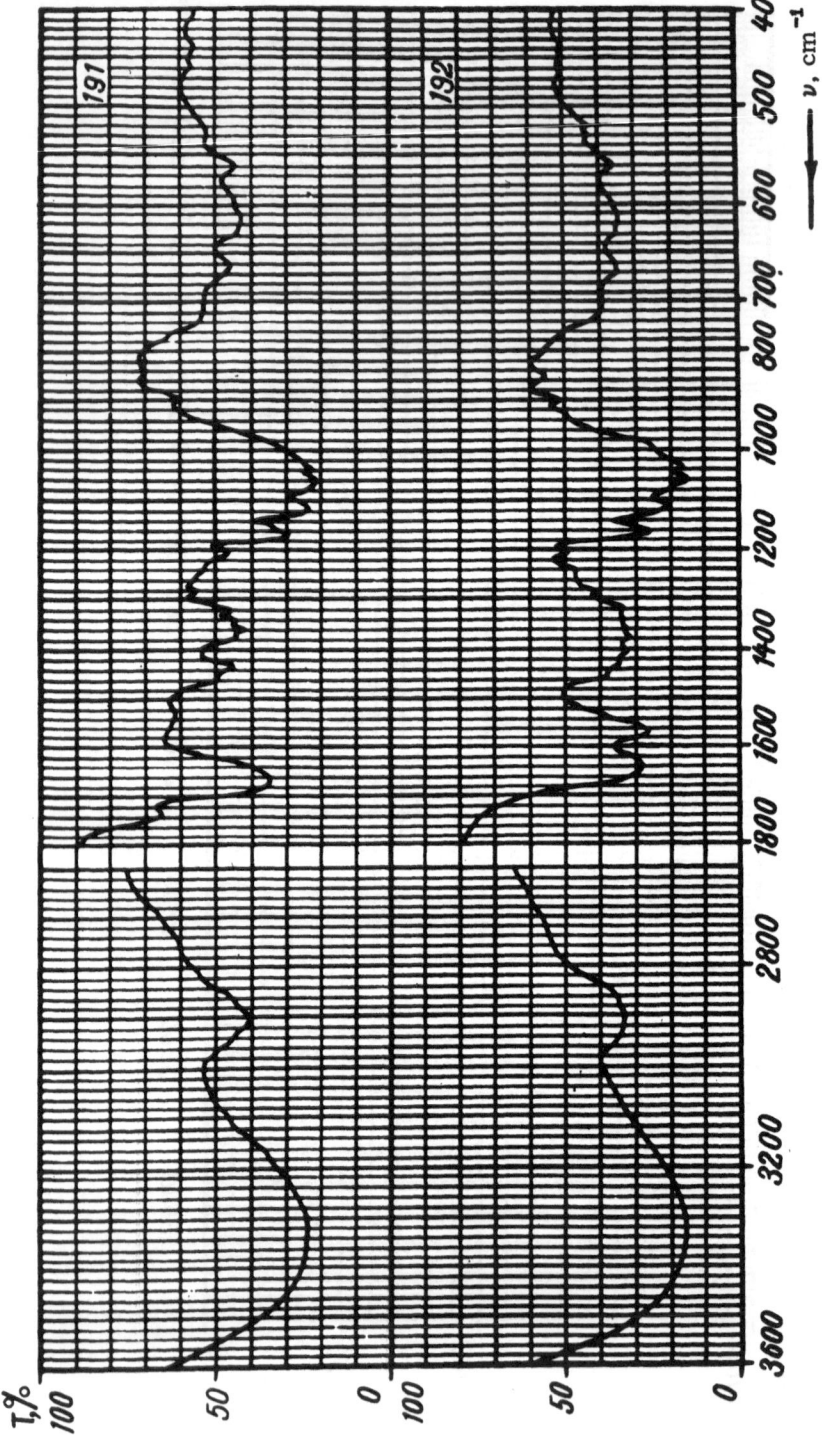

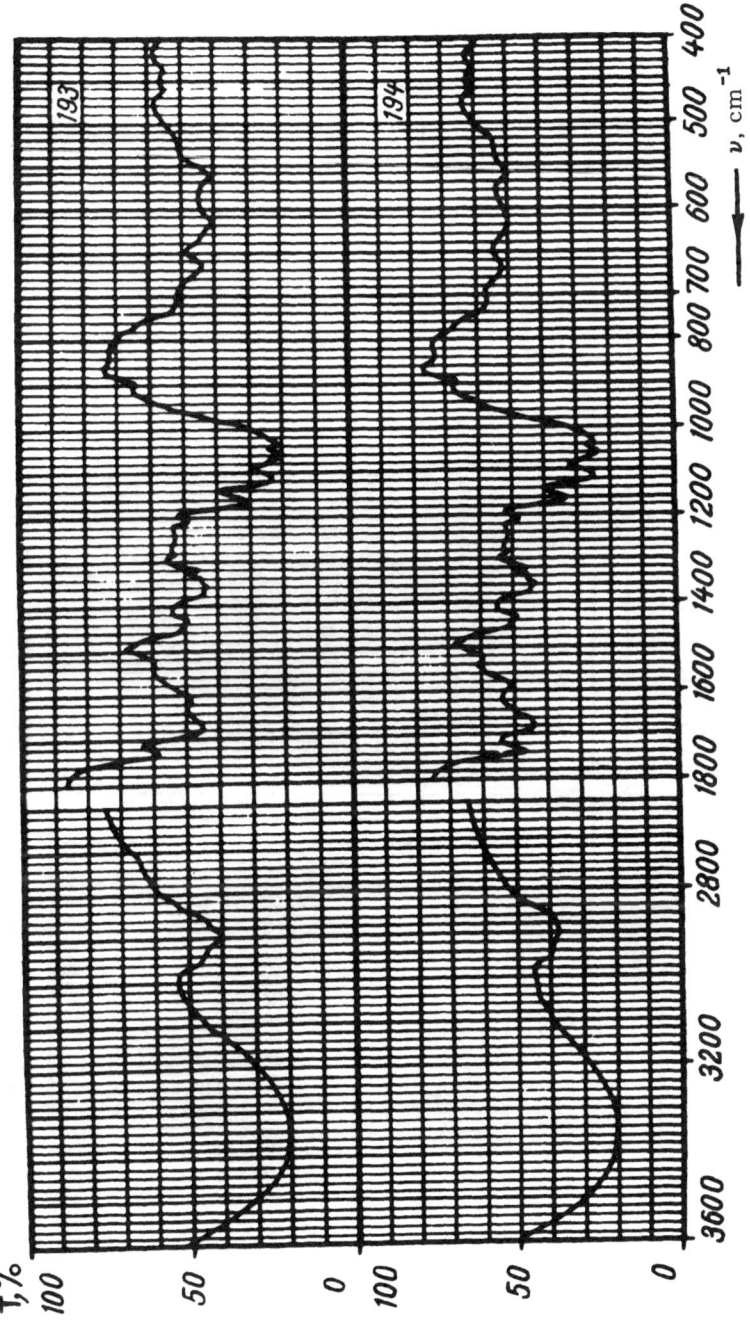

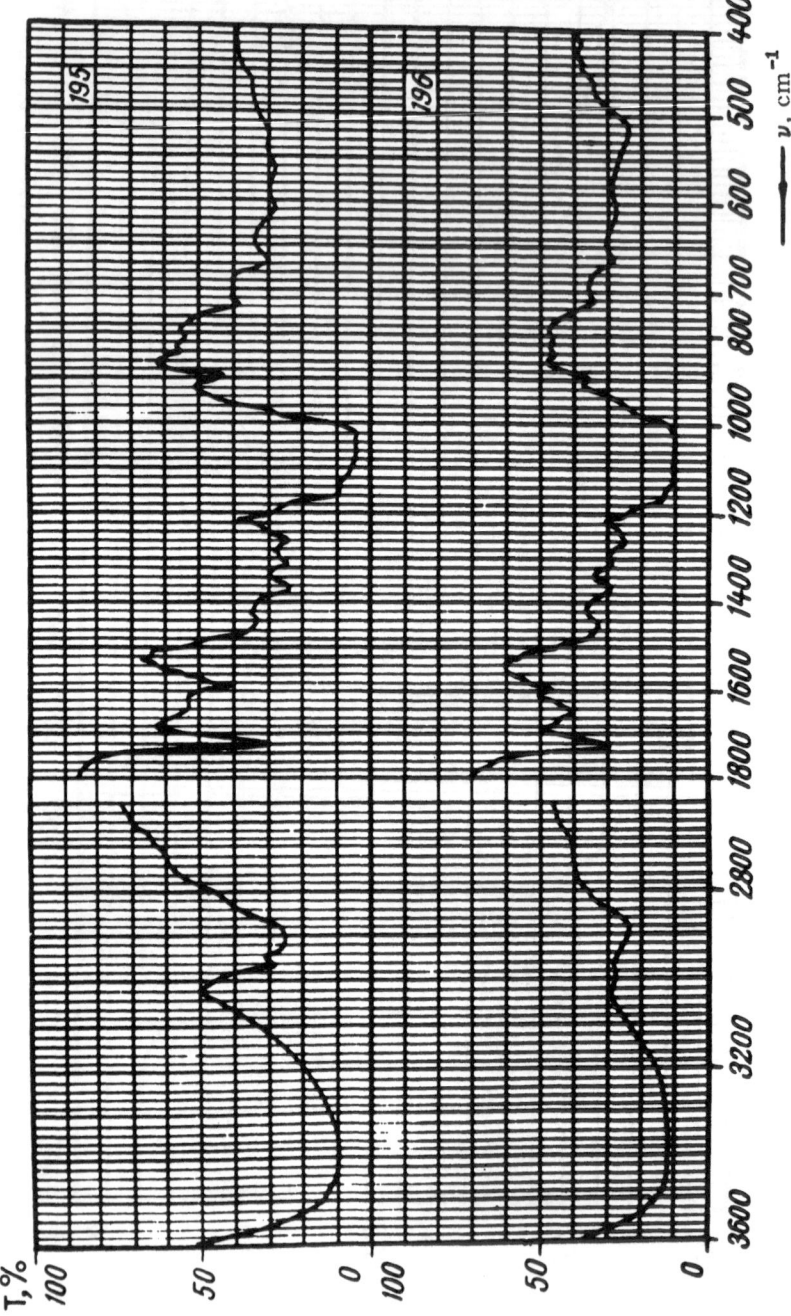

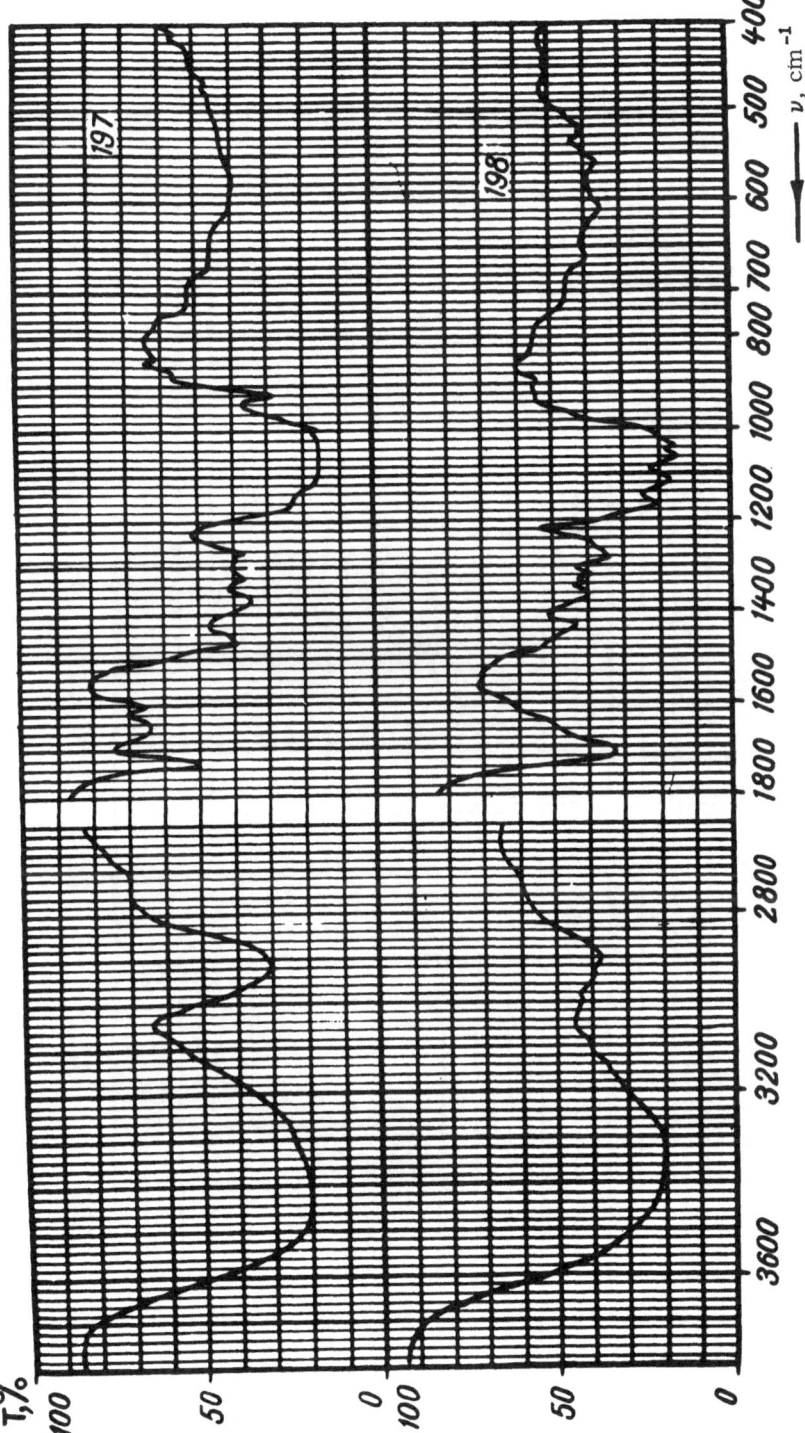

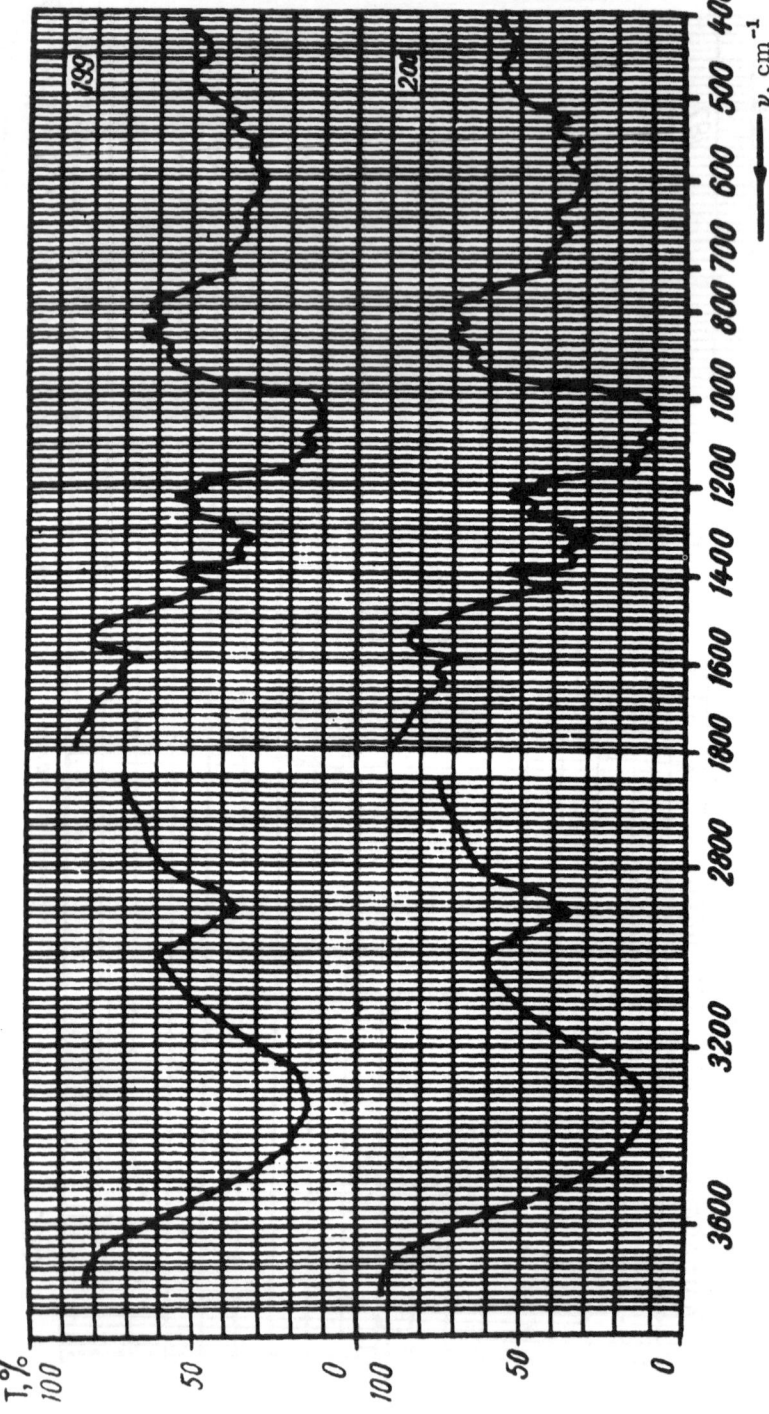

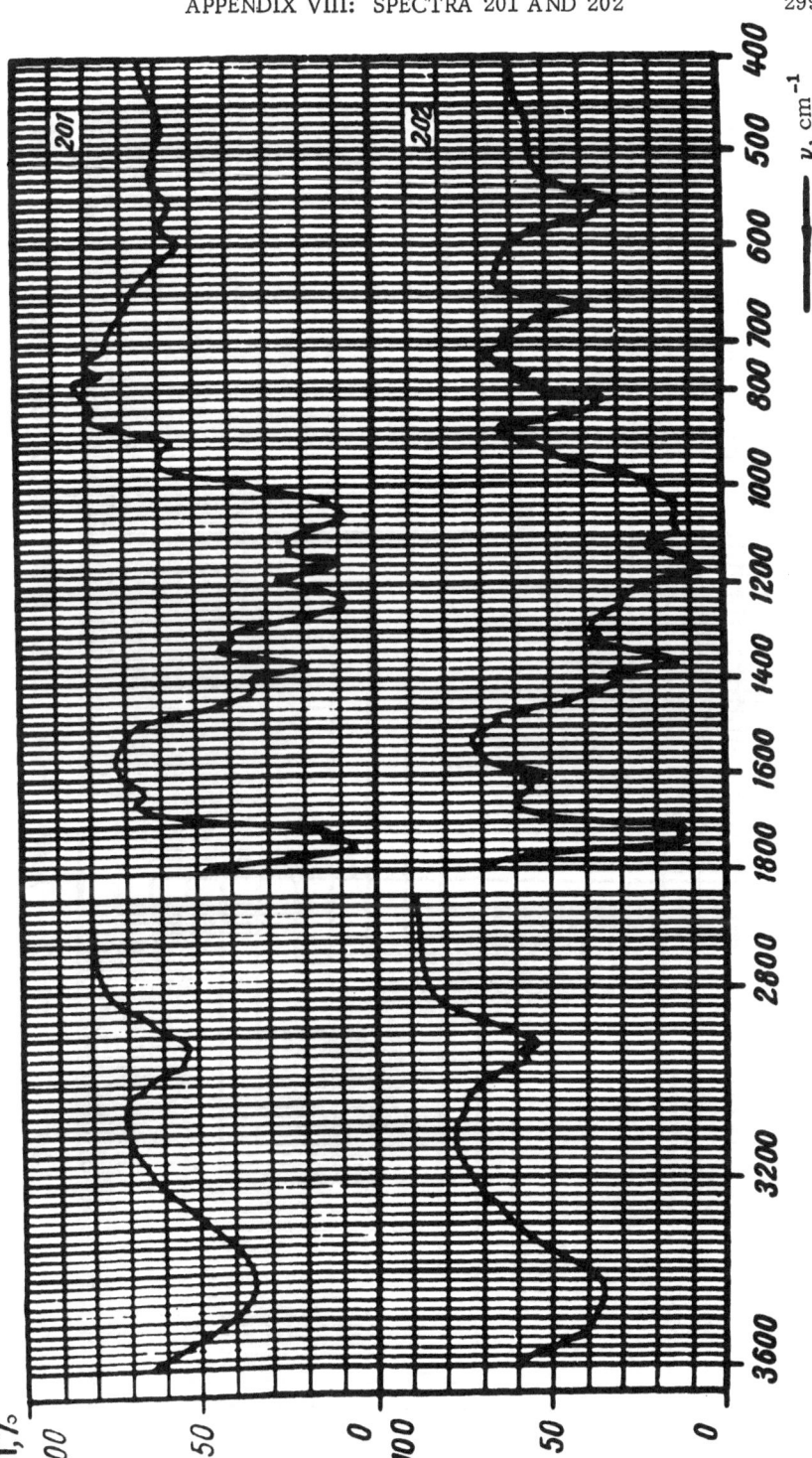

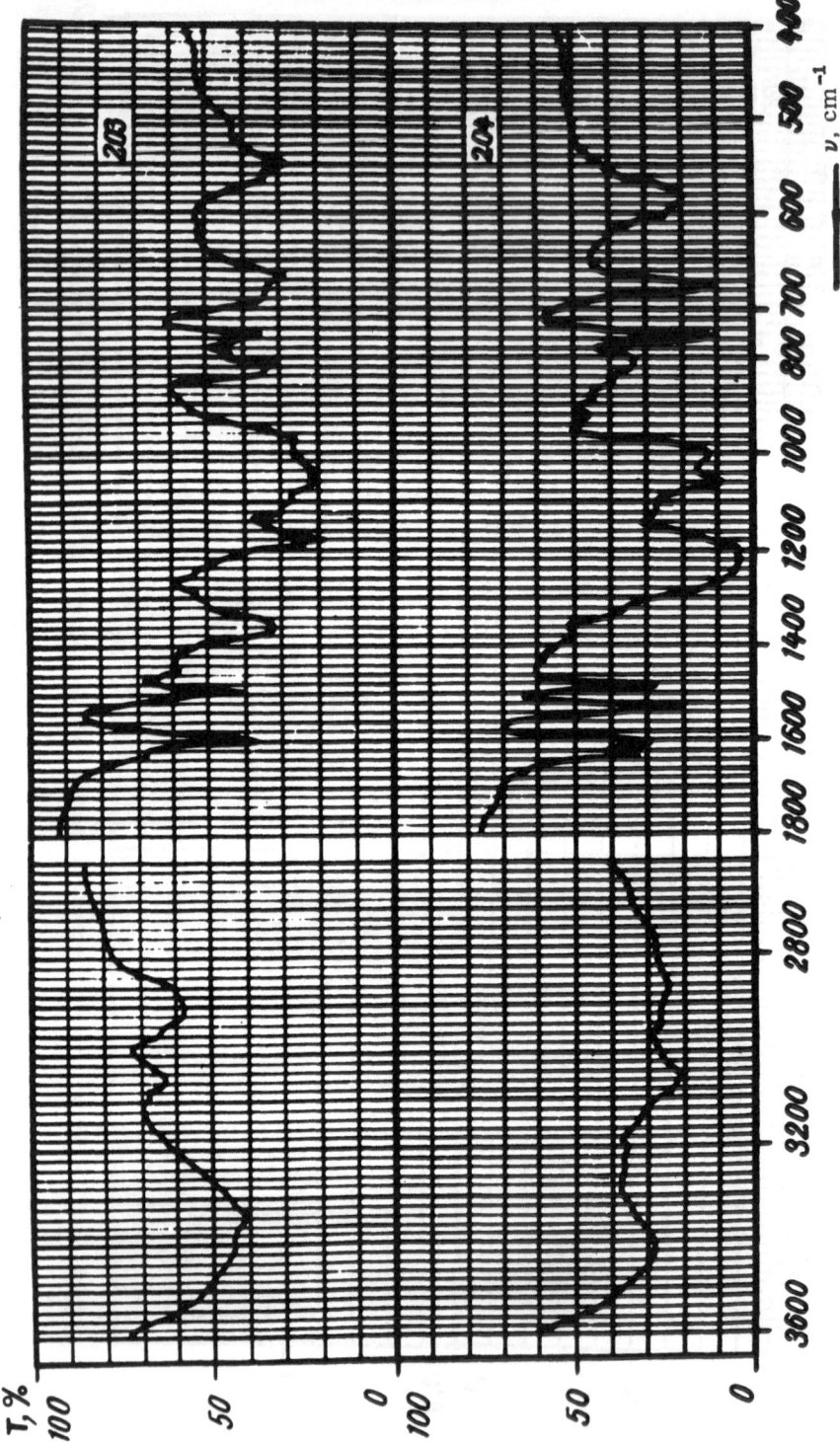

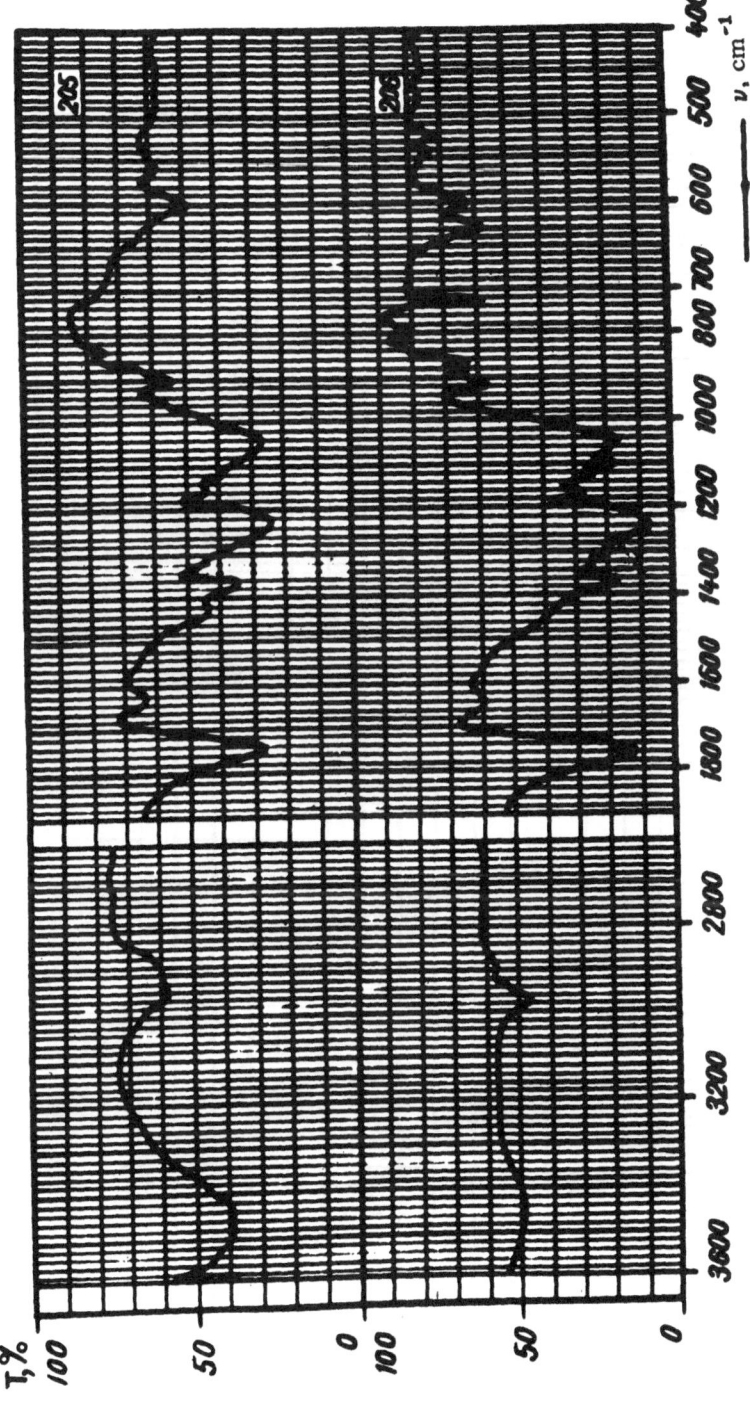

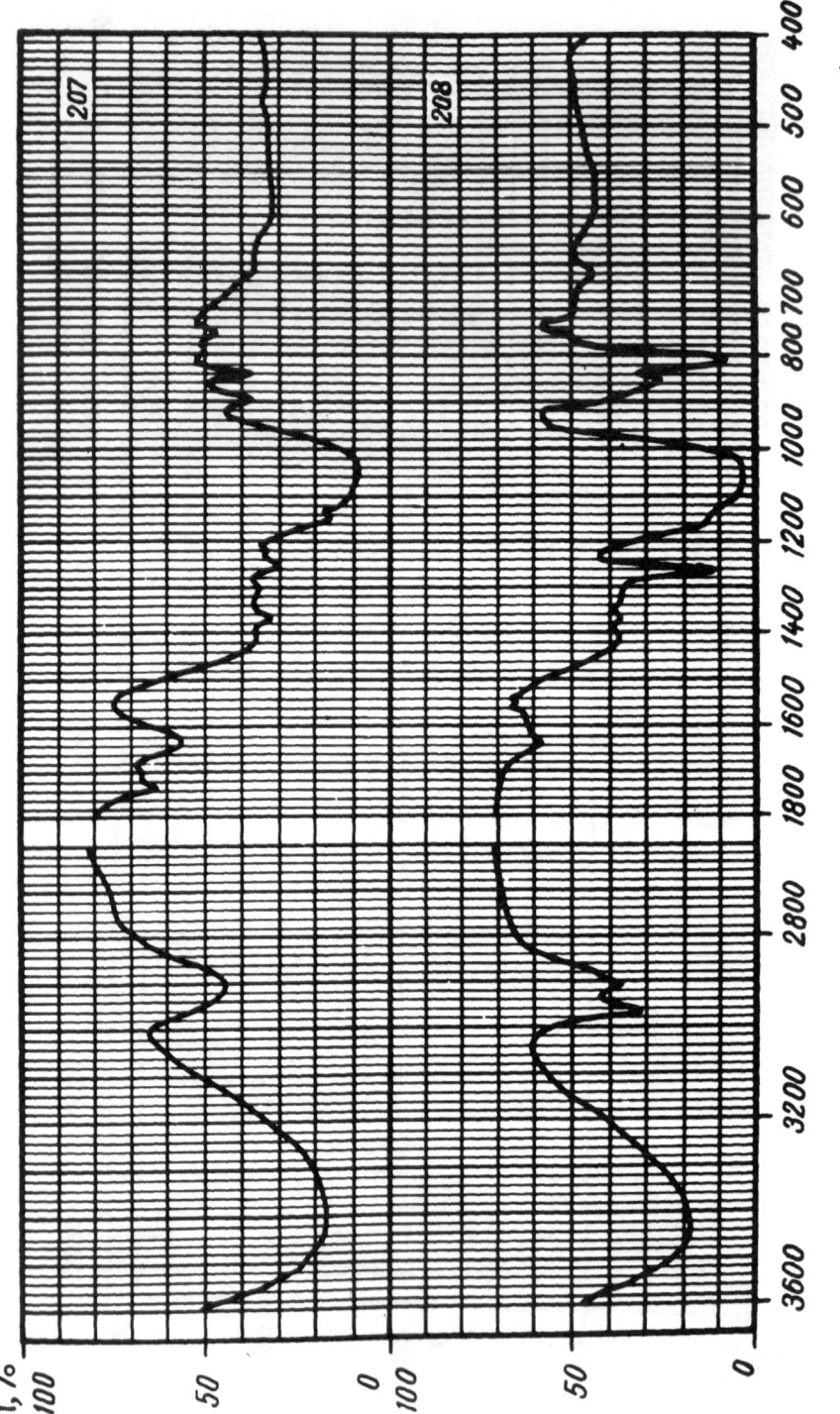

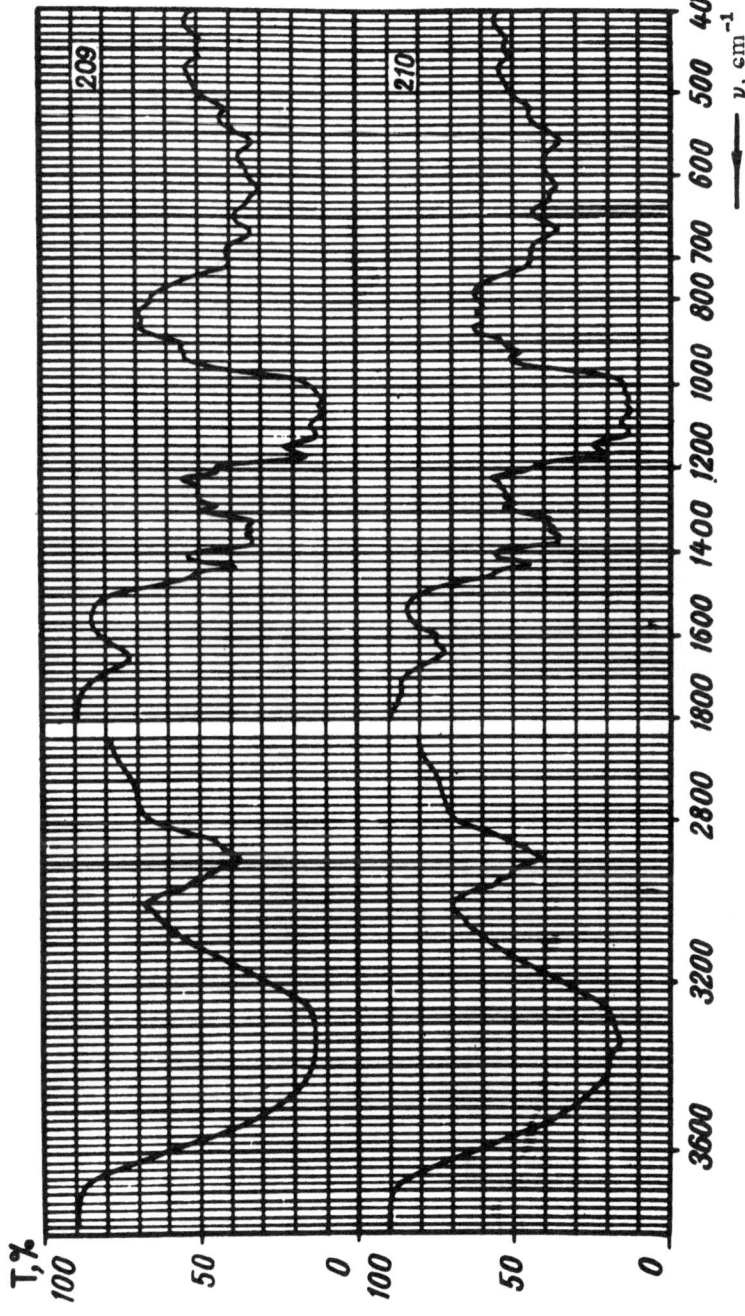

APPENDIX IX

CHARACTERISTIC VIBRATION FREQUENCIES OF GROUPS AND BONDS IN THE MOLECULES OF CELLULOSE
AND ITS DERIVATIVES

No.	Group, bond	Molecule	Frequency,cm^{-1}	Remarks
1	CHOH, CH$_2$OH (OH groups, combined to form hydrogen bonds, stretching vibrations	Natural cellulose	3400 3350 3300	Most clearly visible after partial deuteration of cellulose
		Mercerized cellulose	3480 3450 3350 3170	Most clearly visible after partial deuteration of cellulose
		Doubly oriented film of bacterial cellulose	3405($\perp$) 3375(‖) 3350(‖) 3305($\perp$) 3275(‖) 3245(‖)	Intermolecular hydrogen bonding Intramolecular hydrogen bonding Intermolecular hydrogen bonding [196]
		Doubly oriented film of ramie	3405($\perp$) 3375(‖) 3350(‖)	Intermolecular hydrogen bonding Intramolecular hydrogen bonding

			3305(⊥) 3275(∥)	Intermolecular hydrogen bonding [196]
		Doubly oriented film of mercerized cellulose	3488(∥) 3447(∥) 3350(⊥) 3305(⊥) 3175(⊥)	Intramolecular hydrogen bonding Intermolecular hydrogen bonding [148]
		Highly substituted cellulose ester (excluding cellulose nitrate)	3520-3470	
		Highly substituted cellulose nitrate	3580-3550	
		Alkaline cellulose	3400-3350	Cellulose treated with a solution of NaOD in D$_2$O
2	OH, combined to form very strong hydrogen bonds	Cellulose and its derivatives	3000-2500	Strong hydrogen bonding of the type
		Highly oxidized carboxycellulose	2500	Strong hydrogen bonding of the type

No.	Group, bond	Molecules	Frequency, cm^{-1}	Remarks
3	$>$**CHOH**, $-$**CH$_2$OH** (stretching vibrations)	Natural cellulose	2960 2930 2900 2870	
		Mercerized cellulose	2950 2930 2910 2890 2880	
		Doubly oriented film of bacterial cellulose	2970($\parallel$) 2945($\perp$) 2914($\perp$) 2891($\perp$) 2870($\perp$) 2853($\parallel$)	[196]
		Doubly oriented film of ramie	2970($\parallel$) 2945($\perp$) 2910($\perp$) 2870($\perp$) 2853($\parallel$)	[196]
		Doubly oriented film of mercerized cellulose	2981($\perp$?) 2968($\parallel$)	

No.	Group	Substance	Frequency	[148]
4	CHOD, CH₂OD (OD group involved in hydrogen bonding, stretching vibrations)		$2955(\parallel)$ $2933(\perp\ ?)$ $2904(\perp\ ?)$ $2891(\perp)$ $2874(\perp\ ?)$ $2850(\parallel)$	
		Natural cellulose	2520 2480 2450	Clearly visible after deuteration of the ordered parts of the cellulose
		Mercerized cellulose	2580 2550 2480	Clearly visible after deuteration of the ordered parts of the cellulose
5	$-CH_2-C\equiv N$	Cyanoethylcellulose, cell.$-OCH_2CH_2CN$	2250	Stretching
6	$-CH_2-C\equiv N$	Graft copolymer of acetylcellulose and polyacrylonitrile	2250	Stretching
7	$-CH_2-C\equiv N$	Graft copolymer of cellulose and polyacrylonitrile, cell.$-(CH_2)_2-$ ... $HN-(CH_2-CH_2CN)_n$ ... $-SO_2-$ ... OCH_3	2250	Stretching

No.	Group, bond	Molecules	Frequency, cm^{-1}	Remarks
8	$\bigcirc$—C≡N	Cellulose ester, cell.—O—(CH$_2$)$_2$—SO$_2$—$\bigcirc$—OCH$_3$ / CN	2160	Stretching
9	$>$CH—C≡C—C$<$	Block copolymer of cellulose, —O—(CH$_2$—CH—)$_n$ / C / ≡ / C—OH / CH$_3$, CH$_3$	2230	Stretching
10	—O—CH$_3$	Methylcellulose, cell.—O—CH$_3$	2970 / 2920 / 2870 / 2830 / 2810	Stretching
			1460 / 1380 / 1320 / 950	Deformation
11	—O—CH$_2$CH$_3$	Ethylcellulose, cell.—O—CH$_2$CH$_3$	2970 / 2930	Stretching

No.	Group	Compound	Stretching	Deformation
12	—O—CH₂CH₂CN	Cyanoethylcellulose, cell. -O-CH₂CH₂CN	2900, 2870	1490, 1450, 1410, 1380, 1320, 1280
			2960, 2930, 2890	1480, 1420, 1370, 1330, 1270, 1230, 850 ?
13	—O—CH(CH₃)₂	Isopropyl ether of cellulose, cell. -O-CH(CH₃)₂	2970, 2920, 2880	1460, 1370, 1310

No.	Group, bond	Molecules	Frequency, cm^{-1}	Remarks
14	—O—C(CH₃)₃	Tertiary-butyl ether of cellulose, cell.—O—C(CH₃)₃	2960 2910 2830	Stretching
			1460 1380 950	Deformation
15	—O—(CH₂)₈CH₃	Cellulose ester, cell.—O—(CH₂)₈CH₃	2950 2930 2900 2850	Stretching
			1470 1370 1310 720	Deformation
16	O‖ —O—C—CH₃	Acetylcellulose, cell.—O—C(=O)—CH₃	3000 2960 2940 2890	Stretching
			1430 1370	Deformation
17	O‖ —O—C—C(CH₃)₃	Cellulose 2-dimethylpropionate, cell.—O—C(=O)—C(CH₃)₃	2970 2940 2910 2870	Stretching

No.	Structure	Compound	Wavenumber	Assignment
18	—C(=O)(CH₂)₂CH₃	Cellulose butyrate, cell.—O—C(=O)(CH₂)₂CH₃	1480	Deformation
			1460	
			1400	
			1370	
			1280	
			1230	
			2960	Stretching
			2940	
			2870	
			1460	Deformation
			1420	
			1380	
			1370	
			1310	
			1250	
19	—C(=O)(CH₂)₁₆CH₃	Cellulose stearate, cell.—O—C(=O)(CH₂)₁₆CH₃	2960	Stretching
			2920	
			2850	
			1470	Deformation
			1420	
			1380	
			1360	
			1320	
			720	

Structures:
18: —C(=O)(CH₂)₂CH₃
19: —C(=O)(CH₂)₁₆CH₃

No.	Group, bond	Molecules	Frequency, cm^{-1}	Remarks
20	$\overset{O}{\underset{}{\diagdown}}C\diagup(CH_2)_4Cl$	Cellulose chlorovalerate, cell.—O—C$\overset{O}{\diagup}$(CH$_2$)$_4$Cl	2960 2940 2920 2870	} Stretching
			1460 1420 1390 1360 1320 1300 780 750 720	} Deformation
21	—S—CH$_3$	Cellulose methylxanthate, cell.—O—C$\overset{S}{\diagup}$SCH$_3$	2990 2920	} Stretching
			1420 1320	} Deformation
			970	
22	—S—CH$_2$—C$\overset{O}{\diagdown}$	Diethylacetoamido derivative of cellulose xanthate, cell.—O—C$\overset{S}{\diagup}$SCH$_2$CON(C$_2$H$_5$)$_2$	2970 2920	Stretching
			1450	Internal deformation

No.	Group	Compound	Wavenumbers	Type
23	$O=C-(CH_2)_5CH_3$	Cellulose acetoenanthate, cell. with $-O-C(=O)-CH_3$ and $-O-C(=O)-(CH_2)_5CH_3$	2940, 2860	Stretching
			1460, 720	Deformation
24	$P(=O)-CH_3$	Cellulose methylphosphonate, $(cell.-O)_2P(=O)-CH_3$	3000, 2920	Stretching
			1460, 1310	Deformation
25	$-O-CH_2$ (benzene ring)	Benzylcellulose, $cell.-O-CH_2-C_6H_5$	2900, 2870, 2800	Stretching
			1460 ?, 1370, 1210	Deformation
26	$-CH_2-CCl_2-$	Graft copolymer of cellulose and polyvinylidene chloride, $cell.-O-(CH_2CCl_2-)n$	1410, 1320	Deformation
27	CH of benzene rings	Benzylcellulose, $cell.-O-CH_2-C_6H_5$	3090, 3060, 3030	Stretching
			710, 700	Out of planar deformation

No.	Group, bond	Molecules	Frequency, cm^{-1}	Remarks
28	CH of benzene rings	Cellulose benzoate, cell.—O—C(=O)—C₆H₅	3090, 3060, 3030	Stretching
			710, 690	Out of planar deformation
29	CH of benzene rings	Trityl ether of cellulose, cell.—O—C(C₆H₅)₃	3090, 3070, 3030	Stretching
			770, 750, 700	Out of planar deformation
30	CH of benzene rings	Cellulose nitrophenylxanthate, cell.—O—C(=S)—SC₆H₄NO₂	3100, 3080, 3060, 3030, 3000	Stretching
			740, 730	Out of planar deformation
31	CH of benzene rings	Cellulose phenyl phosphate, cell.—O—P(=O)—OC₆H₅	770, 720, 690	Out of planar deformation

No.	Structure	Compound	Wavenumber	Assignment
32	CH of benzene rings (benzene ring)	Cellulose phenylphophonate, $cell.\!-\!O\!-\!P(\!=\!O)\,O\!-\!C_6H_5$	750 720 690	Out of planar deformation
33	$-C(\!=\!O)OH,\ -C(\!=\!O)H$	Dialdehydocellulose, oxidized by nitrogen oxides (with a high carboxyl group content)	1790 1760 1740	Stretching vibrations
34	$-C(\!=\!O)R$, $R=OH,\ CH_3,\ (CH_2)_nCH_3$	Carboxycellulose, cellulose esters	1750	Stretching vibrations
35	$-C(\!=\!O)C(CH_3)_3$	Cellulose 2-dimethylpropionate, $cell.\!-\!O\!-\!C(\!=\!O)\!-\!C(CH_3)_3$	1740	Stretching vibrations
36	$-C(\!=\!O)\!-\!C_6H_5$	Cellulose tribenzoate, $cell.\!-\!O\!-\!C(\!=\!O)\!-\!C_6H_5$	1730	Stretching vibrations
37	$-C(\!=\!O)CH_2-$	Products of thermal decomposition of cellulose methylxanthate	1720	Stretching vibrations

No.	Group, bond	Molecules	Frequency, cm^{-1}	Remarks
38	—C(=O)—CH$_2$—	Mixed ester of cellulose, cell. $\left[\begin{array}{l}\text{—O—COCH}_3 \\ \text{—O—CO(CH}_2)_2\text{COCH}_3\end{array}\right.$	1720	Stretching vibrations
39	—C(=O)—N(H)—OH	Graft copolymer of cellulose and poly-acrylhydroxamic acid, cell.—O—(CH$_2$—CH—)n C(=O)—NHOH	1670	Stretching vibrations
40	—SCH$_2$C(=O)—N(C$_2$H$_5$)$_2$	Cellulose diethylacetamidoxanthate cell.—O—C(=S)—SCH$_2$CON(C$_2$H$_5$)$_2$	1650	Stretching vibrations
41	(carboxylate O⋯C⋯O)	Salts of oxidized cellulose	1640–1580 1450–1350	Asymmetric stretching vibrations Symmetric stretching vibrations
42	—O—CH$_2$C(=O)—ONa	Sodium salt of highly substituted carboxymethylcellulose	1620	Stretching vibrations
43	—C(=O)—N(H)—O cat.	Products of ion exchange of graft co-polymer of cellulose and polyacryl-hydroxamic acid	1600–1580	Stretching vibrations

	Structure	Compound	Frequency, cm⁻¹	Assignment
44	$\rangle C{=}C\langle$ $_{OH}$ ^{H}	Products of thermal decomposition of cellulose methylxanthate	1650	Stretching
45	H_2O	Cellulose and its derivatives	1640	Adsorbed water
46	(N with two H)	Cellulose ether of 4-β-hydroxyethyl-sulfonyl-2-aminoanisole, cell.$-$O$-$CH$_2$CH$_2$SO$_2-$⟨⟩$-$OCH$_3$, NH$_2$	1630	Internal deformation
47	$-O-N\!\!\begin{smallmatrix}O\\\\O\end{smallmatrix}$	Cellulose nitrate, cell.$-$O$-$NO$_2$	1650 1280 840	Stretching
48	(S ring N with two O)	Cellulose nitrophenylxanthate, cell.$-$O$-$C$\begin{smallmatrix}S\\\\SC_6H_4NO_2\end{smallmatrix}$	1520 1350	Stretching
49	Aromatic impurities in cellulose (lignin)	Cellulose and its derivatives	1600 1520	Stretching vibrations of C=C in aromatic rings
50	$-$CH$_2-$ (benzene ring)	Benzylcellulose, cell.$-$CH$_2-$C$_6$H$_5$	1610 1590 1500 1460 ?	Stretching vibrations of C=C in aromatic rings
51	$O{=}C$ (benzene ring)	Cellulose benzoate	1610 1590 1500 1460	Stretching vibrations of C=C in aromatic rings

No.	Group, bond	Molecules	Frequency, cm^{-1}	Remarks
52	(triphenylmethyl structure)	Trityl ether of cellulose, cell.$-$O$-$C$-$(C$_6$H$_5$)$_3$	1600, 1500, 1450	Stretching vibrations of C=C in aromatic rings
53	NH$_2$ OCH$_3$, $-$SO$_2-$		1700, 1600, 1490	Stretching vibrations of C=C in aromatic rings
54	NO$_2$, $-$S$-$	Cellulose nitrophenylxanthate, cell.$-$O$-$C$\overset{S}{\diagdown}$SC$_6$H$_4$NO$_2$	1600, 1580	Stretching vibrations of C=C in aromatic rings
55	$-$S$-$NH$-$	Cellulose phenylthiourethane, cell.$-$O$-$C$\overset{S}{\diagdown}$SNHC$_6$H$_5$	1600, 1500	Stretching vibrations of C=C in aromatic rings
56	$\overset{O}{\underset{}{=}}P-O-$	Ester of cellulose with phosphorus-containing acids	1590, 1490	Stretching vibrations of C=C in aromatic rings

	Structure	Substance	Frequency, cm⁻¹	Type of vibration
57	—O—C(=S)	Cellulose xanthate cell.—O—C(=S)—SR	1380 ?	Stretching
58	—O—P=O	Cellulose esters of phosphorus-containing acids	1290–1200	Stretching
	P = O group unassociated	Cellulose esters of phosphorus-containing acids	1290–1280	
	P = O group associated	Cellulose esters of phosphorus-containing acids	1220–1200	
59	O=P	Cellulose phenylphosphonate, cell.(O)(O)P(O)C₆H₅	1440	Stretching
60	>C(H)(H)	Cellulose and its derivatives	1460–1420	Internal deformation (scissoring)
61	C—C—C with H H	Cellulose and its derivatives	1300–1200	External deformation (wagging)
62	C—C—C with H H	Cellulose and its derivatives	950–800	External deformation (rocking)

No.	Group, bond	Molecules	Frequency, cm^{-1}	Remarks
63	$\diagup\!\!\diagdown C - O \diagdown^{H}$ $\diagup\!\!\diagdown C_{(6)} - O \diagdown^{H}$	Cellulose and its derivatives Cellulose and its derivatives	1400–1200 1400–1300	} Planar deformation
64	$\left.\begin{array}{l} -C-O-H \\ -C-O-C \end{array}\right\}$ $C - O$ (ring)	Cellulose and its derivatives	1200–1100 1150–1000	Stretching
65	$O\!\!\diagdown\!\!\underset{O}{\overset{O}{S}}\!\!\diagup$	Cellulose ethers containing sulfonyl group $O-(CH_2)_2-SO_2-\!\!\langle\text{NH}_2\rangle\!-OCH_3$ $O-(CH_2)_2-SO_2-\!\!\langle\text{SCN}\rangle\!-OCH_3$	1300–1250 1320 1280 1260	Stretching Stretching
66	$-C\diagup^{CF_3}_{\diagdown CF_3}$	Cellulose mixed ester with acetic and hexafluoroisobutyric acids, cell. $\left[\begin{array}{l} -O-\overset{O}{\overset{\|}{C}}-CH_3 \\ -O-\overset{O}{\overset{\|}{C}}-CH(CF_3)_2 \end{array}\right.$	1300 1100 730	} Stretching Deformation

67	$\mathrm{C-O-C\!\!=\!\!O}$; R	Cellulose ester, cell.—O—C(=O)R	1280-1100	Stretching
68	$\mathrm{C-O-C\!\!=\!\!O}$; CH_3	Acetylcellulose, cell.—O—C(=O)CH_3	1230 1050	Stretching
69	$\mathrm{C-O-C\!\!=\!\!O}$; CH_2CH_3	Cellulose propionate, cell.—O—C(=O)CH_2CH_3	1170 1080-1070	Stretching
70	$\mathrm{C-O-C\!\!=\!\!O}$; $CH_2CH_2CH_3$	Cellulose butyrate, cell.—O—C(=O)$CH_2CH_2CH_3$	1170 1080-1060	Bands can also be observed in the regions 1110, 1040, and 1010 cm^{-1}; these are evidently attributable to structural factors
71	$\mathrm{C-O-C\!\!=\!\!O}$; $(CH_2)_{16}CH_3$	Cellulose stearate, cell.—O—C(=O)$(CH_2)_{16}CH_3$	1160 1070	Bands can also be observed at 1120 and 1080 cm^{-1}; these are evidently attributable to structural factors

No.	Group, bond	Molecules	Frequency, cm^{-1}	Remarks
72	$-C-O-C(=O)-C(CH_3)_3$	Cellulose 2-dimethylpropionate, cell.$-O-C(=O)-C(CH_3)_3$	1150 1070 1040	Stretching
73	$-C-O-C(=O)-(CH_2)_4Cl$	Cellulose chlorovalerate, cell.$-O-C(=O)-(CH_2)_4Cl$	1170 1130 1060	Stretching
74	$-C-O-C(=O)-C_6H_5$	Cellulose benzoate, cell.$-O-C(=O)-C_6H_5$	1270 1090	Stretching
75	$-C-O-C(=S)-SCH_3$	Cellulose methylxanthate, cell.$-O-C(=S)-SCH_3$	1220 1060	Stretching
76	$-C-O-N(=O)$	Cellulose nitrate, cell.$-ONO_2$	1070	Stretching
77	β-Glucopyranose unit in 1C conformation	3,6-Anhydrocellulose	840	Deformation vibration C_1H?

No.	Structure	Group	Frequency	Remark
78	(bicyclic structure)	3,6-Anhydrocellulose	920 ? 800	Vibrations of hydrofuranol ring?
79	$-C(=S)SR$	Cellulose xanthate, $\text{cell.}-O-C(=S)SR$	750–700	Stretching
80	$>C-Cl$	Ester of cellulose and chlorinated fatty acid $\text{cell.}-O-C(=O)(CH_2)_nCl$	650	Stretching
81	$-C-O-C(=O)CH_3$	Acetylcellulose, $\text{cell.}-O-C(=O)CH_3$	600	
82	$>C(H)O$	Cellulose and its derivatives	700–400	Out-of-planar deformation

N o t e . The heavy type and stars denote the groups of atoms whose vibrations are mainly responsible for the appearance of the corresponding bands. The literature citations in the remarks show the sources from which the values of the frequencies are taken.

LITERATURE CITED

1. V. I. Nikitin. Vestn. Leningr. Gos. Univ. 3: 33 (1950).
2. A. Elliot. "The infrared spectra of polymers," in: Advances in Spectroscopy, H. W. Thompson (ed.), Interscience Publishers, Inc., New York (1960)[Russian translation], IL, Moscow (1963).
3. A. A. Boldin and R. F. Vasil'eva. Zavodsk. Lab. 27(7): 819 (1961).
4. N.I.Makarevich and N. A. Borisevich. Zavodsk. Lab. 29(6) (1962).
5. I. N. Ermolenko. Doctoral Dissertation, Tashkent (1963).
6. R. G. Zhbankov and I. N. Ermolenko. Izv. Akad. Nauk BSSR, Ser. Fiz.-Tekhn. 1: 15 (1956).
7. R. G. Zhbankov. Dissertation, Minsk (1958).
8. I. N. Ermolenko and S. S Gusev. Vysokomolekul. Soedin. 1(3): 466 (1959).
9. B. I. Stepanov, R. G. Zhbankov, and I. N. Ermolenko. Izv. Akad. Nauk. SSSR, Ser. Fiz. 23(10): 1222 (1959).
10. R. G. Zhbankov. Opt. i Spektroskopiya 4(3): 318 (1958).
11. R. G. Zhbankov. Collected Proceedings of the First Conference on Molecular Spectral Analysis, Akad. Nauk Belorussk.SSR, Minsk (1958).
12. I. N. Ermolenko. The Spectroscopy and Chemistry of Oxidized Cellulose, Minsk (1959).
13. B. I. Stepanov, R. G. Zhbankov, E. N. Bolkov, A. I. Skrigan, A. M. Shishko, and A. Ya. Rozenberg. "A method for determining the degree of mercerization of cellulose," Author's Certificate No. 119,303, with priority from August 21, 1958.
14. B. I. Stepanov, R. G. Zhbankov, and A. Ya. Rozenberg, Zh. Fiz. Khim. 33(9): 1907 (1959).
15. B. I. Stepanov, A. I Skrigan, A. M. Shishko, and R. G. Zhbankov, Dokl. Akad. Nauk SSSR 135: 3 (1960).
16. R. G. Zhbankov, N. V. Ivanova, and A. Ya. Rozenberg, Zavodsk. Lab., No. 11: 1324 (1962).
17. B. I. Stepanov, R. G. Zhbankov, and R. Marupov. Vysokomolekul. Soedin. 3(11): 1633 (1961).

18. R. G. Zhbankov. Zavodsk. Lab., No. 12:1438 (1963).

19. 7 A. Rogovin and N. N. Shorygina. The Chemistry of Cellulose and Related Compounds, Moscow-Leningrad (1953).

20. I. R. Petrov and A. N. Filatov. Plasma-Substitute Solutions, Medgiz (1958).

21. R. G. Zhbankov, N. P. Krivoshheev, and G. V. Reutovich. Dokl. Akad. Nauk Belorussk.SSR 6(9): 592 (1962).

22. A. A. Konkin, D. N. Shigorin, and L. I. Novikova. Zh. Fiz. Khim. 32: 894 (1958).

23. D. N. Sokolov, Usp. Fiz. Nauk 57(2): 205 (1955).

24. L. Bellamy. The Infrared Spectra of Complex Molecules [Russian translation], IL, Moscow (1963).

25. R. G. Zhbankov, R. V. Zueva, P. V. Kozlov, L. V. Savel'eva. Vysokomolekul. Soedin. 2(8):1270 (1960).

26. R. G. Zhbankov, N. V. Ivanova, and Z. A. Rogovin. Vysokomolekul. Soedin. 4(6): 901 (1962).

27. R. G. Zhbankov, V. I. Nepochatykh, R. Marupov, and Z. A. Rogovin. Vysokomolekul. Soedin. 4(11): 1696 (1962).

28. B. I. Stepanov, R. G. Zhbankov, A. I. Skrigan, and A. M. Shishko. Izv. Akad. Nauk Belorussk.SSR, Ser. Fiz.-Tekhn. 4: 105 (1957).

29. A. I. Skrigan, A. M. Shishko, and R. G. Zhbankov, Dokl. Akad. Nauk. Belorussk.SSR 1: 17 (1957).

30. R. G. Zhbankov, R. Marupov, N. I. Garbuz, A. I. Skrigan, and A. M. Shishko. Izv. Akad. Nauk Belorussk. SSR, Ser. Fiz.-Tekhn., No.4: 65 (1963).

31. F. F. Derbentsev, A. M. Shishko, and N. A. Derbentseva. Izv. Akad. Nauk Belorussk.SSR, Ser. Fiz.-Tekhn., No. 3 (1960).

32. B. I. Stepanov and R. G. Zhbankov. Zavodsk. Lab., No. 6: 696 (1963).

33. A. I. Skrigan, A. M. Shishko, and R. G. Zhbankov. Izv. Akad. Nauk Belorussk.SSR, Ser. Fiz.-Tekhn. 1: 29 (1957).

34. A. I. Skrigan, A. M. Shishko, and R. G. Zhbankov, Dokl. Akad. Nauk SSSR 115(1): 114 (1957).

35. A. I. Skrigan, A. M. Shishko, and R. G. Zhbankov. Tr. Inst. Fiz. Khim., Akad. Nauk Belorussk.SSR, Vol. 7 (1959).

36. A. I. Skrigan and A. M. Shishko. Izv. Akad. Nauk Belorussk.SSR, Ser. Fiz.-Tekhn. 2: 56 (1959).

37. N. I. Nikitin. The Chemistry of Wood and Cellulose, Akad. Nauk SSSR, Moscow-Leningrad (1962), p. 209.

38. B. I. Stepanov, R. G. Zhbankov, and A. Ya. Rozenberg. Dokl. Akad. Nauk Belorussk.SSR 1(3):92 (1957).

39. R. G. Zhbankov and A. Ya. Rozenberg. Collected Proceedings of the First Conference on Molecular Spectral Analysis, Minsk (1958), p. 112.

40. B. I. Stepanov, R. G. Zhbankov, A. M. Shishko, A. I. Skrigan, and A. Ya. Rozenberg, Author's certificate No. 138,408, with priority from August 17, 1959.

41. M. Mizushima. The Structure of Molecules and Internal Rotation, Academic Press, Inc., New York; Russian translation: IL, Moscow (1957).

42. V. M. Vol'kenshtein. The Configuration Statistics of Polymer Chains, Akad. Nauk SSSR, Moscow-Leningrad (1959).

43. V. A. Kargin, N. V. Mikhailov, and V. I. Elinke. In: Investigations in the Field of High Molecular Compounds, Akad. Nauk SSSR (1949).

44. K. Gettse (ed.). The Production of Viscose Fiber, Gizleprom (1958), p. 15.

45. O. Orlova and D. Fedorov. Zh. Tekhn. Fiz. 3(7): 1124 (1933).

46. V. N. Nikitin. Zh. Fiz. Khim. 23: 775, 786 (1949); The Chemistry of Wood and Cellulose, Vol. 3. Akad. Nauk SSSR, Moscow-Leningrad (1962), pp. 58-59.

47. M. A. Katibnikov, I. N. Ermolenko, A. I. Somova, O. G. Efremova, and S. A. Glikman. Vysokomolekul. Soedin. 2(12): 1805 (1960).

48. V. I. Kurlyankina, A. B. Polyak, and O. P. Koz'mina. Vysokomolekul. Soedin. 2(12): 1850 (1960).

49. R. V. Zueva, R. G. Zhbankov, N. V. Ivanova, P. V. Kozlov, and E. K. Podgorodetskii. In: Cellulose and Its Derivatives, Akad. Nauk SSSR (1963), pp. 118-123, 124-130.

50. P. V. Kozlov, R. G. Zhbankov, R. V. Zueva, N. V. Ivanova, and E. K. Podgorodetskii. In: Cellulose and Its Derivatives, Akad. Nauk SSSR (1963), pp. 131-138.

51. P. V. Kozlov and F. S. Sherman. Zh. Prikl. Khim. 25: 384 (1952).

52. E. K. Podgorodetskii and V. A. Savitskaya. Tr. Nauch.-Issledovatel. Kino-Foto-Inst., No. 4: 23 (1958).

53. W. West (ed.). Chemical Applications of Spectroscopy (Technique of the Uses of Spectroscopy in Chemistry. Organic Chemistry, Vol. 9), Interscience Publishers, Inc. (1956) [Russian translation, IL (1959)].

54. I. N. Ermolenko, R. G. Zhbankov, V I. Ivanov, N. Ya. Lenshina, and V. S. Ivanova. Izv. Akad. Nauk SSSR, Otd. Khim. Nauk 2: 249 (1958); Izv. Akad. Nauk SSSR 12: 1495 (1958).

55. I. N. Ermolenko and R. G. Zhbankov. Izv. Akad. Nauk Belorussk. SSR, Ser. Fiz.-Tekhn. 1: 35 (1958).

56. I. N. Ermolenko and S. S. Gusev. Vysokomolekul. Soedin. 1(10): 1462 (1959).

57. A. B. Polyak. Trudy Lesotekhnicheskoi Akad. im. Kirova, Vol. 9 (1960).

58. I. N. Ermolenko and R. G. Zhbankov. Zh. Fiz. Khim. 33(6): 1191 (1959); Kolloidn. Zh. 20(4): 429 (1958); Dokl. Akad. Nauk Belorussk. SSR 3(5): 202 (1959).

59. N. Ya. Lenshina, V. S. Ivanova, and V. I. Ivanov. Izv. Akad. Nauk SSSR, Otd. Khim. Nauk (1960), p. 1894.

60. V. I. Ivanov, N. Ya. Lenshina, and V. S. Ivanova. Dokl. Akad. Nauk SSSR 129: 323 (1959).

61. N. Ya. Lenshina, V. S. Ivanova, and V. I. Ivanov. Izv. Akad. Nauk SSSR, Otd. Khim. Nauk 2: 519 (1961).

62. N. Ya. Lenshina and V. I. Ivanova. Vysokomolekul. Soedin. 4(11): 1647 (1962).

63. N. Ya. Lenshina. Dissertation, Moscow (1964).

64. R. G. Zhbankov and I. N. Ermolenko. Author's certificate No. 113,571, with priority from September 18, 1957.

65. I. N. Ermolenko, R. G. Zhbankov, and A. Ya. Rozenberg. Izv. Akad. Nauk Belorussk. SSR, Ser. Fiz.-Tekhn. 3: 25 (1960).

66. S. S. Gusev and I. N. Ermolenko. Kolloidn. Zh. 24(3): 278-282 (1962).

67. S. S. Gusev, M. A. Katibnikov, and I. N. Ermolenko. Kolloidn. Zh., 23(2): 140-144 (1961).

68. I. N. Ermolenko and R. G. Zhbankov. Inzh.-Fiz. Zh. 1(2): 94-98 (1958).

69. V. I. Ivanov and N. Ya. Lenshina. Izv. Akad. Nauk SSSR, Otd. Khim. Nauk 4: 506 (1956).

70. G. Herzberg. Vibrational and Rotational Spectra of Polyatomic Molecules [Russian translation], IL, Moscow (1952).

71. R. G. Zhbankov, R. Marupov, Wu Mei-Yen, M. A. Tyuganova, and Z. A. Rogovin, Vysokomolekul. Soedin. 5(9): 1292 (1963).

72. Z. A. Rogovin. Khim. i Tekhnol. Polimerov 7-8: 174 (1960).

73. Wu Mei-Yen, T. Zharova, and Z. A. Rogovin. Zh. Prikl. Khim. 35(8): 1820 (1962).

74. Z. A. Rogovin, Wu Mei-Yen, M. A. Tyuganova, T. Ya. Zharova, and E. L. Gefter. Vysokomolekul. Soedin. 5(4): 506 (1963).

75. A. E. Arguzov, M. I. Batuev, and V. S. Vinogradova. Dokl. Akad. Nauk 54: 603 (1946).

76. E. M. Popov, I. Kabanchik, and L. S. Mayants. Usp. Khim. 30(7): 846 (1961).

77. M. I. Kabanchik, T. A. Mastryukova, N. P. Radionova, and E. M. Popov. Zh. Obshch. Khim. 26: 120 (1956).

78. M. I. Kabanchik, N. I. Kurochkin, T. A. Mastryukova, S. T. Ioffe, E. M. Popov, and N. P. Radionova. Dokl. Akad. Nauk SSSR 104: 861 (1955).

79. L. S. Mayants, E. M. Popov, and M. I. Kabanchik. Opt. i Spektroskopiya 7: 170 (1959).

80. Wu-jeng Wei-ch'ang and Z. A. Rogovin, Vysokomolekul. Soedin. 2: 456 (1960).

81. J. Lecomte. Infrared Radiation [Russian translation], Gosizdat Fiz.-Mat. Lit., Moscow (1958).

82. V. I. Nepochatykh and Z. A. Rogovin. Khim. Volokna 1 : 40 (1960).

83. L. Aleksandru and Z. A. Rogovin. Zh. Obshch. Khim., No. 23: 1199, 1203 (1953).

84. A. I. Polyakov, V. A. Derevitskaya, and Z. A. Rogovin, Vysokomolekul. Soedin. 2(3): 386 (1960).

85. E. D. Kaverzneva, V. I. Ivanov, and A. S. Salova. Izv. Akad. Nauk SSSR, Otd. Khim. Nauk (1949), p. 369.

86. L. A. Chugaev. Investigations in the Field of Terpenes and Camphor, Dissertation (1903); Selected Works, Vol. 2, Akad. Nauk SSSR (1955).

87. M. D. Balabaeva, M. O. Lishevskaya, A. D. Virnik, Z. A. Rogovin, and R. G. Zhbankov. In: Cellulose and Its Derivatives, Akad. Nauk SSSR (1963), p. 57.

88. R. G. Zhbankov, R. Marupov, M. D. Balabaeva, O. Lishevskaya, and M. A. Tyuganova. Izv. Akad. Nauk Belorussk.SSR, Ser. Fiz.-Tekhn., No. 2 (1963).

89. M. O. Lishevskaya, A. D. Virnik, and Z. A. Rogovin. In: Cellulose and Its Derivatives, Akad. Nauk SSSR (1963), p. 32.

90. Yu. G. Kryazhev and Z. A. Rogovin, Vysokomolekul. Soedin. 3: 1847 (1961).

91. Yu. G. Kryazhev, Z. A. Rogovin, and V. V. Chernaya. In: Cellulose and Its Derivatives, Akad. Nauk SSSR (1963), p. 94.

92. R. Marupov, R. G. Zhbankov, Yu. G. Kryazhev, and Z. A. Rogovin. In: Cellulose and Its Derivatives, Akad. Nauk SSSR (1963), p. 150.

93. R. M. Livshits and Z. A. Rogovin. In: Cellulose and Its Derivatives, Akad. Nauk SSSR (1963), p. 12.

94. R. M. Livshits, A. A. Predvoditelev, and Z. A. Rogovin. In: Cellulose and Its Derivatives, Akad. Nauk SSSR (1963), p. 60.

95. R. M. Livshits, R. Marupov, R. G. Zhbankov, and Z. A. Rogovin. In: Cellulose and Its Derivatives, Akad. Nauk SSSR (1963), p. 65.

96. Z. A. Rogovin, Sun T'ung, A. D. Virnik, and N. M. Khvostenko. Vysokomolekul. Soedin. 4: 571 (1962).

97. V. N. Nikitin and N. V. Mikhailova. In: Cellulose and Its Derivatives, Akad. Nauk SSSR (1963), p. 40.

98. N. V. Ivanova and R. G. Zhbankov. The Hydrogen Bond, Collected Papers, Nauka, Moscow (1964), p. 149.

99. R. G. Zhbankov, R. Marupov, N. V. Ivanova, and A. M. Prima. Proceedings of the Fifteenth All-Union Conference on Spectroscopy, Minsk (1963).

100. A. I. Skrigan, A. M. Shishko, and R. G. Zhbankov. Collected Proceedings of the First Conference on Molecular Spectral Analysis, Minsk (1958), p. 122.

101. B. I. Stepanov, Zh. Fiz. Khim. 20:917 (1946).
102. E. P. Volkov. Collected Proceedings of the First Conference on Molecular Spectral Analysis, Minsk (1958), p. 130.
103. S. S. Gusev and I. N. Ermolenko. Zavodsk. Lab., No. 2:181 (1964).
104. G. L. Elina, S. S. Gusev, and I. N. Ermolenko, Dokl. Akad. Nauk Belorussk.SSR 8(2):104 (1964).
105. R. Holiday, Nature 163:602 (1949).
106. C. Ruscher and R. Schmolke. Faserforschung und Textiltechnik 8:11 (1960).
107. E. R. Blout and G. R. Bird. J. Opt. Soc. Am. 41:547 (1951).
108. C. R. Barer, A. R. Cole, and H. W. Thompson. Nature 163:198 (1949).
109. K. P. Norris. J. Sci. Instr. 31:284 (1954).
110. E. R. Blout and M. J. Abbate. J. Opt. Soc. Am. 45:1028 (1955).
111. F. H. Forziati and I. W. Rowen. J. Res. Natl. Bur. St. 46(1):38 (1951).
112. R. T. O'Connor, E. T. Du Pre, and E. K. McCall. Anal. Chem. 29(7):998 (1957).
113. H. G. Higgins. Australian J. Chem. 10:496 (1957).
114. M. A. Ford and Wilkinson. J. Sci. Instr. 31:338 (1954).
115. R. O. French, M. E. Wadsworth, M. A. Cook, and I. B. Gutler. J. Phys. Chem. 58(10):805 (1954).
116. A. W. Beker. J. Phys. Chem. (1957), p. 450.
117. V. C. Farmer. Spectrochim. Acta 8(6):374 (1957).
118. V. W. Meloche and G. E. Kalbus. J. Inorg. Nucl. Chem. 6:104 (1958).
119. F. Vrathy. J. Inorg. Nucl. Chem. 10:328 (1959).
120. H. Röpke and W. Nendert. J. Anal. Chem. 170:78 (1959).
121. S. Burgess and H. Spedding. Chem. Ind. 29:1161 (1961).
122. Olli Ant-Wuorinen and Asko Visapa. Paper and Timber 42(6):367 (1960).
123. S. Krimm, C. Liang, and G. Sutherland. J. Polymer Sci. 22:227 (1956).
124. L. R. Kuhn. Anal. Chem. 22(2):276 (1950).
125. S. A. Barker, E. J. Bourne, M. Stasey, and D. H. Whiffen. J. Chem. Soc. (1954), p. 171; Chem. Ind. (1953), p. 196.
126. S. C. Burket and R. M. Badger. J. Am. Chem. Soc. 72:4397 (1950).
127. S. A. Barker, E. J. Bourne, R. Stephens, and D. H. Whiffen. J. Chem. Soc. (1954), p. 3468.
128. R. L. Whistler and L. R. House. Anal. Chem. 25:1463 (1953).
129. T. Urbanski, W. Hofman, and M. Witanowski. Bull. Accad. Polon. Sci., Ser. Sci., Chim., Geol. Geograph. 7:619 (1959).
130. R. L. Tipson and H. S. Isbell. J. Res. Natl. Bur. Std. 64A:239 (1960).
131. R. L. Tipson and H. S. Isbell. J. Res. Natl. Bur. Std. 64A(5):405 (1960); 65A(1):31 (1961); 66A(1):31 (1962).

132. J. D. Goulden. Spectrochim. Acta 9: 657 (1959).

133. H. Higgins, C. M. Steward, and K. J. Harrington. J. Polymer Sci. 51: 59 (1961).

134. L. Legal, R. T. O'Connor, and F. V. Eggerton. J. Am. Chem. Soc. 82: 2807 (1960).

135. M. M. Davies. J. Chem. Phys. 16: 267 (1948).

136. A. Fürst, H. H. Kuhn, R. Scotini, and H. H. Günthard. Helv. Chim. Acta 35: 951 (1952).

137. C. Y. Liang and R. H. Marchessault. J. Polymer Sci. 39: 135-269 (1959).

138. M. Falk and E. Whally. J. Chem. Phys. 34: 1554 (1961).

139. A. Miyake. J. Am. Chem. Soc. 82: 12 (1960).

140. S. A. Barker and R. Stephens. J. Chem. Soc. (1954), p. 4550.

141. R. Giray-Deponthière. Bull. Soc. Chim. Belges. 69(3): 169 (1960).

142. R. Reeves. Advan. Carbohydr. Chem. 6: 107 (1951).

143. M. Tsuboi. J. Polymer Sci. 25: 159 (1957).

144. R. T. O'Connor, E. DuPre, and D. Mitcham. Textile Res. J. 28(5): 382 (1958).

145. A. W. McKenzie and H. G. Higgins. Svensk. Paperstid. 61: 20, 893 (1958).

146. C. Y. Liang, K. H. Bosset, E. A. McGinnes, and R. H. Marchessault. Tappi 43(12): 1017 (1960).

147. H. I. Marrinan and I. Mann. J. Polymer Sci. 21: 301 (1956).

148. R. H. Marchessault and C. Y. Liang. J. Polymer Sci. 43: 71 (1960).

149. F. G. Hurtübise and U. Krässig. Anal. Chem. 32: 177 (1960).

150. T. Kubo. Kolloid. Z. 93: 338 (1940); 96: 41 (1941).

151. H. I. Marrinan and I. Mann. J. Appl. Chem. 4: 204 (1954).

152. A. I. Wells. J. Appl. Phys. 11(2): 137 (1940).

153. R. Stelle and E. Pacsu. Textile Res. J. 19(12): 790 (1949).

154. L. W. Brown, P. Holiday, and F. Trotter. J. Chem. Soc. (1951), p. 1532.

155. R. Gerbaux. Bull. Soc. Chim. Belges. 66(5-6): 382 (1957).

156. M. Bouriot. Bull. Inst. Textile France (ITF) 94: 7-19 (1961).

157. F. G. Hürtubise. Tappi 45(6): 460 (1962).

158. H. L Barthelemy. Chemie et Industrie 25: 819 (1931).

159. F. B. Cramer and C. B. Purves. J. Am. Chem. Soc. 61: 3458 (1939).

160. T. S. Gardner and C. B. Purves. J. Chem. Soc. 64: 1539 (1942).

161. R. Gerbaux. Bull. Soc. Chim. Belges. 65: 270-290 (1956).

162. H. W. Thompson and P. Torkington. J. Chem. Soc. (1945), p. 640.

163. N. B. Colthup. J. Opt. Soc. Am. 40: 397 (1950).

164. I. W. Rowen, C. M. Hunt, and E. K. Plyler. Textile Res. J. 17(9): 504 (1947).

165. I. W. Rowen and E. J. Plyler. J. Res. Natl. Bur. Std. 44(3): 313 (1950).

166. I. W. Rowen, F. H. Forziati, and R. S. Reeves. J. Am. Chem. Soc. 73(9): 4484 (1951).

167. F. H. Forziati, I. W. Rowen, and E. K. Plyler. J. Res. Natl. Bur. Std. 46: 288 (1951).

168. H. G. Higgins and A. W. McKenzie. Australian J. Appl. Sci. 9(2): 167 (1958).

169. H. Spedding. J. Chem. Soc. (July, 1960), pp. 3147-3152.

170. F. S. H. Head. J. Textile Institute 46: 400 (1955).

171. E. Ott et al. Cellulose and Cellulose Derivatives, Second edition, three parts, Interscience Publishers, Inc. (1954).

172. Lecomte. Rev. Optique 28: 353 (1949).

173. R. L. Favre. J. Microchim. Acta 2-3: 517 (1955).

174. L. W. Daasch and D. C. Smith. Anal. Chem. 23: 853 (1951).

175. E. D. Bergmann, U. L. Litauer, and Pinchas. J. Chem. Soc. (March, 1952), pp. 847-849.

176. R. A. McIvor, Y. A. Grant, and C. E. Hubly. Can. J. Chem. 3: 11 (1956).

177. L. J. Bellamy and L. Beecher. J. Chem. Soc. (1952), p. 476; (1952), p. 1701; (1953), p. 729.

178. C. D. Miller, R. C. Miller, and J. W. R. Rogers. J. Am. Chem. Soc. 80: 1562 (1958).

179. H. Gerding and J. W. Maarsen. Rec. Trav. Chim. 76: 6 (1957).

180. A. Pojefcky and W. D. Coggeshall. Anal. Chem. 23: 1611 (1951).

181. Landolt-Börnstein. Atom- und Molecular-Physik, Second Part, Berlin (1951).

182. I. I. Fox and A. E. Martin. Proc. Roy. Soc. (London), Ser. A, 175: 208 (1940).

183. H. W. Thompson, D. L. Nichelson, and L. W. Sort. Discussions Faraday Soc. 9: 222 (1950).

184. P. F. Gardiner and R. B. Stasiak. Appl. Spectry. 12: 116 (1958).

185. E. Klein, J. Bosarge, and J. Norman. J. Phys. Chem. 64: 1666 (1960).

186. L. Sittle and J. Leja. Electr. Phenomena and Solid-Liquid Interface. Butterworths,Scientific Publishers, London (1957), pp. 261-266.

187. J. C. Evans. Spectrochim. Acta 16(4): 428 (1960).

188. M. Tsuboi. Spectrochim. Acta 16(4): 505 (1960).

189. K. Frendenberg and A. Wolf. Ber. 60: 232 (1927).

190. E. A. Coulson, I. I. Hales, and E. F. G. Herington. J. Chem. Soc. (1951), p. 2125.

191. L. Mirion, D. A. Ranisay, and R. N. Jones. J. Am. Chem. Soc. 73: 305 (1951).

192. The Chemistry of Penicillin, Princeton University Press, Princeton, New Jersey (1949), p. 390.

193. R. E. Richards and H. W. Thompson. J. Chem. Soc. (1947), p. 1248.

194. Sakai Takenski. Ann. Rept. Fac. Pharm. Kanazawa Univ. 10:1-2 (1960).

195. J. Mann and H. J. Marrinan. J. Polymer Sci. 27:595 (1958).

196. C. Y. Liang and R. H. Marchessault. J. Polymer Sci. 35:529 (1959); J. Polymer Sci. 37:385 (1959).

197. K. H. Meyer and L. Misch. Helv. Chim. Acta 20:232 (1937).

198. A. Frey-Wyssling. Biochim. et Biophys. Acta 18:166 (1955).

199. J. Mann and H. J. Marrinan. Trans. Faraday Soc. 52:481, 487, 492 (1956).

200. H. Brock-Neely. Advan. Carbohydr. Chem. 12:13 (1957).

201. H. Isbell and P. Smith. J. Res. Natl. Bur. Std. 59:141 (1957).

202. B. Schneider and J. Vodnansky. Reprinted from: Collection Czech. Chem. Commun. 28:2080-2088 (1963).

203. M. L. Nelson and R. T. O'Connor. J. Appl. Polymer Sci. 8:1311-1341 (1964).

204. R. Marupov, R. G. Zhbankov, A. I. Polyakov, and Z. A. Rogovin. In: Cellulose and Its Derivatives, Akad. Nauk SSSR (1963), p. 196.